Calculus

GEOFFREY MATTHEWS, M.A., Ph.D.

Shell Professor of Mathematics Education
at the Centre for Science Education
Chelsea College, University of London

Foreword by
PROFESSOR T. A. A. BROADBENT
Royal Naval College, Greenwich

John Murray

FIFTY ALBEMARLE STREET LONDON

First published 1964
Second edition 1980

Reproduced and printed by photolithography and
bound in Great Britain at The Pitman Press, Bath

ISBN: 0 7195 3734 7

Foreword

Fittingly, possibly inevitably, the calculus, dealing with problems of change and growth, was devised during the seventeenth century, in which the climate of thought was changing from the view of the universe, and man's place in it, as static and immutable, to that of a dynamic and developing world. But in spite of the power and inherent simplicity of the new doctrine, the calculus was recognized as a school subject only towards the end of the nineteenth century, and even then it was often presented as an elaborate and sometimes tedious technique. In the last fifty years, a number of excellent texts have gone far to remedy this state of affairs, by stressing the immediate applicability of calculus methods before an attempt is made to deal with the differentiation and integration of complicated functions, and by allowing problems to suggest methods of attack. It may well be that in this age of the car and the aeroplane, correct concepts about rates of change are more readily understood and their importance more easily appreciated.

Those who know of Dr Matthews' exceptional skill in rendering classroom mathematics exciting and stimulating will recognize these qualities in the present book. Not even the dullest reader can fail to catch something of the spirit of an intellectual adventure, to sense the power provided by calculus methods. Naturally, in following such a well-trodden path, there is little room for innovation; the book must be judged mainly on the use and arrangement of fairly familiar material. Here three points call for comment.

Firstly, Dr Matthews tells us the truth, if not always the whole truth; he warns us that what is 'obvious' may, at a later stage, call for proof. Though he is thinking chiefly of the intending scientist or engineer, he does not make this an excuse for slipshod reasoning; and if the intending mathematician uses this book at school, he will of course have a great deal to learn at the university, but little if anything to unlearn. Secondly, steps forward in the theory are, where possible, shown as natural responses to natural questions, theoretical or practical; and examples to show the power of the methods are selected from a wide field. Thirdly, the active cooperation of the reader is demanded. He is not being drilled (though plenty of drill examples are available), but being asked to think. The text is devised to provoke thought; while some of the exercises are very simple, others may seem to be too difficult for basic A-level work, but Dr Matthews clearly believes that more harm can be done by under-estimating a pupil's capacity than by over-estimating it, that the boy (or girl) should be asked to put the brain to full stretch rather than let it always work at less than full power. Much of the provocative enthusiasm which, so his pupils have assured me, Dr Matthews displays in the classroom has come through into the more restrictive medium of the printed page, and so, we may hope, can be transmitted back into many other classrooms.

T. A. A. BROADBENT

Royal Naval College,
Greenwich

January 1964

Preface to First Edition

This book is an introduction to calculus; it covers the requirements at school of future scientists and engineers. Chapters 1 to 6 provide adequate preparation for the various additional and alternative O papers of the GCE. Chapters 7 to 11 complete a basic A-level course; the remaining chapters cover the extra topics required by certain boards and complete the course to (scientists') open scholarship standard.

There is no lack of traditional textbooks covering just this ground—nor of university lecturers who complain that their first-year students are ignorant of the very first ideas on the subject. I believe this is due to the 'drilling' of pupils on one topic after another, glossing over difficulties and not giving any sense of purpose. This book attempts a more humane and contemporary approach. It is my experience, and that of a growing number of other teachers, that the so-called 'duller' pupil (taking an A-level course just the same!) will respond readily to a challenge which seems worthwhile; he likes to feel he is really getting down to the subject instead of learning a set of tricks which might one day be useful. It is accordingly possible to penetrate the subject right from the start more deeply than is customary.

Students should, in fact, be able to read a good deal of the book with a minimum of supervision, thus getting used to the idea of independent study. When it is necessary to forego rigour in the interests of brevity or comprehension a clear indication is given, so that the future mathematician can also start on this book without having to unlearn anything later. In this connexion, I have introduced the notion of function from the start in terms of 'many-one' correspondence, which incidentally simplifies the treatment of inverse functions later.

As already indicated, emphasis has been placed on 'motivation' and understanding rather than mechanical learning. Thus, for example, the idea of gradient is first shown to be useful, and later the underlying concept of limit is discussed in more than customary detail. Again, the exponential function and Taylor's theorem are introduced most informally, more rigorous treatment being deferred until after their importance has been justified.

On the other hand, much practice is necessary in differentiating and integrating, and there are numerous exercises for the reader. These include many from past examinations, and I have to thank the authorities of the Oxford and Cambridge Schools Examination Board, the University of London, the Joint Matriculation Board, the Welsh Joint Education Committee, the University of Durham School Examination Board, the Southern Universities'

Joint Board and the Delegates for Local Examinations, University of Oxford for permission to include questions from papers set by them.

A few questions are marked (D) for class-discussion and others (P) for 'project'; it is hoped that the reader will be tempted to use a library for research on the latter.

I have found much useful guidance in three reports: *The Teaching of Calculus in Schools* (Mathematical Association), *Teaching of Mathematics* (IAAM), and *Teaching Mathematics in Secondary Schools* (Ministry of Education).

I have not, however, felt bound to follow all their recommendations slavishly. For example, the first-named report, advocating numerical work on gradients of particular curves, advises: 'Work of this kind can easily and profitably occupy half a dozen lessons.' This is directly contrary to my experience; after two or three examples of this sort, any pupil who is going to stay the course at all is eager to seek some sort of generalization.

I thank Professor T. A. A. Broadbent for generously writing a Foreword and for providing a number of helpful suggestions. I am indebted to Mr A. P. Rollett for many useful comments and much encouragement as well as for reading the proofs. I also express my thanks to Professor D. B. Scott of the University of Sussex for some useful remarks, to Miss A. R. Russell and Messrs C. A. R. Bailey and F. B. Lovis for much forthright and constructive criticism, firmly based on their experience as fellow-teachers, and to Messrs D. J. Andrew and E. D. Cooke for willing help in checking the answers. Thanks are also due to the Oxford University Press for permission to reproduce the passage on p. 143 from *Nuclear Physics in Photographs* by Powell and Occhialini.

GEOFFREY MATTHEWS

Preface to Second Edition

In order to bring this book into line with current practice, the major changes incorporated are those concerned with units and notation. These now follow, in the main, the guidelines set out by the Teaching Committee of the Mathematical Association in the December 1977 issue of the *Mathematical Gazette*.

Contents

1
Gradients

Shall I refuse my dinner because I do not fully understand the process of digestion?
                                                                    HEAVISIDE

Introduction

The calculus was invented in the first place to deal with the slopes of curves and the areas 'under' them, Fig. 1.1. It has since turned out to be probably the most fruitful branch of mathematics, certainly in its application to science and engineering.

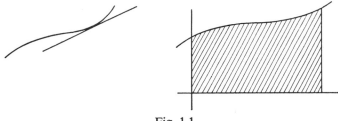

Fig. 1.1

The chief credit for its discovery is due to two seventeenth-century mathematicians, Sir Isaac Newton and Baron Gottfried von Leibniz, but some of the problems which the subject has helped to solve had been tackled already in classical times.*

We shall first enquire why problems connected with curves should still be of interest to-day.

EXAMPLE 1.1 *A man, walking with a steady speed, covers 15 kilometres in 3 hours. To draw the graphs of* (i) *the distance, s, against time, t, and* (ii) *velocity, v, against t.*

The distance covered divided by the time taken gives the man's velocity, 5 km h⁻¹. If the unit scales for s and t are taken to be equal (e.g. 1 cm to represent 1 hour, and also to represent 1 km), then v is the *gradient* of the line in the $s(t)$ graph, Fig. 1.2(i), that is, the tangent of the angle it makes

---

* The word *calculus* is Latin for a pebble. Calculating 2000 years ago was largely a matter of counting pebbles; some cricket umpires still use this method to check the number of balls in an over.

with the *t*-axis. *In what follows, we shall always take the scales along the two axes of a graph to be equal.*

The $v(t)$ graph, Fig. 1.2(ii), is a horizontal straight line, since v remains equal to 5 throughout the journey. The area $OABC$ under this graph is 15, the number of kilometres travelled.

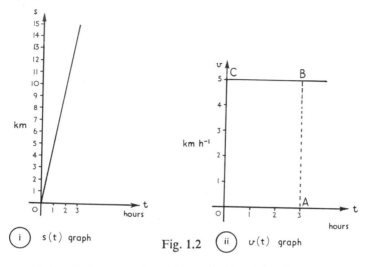

(i) $s(t)$ graph   **Fig. 1.2**   (ii) $v(t)$ graph

EXAMPLE 1.2   *To interpret the $v(t)$ graph, Fig. 1.3, of a particle moving in a straight line, and sketch the shape of the corresponding $s(t)$ graph.*

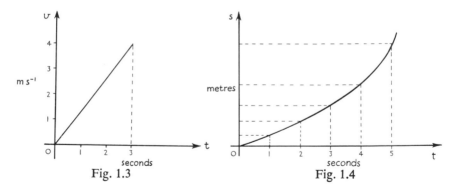

Fig. 1.3   Fig. 1.4

Here the particle starts 'from rest', and its pace increases steadily, so that after 3 seconds it has acquired a velocity of 4 m s⁻¹. It goes faster, or *accelerates*, at the rate of $\frac{4}{3}$ metres per second every second, which can be written as $\frac{4}{3}$ m s⁻². This acceleration is given by the gradient of the graph.

Now, as the particle accelerates, the distance travelled in successive seconds will increase, so that the $s(t)$ graph will look something like Fig. 1.4.

EXERCISE I

1. **(D)** Guess the form of equation of the $s(t)$ graph in Example 1.2.
2. **(D,** *but each should draw his own graph.*)
   A particle moves in a straight line so that the distance s metres it has travelled after t seconds is given by $s = \frac{1}{2}t^2$. Plot accurately the graph of s against t for values of t from 0 to 4.
     (i) Read off from the graph the total distance covered (*a*) between $t = 0$ and $t = 3$; (*b*) between $t = 0$ and $t = 2$.
    (ii) What is the distance travelled in the third second (i.e. between $t = 2$ and $t = 3$)?
   (iii) What is the average velocity between $t = 2$ and $t = 3$?
    (iv) Find the average velocity (*a*) between $t = 2$ and $t = 2\cdot5$; (*b*) between $t = 2$ and $t = 2\cdot1$; (*c*) between $t = 1\cdot7$ and $t = 2\cdot3$.
     (v) Estimate the velocity when $t = 2$.
    (vi) Draw by eye, as accurately as possible, the straight line touching the graph at the point for which $t = 2$, and calculate its gradient (the tangent of the angle it makes with the t-axis).
   (vii) What do you infer by comparing the answers to (v) and (vi)?

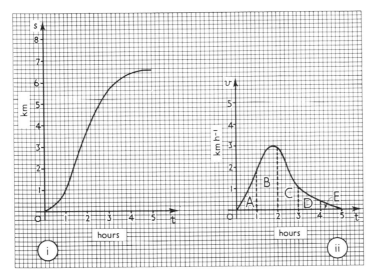

Fig. 1.5

3. **(D)** The graphs in Fig. 1.5 are those of (i) distance against time, (ii) velocity against time, for a certain journey.
   (*a*) By counting squares, estimate the areas marked A, B, C, D and E.
   (*b*) Look for relationships between these and the values of s for different values of t.
4. A particle moves in a straight line in such a way that the distance s metres it has travelled after t seconds is given by $s = \frac{1}{4}(t^2 + t)$. Plot accurately the $s(t)$ graph for values of t from 0 to 4.

Draw by eye straight lines 'touching' this graph at the points where $t = 0$, 1, 2, 3 and 4, and calculate their gradients.

Guess what these gradients represent and, on a fresh piece of graph paper, plot their values against t; estimate the area under this new graph, and compare it with the value of s when $t = 4$.

The gradient of a straight line which 'touches' a curve we provisionally call the 'gradient of the curve' at its point of contact. Later we shall give formal definitions of both *gradient* and *tangent*.

By this time, it is possible to make three fair guesses about motion in a straight line:

    (i) The gradient of *any* $s(t)$ curve at any point gives the velocity at that instant.

    (ii) The gradient of any $v(t)$ graph gives the acceleration (cf. Example 1.2).

    (iii) The area under any section of a $v(t)$ graph gives the distance travelled in the corresponding time.

We shall presently show these conjectures to be true, and also find a way of *calculating* the gradients and areas without having to draw 'accurate' graphs. The resulting theory can be applied alike to men, motor-cars and missiles—and this gives an indication of only one of the many applications of the 'calculus'.

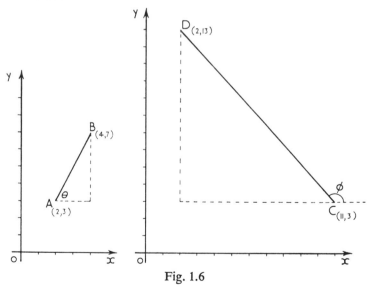

Fig. 1.6

Gradients of Straight Lines

Referring to rectangular axes Ox, Oy, in Fig. 1.6, the gradient of a straight line is the tangent of the angle which it makes with the positive x-axis. For example, the gradient of the line joining A (2, 3) to B (4, 7) is given by

$\tan \theta = \dfrac{7-3}{4-2}$ which equals 2. Similarly, the gradient of the line joining $C\,(11, 3)$ to $D\,(2, 13)$ is $\dfrac{3-13}{11-2}$, which equals $-\dfrac{10}{9}$, the tangent of the (obtuse) angle ϕ which CD makes with the *positive x*-axis.

More generally, Fig. 1.7, we consider the line joining the points $P_1(x_1, y_1)$ and $P_2(x_2, y_2)$. We denote the difference in the y-coordinates of the two points by δy ('delta y'), so that $\delta y = y_2 - y_1$; similarly, for the difference in the

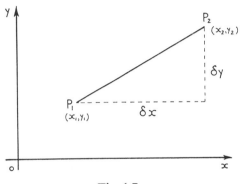

Fig. 1.7

x-coordinates, we write $\delta x = x_2 - x_1$. Then the gradient of the line P_1P_2 is $\dfrac{\delta y}{\delta x}$, i.e. $\dfrac{y_2 - y_1}{x_2 - x_1}$. $\left(\text{We could just as well have taken the difference } \delta y \text{ as } y_1 - y_2 \right.$ provided that we also took the x-coordinates in the same order, i.e. set $\delta x = x_1 - x_2$. The quotient $\dfrac{\delta y}{\delta x}$ would then have been $\dfrac{y_1 - y_2}{x_1 - x_2}$, which equals $\dfrac{y_2 - y_1}{x_2 - x_1}$ as before.$\left.\vphantom{\dfrac{y_1}{x_1}}\right)$

EXAMPLE 1.3 *To find the gradients of the lines joining the following pairs of points; (a) (5, 2) and (12, 7): (b) (3, −2) and (7, 10): (c) (4, 7) and (6, −2).*

(a) For (5, 2) and (12, 7), the difference in the y-coordinates is given by $\delta y = 7 - 2 = 5$, and the difference in the corresponding x-coordinates is $\delta x = 12 - 5 = 7$. Thus the gradient is given by $\dfrac{\delta y}{\delta x} = \dfrac{5}{7}$.

(b) The gradient is given by $\dfrac{\delta y}{\delta x} = \dfrac{10 - (-2)}{7 - 3} = \dfrac{12}{4} = 3$.

(c) The gradient is $\dfrac{\delta y}{\delta x} = \dfrac{-2 - 7}{6 - 4} = -\dfrac{9}{2}$.

EXERCISE II

1. (D) Using the 'delta' notation, δy, δx, find the gradients of the lines joining the following pairs of points: (a) (7, 12), (4, 9); (b) (−3, 2), (4, 1); (c) (−2, −7), (6, −3); (d) (0, 6), (4, 0).

The remainder of this exercise can be omitted if the class is already familiar with gradients of straight lines. Otherwise, questions 3 to 8 are suitable for class-discussion.

2. Find the gradients of the lines joining the following pairs of points: (a) (5, 4) (8, 9); (b) (6, 3), (10, 1); (c) (−1, 2), (7, 4); (d) (−4, −6), (4, 3); (e) (−5, −1), (−14, 3).

3. On the same sheet of graph paper, draw accurately the graphs of the straight lines $y = x$, $y = 2x$, $y = 5x$, $y = \frac{1}{3}x$, $y = -3x$, and find their gradients.

4. Sketch the graphs of $y = 2x$, $y = 2x + 1$, $y = 2x - 3$.

5. If m, c are any constants, what is represented by $y = mx + c$?

6. Verify that $y - 7 = 3(x - 2)$ represents a straight line passing through the point (2, 7). What is its gradient?

7. Write down the equations of the straight lines through the given points with the given gradients:
   (a) through (6, 2), gradient 4;   (b) through (−2, 4), gradient 7;
   (c) through (2, −3), gradient $-\frac{1}{2}$.

8. Verify that the equation $y - y_1 = m(x - x_1)$ represents a straight line which passes through the point (x_1, y_1) and whose gradient is m.

9. Find the gradients of the lines joining the following pairs of points: (a) (3, 4) (8, 16); (b) (2, −4), (−1, 3); (c) (0, −3), (−4, −7); (d) (p, q), (r, s).

10. Sketch the graphs of the following on the same sheet of paper, and give their gradients: (a) $y = 2x + 1$; (b) $y = 2x - 7$; (c) $y = \frac{1}{2}x - 3$; (d) $y = -\frac{1}{2}x + 1$.

11. Write down the equations of the straight lines through the given points with the given gradients:
   (a) through (3, 4), gradient 2;   (b) through (−4, 7), gradient 1/3;
   (c) through (3, −1), gradient −2/5.

Gradients of Curves

Drawing a line 'by eye' to touch a curve (as in Exercise I) is all right for guess-work, but clearly something more precise is required for a theory. We shall require a proper definition of the *tangent* to a curve at a given point.

We start with the familiar parabola $y = x^2$ (Fig. 1.8). Let P be the point (2, 4) on the curve, and Q_1 the point (5, 25). The gradient of the chord PQ_1 is $\dfrac{NQ_1}{PN} = \dfrac{25 - 4}{5 - 2}$, $= 7$. We now take successive points Q_2, Q_3 etc. (Fig. 1.9), nearer and nearer to P:

(a) $Q_2(4, 16)$; gradient of $PQ_2 = \dfrac{16 - 4}{4 - 2} = 6,$

(b) $Q_3(3, 9)$; gradient of $PQ_3 = \dfrac{9 - 4}{3 - 2} = 5,$

(c) $Q_4(2{\cdot}5, 6{\cdot}25)$; gradient of $PQ_4 = 4{\cdot}5,$

(d) $Q_5(2{\cdot}1, 4{\cdot}41)$; gradient of $PQ_5 = 4{\cdot}1,$

(e) $Q_6(2{\cdot}0001, 4{\cdot}00040001)$; gradient of $PQ_6 = 4{\cdot}0001,$

(f) $Q_7(2{\cdot}0000001, 4{\cdot}00000040000001)$; gradient of $PQ_7 = 4{\cdot}0000001.$

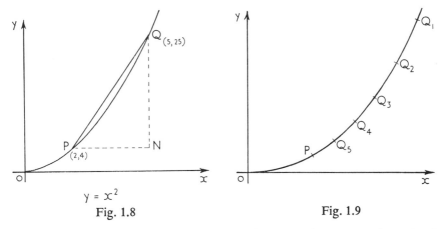

Fig. 1.8   Fig. 1.9

We conclude that, the nearer we take a point Q on the curve to the point P (2, 4), the closer the gradient of the chord PQ gets to the 'limiting value' 4, which could fairly be described as the gradient 'at P'. This suggests that we should define the *gradient* at a point P of a curve as the limit of the gradient of a chord PQ as Q moves along the curve towards P, and the *tangent* at P as the line through P with this gradient.

So far, we have found that the gradient of the curve $y = x^2$ at (2, 4) is 4. This was a haphazard choice; it would be far more useful to find a general recipe for the gradient at *any* point on the parabola.*

Instead of (2, 4), we shall take the coordinates of P as (x, y); and instead of taking in turn the points $Q_1, Q_2, \ldots$, we shall let the coordinates of a neighbouring point Q on the curve be $(x + \delta x, y + \delta y)$. Thus, δx is the difference between the x-coordinates of P and Q, so that, for example, when we had $x = 2$ above, δx for PQ_6 was 0·0001. δy is the corresponding small increment of y (e.g. for PQ_6, $\delta y = 0{\cdot}00040001$).

If P is any point (x, y) on the parabola $y = x^2$ (Fig. 1.10) and Q is a neighbouring point $(x + \delta x, y + \delta y)$ also on the curve, then the gradient of the chord PQ is $\dfrac{\delta y}{\delta x}$.

* It may be helpful to work Exercise III, questions 4 and 5 before proceeding to generalize.

Since both points lie on the parabola,

$$y = x^2$$

and

$$y + \delta y = (x + \delta x)^2$$

Hence

$$\delta y = (x + \delta x)^2 - x^2$$

$$= 2x\,\delta x + (\delta x)^2$$

and

$$\frac{\delta y}{\delta x} = 2x + \delta x.$$

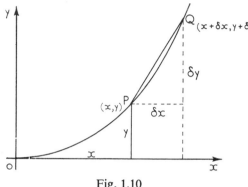

Fig. 1.10

Now, as Q moves towards P, $\delta x \to 0$ (this reads 'δx tends to zero'); that is, by taking Q close enough to P, we can make δx as small as we please. Thus the limit of this gradient is $2x$, so that *the gradient at any point P (x, y) on*

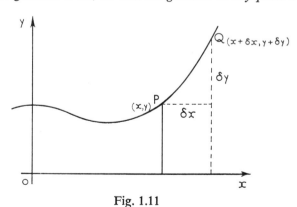

Fig. 1.11

the parabola $y = x^2$ is given by $2x$, that is, twice the x-coordinate of the point.
This method can, of course, be applied to other curves (Fig. 1.11). The gradient at $P(x, y)$ is again the limit as $\delta x \to 0$ of $\dfrac{\delta y}{\delta x}$, where $Q(x + \delta x, y + \delta y)$

is a neighbouring point on the curve. This may be written more shortly as 'gradient $= \lim\limits_{\delta x \to 0} \dfrac{\delta y}{\delta x}$', but this is still rather clumsy, and we write

$$\frac{dy}{dx} \text{ as short for } \lim_{\delta x \to 0} \frac{\delta y}{\delta x}.$$

$\dfrac{dy}{dx}$ is simply a notation, signifying the gradient of the curve in question. It is *not* to be considered here as a ratio, as $\dfrac{\delta y}{\delta x}$ is, but just as a handy way of expressing 'the limit as $\delta x \to 0$ of $\dfrac{\delta y}{\delta x}$'. Thus if $y = x^2$, $\dfrac{dy}{dx} = 2x$. Alternatively, this can be written

$$\frac{d}{dx}(x^2) = 2x,$$

or *the gradient of x^2 is $2x$.*

EXAMPLE 1.4   *To find the gradient of the curve $y = 2x^2 - 7$ at the point $(2, 1)$.*

Method 1. Let a neighbouring point on the curve be $(2 + \delta x, 1 + \delta y)$. Then $1 + \delta y = 2(2 + \delta x)^2 - 7$, so that $\delta y = 2\{4 + 4\delta x + (\delta x)^2\} - 8 = 8\delta x + 2(\delta x)^2$.

The gradient at $(2, 1)$ is given by

$$\frac{dy}{dx} = \lim_{\delta x \to 0} \frac{\delta y}{\delta x} = \lim_{\delta x \to 0}(8 + 2\delta x) = 8.$$

Method 2. Let $P(x, y)$ be any point on the curve $y = 2x^2 - 7$, and let $Q(x + \delta x, y + \delta y)$ be another point on the same curve. Then we have

$$y + \delta y = 2(x + \delta x)^2 - 7$$

as well as
$$y = 2x^2 - 7$$

so that
$$\delta y = \{2(x + \delta x)^2 - 7\} - \{2x^2 - 7\}$$
$$= 4x : \delta x + 2(\delta x)^2.$$

The gradient at P is

$$\frac{dy}{dx} = \lim_{\delta x \to 0} \frac{\delta y}{\delta x} = \lim_{\delta x \to 0}(4x + 2\delta x) = 4x.$$

The gradient at $(2, 1)$ is obtained by putting $x = 2$ in this formula, i.e. $4 \times 2 = 8$.

EXAMPLE 1.5   *To find the gradient of $y = -3x + 6$.*

If $P(x, y)$ and $Q(x + \delta x, y + \delta y)$ lie on the curve, we have $y = -3x + 6$, $y + \delta y = -3(x + \delta x) + 6$, so $\delta y = -3\delta x$ and $\dfrac{dy}{dx} = \lim\limits_{\delta x \to 0} \dfrac{\delta y}{\delta x} = -3$. The

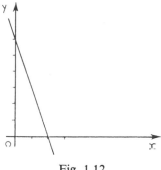

Fig. 1.12

gradient is therefore the same at all points on the 'curve' (Fig. 1.12), which in fact is a straight line.*

EXAMPLE 1.6   *To find the gradient of* $4x^2 - 7x + 3$.

Let $y = 4x^2 - 7x + 3$. Then, proceeding as before,
$$y + \delta y = 4(x + \delta x)^2 - 7(x + \delta x) + 3,$$
so that $\delta y = \{4(x + \delta x)^2 - 7(x + \delta x) + 3\} - \{4x^2 - 7x + 3\}$
$$= 8x \cdot \delta x + 4(\delta x)^2 - 7\delta x$$
$$= (8x - 7)\delta x + 4(\delta x)^2$$

and $\dfrac{dy}{dx} = \lim\limits_{\delta x \to 0} \dfrac{\delta y}{\delta x} = 8x - 7$, which is the required gradient.

EXAMPLE 1.7†   *To sketch the graph of* $y = \dfrac{1}{x^2}$, *and find an expression for* $\dfrac{dy}{dx}$ *at any point on the curve.*

(i) y cannot be negative $\left(\text{being the square of } \dfrac{1}{x}\right)$.

(ii) As $x \to \infty$, $y \to 0$ ('as x tends to infinity, y tends to zero'). There is no number 'infinity'; this is simply a short way of saying that however small a number you care to mention, I can find N such that y is numerically smaller than your number whenever $x > N$.‡ $\left(\text{e.g. if you choose } 10^{-6}, \text{I put } N = 1000,\right.$ and for *all* $x > 1000$, $y < \dfrac{1}{10^6}\Big)$.

---

* The reader will have realized this; but it is perhaps reassuring that the general method we are building up 'works' in this familiar circumstance.
† The reader should not be dismayed if this example proves indigestible; he can return to it later (cf. p. 29).
‡ If this sounds unnecessarily complicated, cf. Exercise III, No. 25, p. 13.

(iii) As $x \to 0, y \to \infty.$ $\left(\text{If } x = \dfrac{1}{100}, y = 10^4, \text{and for all } x < \dfrac{1}{100}, y > 10^4. \right)$

(iv) If we replace x by $-x$, the value of y is unaltered. For example, $(4, \frac{1}{16})$ is on the curve and so is $(-4, \frac{1}{16})$. Thus the curve is symmetrical about the y-axis.

By this time, we have enough 'clues' to give a rough sketch of the curve (Fig. 1.13).

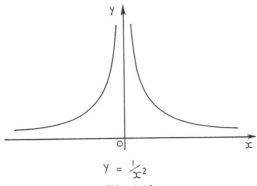

$$y = \tfrac{1}{x^2}$$

Fig. 1.13

To find $\dfrac{dy}{dx}$, we have $y = \dfrac{1}{x^2}, y + \delta y = \dfrac{1}{(x + \delta x)^2}$,

so $\quad \delta y = \dfrac{1}{(x + \delta x)^2} - \dfrac{1}{x^2} = \dfrac{x^2 - (x + \delta x)^2}{x^2(x + \delta x)^2} = -\dfrac{(2x + \delta x)\,\delta x}{x^2(x + \delta x)^2}$,

and

$$\frac{dy}{dx} = \lim_{\delta x \to 0} \frac{\delta y}{\delta x} = \lim_{\delta x \to 0} \left[-\frac{2x + \delta x}{x^2(x + \delta x)^2} \right] = \lim_{\delta x \to 0} -\frac{2x\left(1 + \dfrac{\delta x}{2x}\right)}{x^4\left(1 + \dfrac{\delta x}{x}\right)^2} = -\frac{2}{x^3}.$$

When $x > 0, \dfrac{dy}{dx} = -\dfrac{2}{x^3} < 0$ and as $x \searrow 0$ (that is approaches zero 'from above', i.e. through positive values), $\dfrac{dy}{dx} \to -\infty$*, while if $x \to \infty, \dfrac{dy}{dx} \to 0.$

These facts provide a good check for our sketch.

EXERCISE III

1. Give a rough sketch of the $s(t)$ and $v(t)$ graphs of a school-boy who starts slowly for school, panics because he thinks he's late and then dawdles again as he approaches the school gate.

---

* The reader should write a sentence, as in (ii) above, to satisfy himself that he knows clearly the meaning of 'tending to minus infinity'.

2. Treat similarly a more complicated journey (e.g. he could set out and have to return for a forgotten book, jump on a bus, etc.).

3. Using the same axes, sketch the graphs of the straight lines $y = x$, $y = -x$, $y = 3x + 1$, $y = 4 - 7x$. State the gradient of each.

4. P is the point $(1, 4)$ on the curve $y = 3x^2 + 1$ (Fig. 1.14). Find the gradient of the chord PQ, when the x-coordinate of Q has the following values: (i) 4, (ii) 3, (iii) 2, (iv) 1·5, (v) 1·2, (vi) 1·1, (vii) 1·01. What is the gradient of the curve at P?

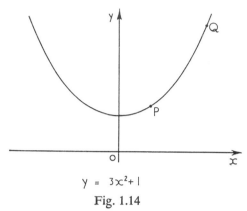

$$y = 3x^2 + 1$$

Fig. 1.14

5. P is the point $(2, 10)$ on the curve $y = x^3 + x$. Find the gradient of the chord PQ when the x-coordinate of Q has the values (i) 2·5, (ii) 2·1, (iii) 2·01, (iv) 2·001. What is the gradient of the curve at P?

6. Find the gradient of (a) $2x$, (b) $3x + 5$, (c) $7x^2 - 3$, (d) $5x^2 - 4x + 2$. [First guess the answers and then justify them.]

7. Find the gradient of (a) $y = 8x + 17$, (b) $y = 17$.

8. Find (i) where $y = 3x^2 - 12x$ meets the x-axis (i.e where $y = 0$), (ii) the gradient of the curve at $(2, -12)$. Give a rough sketch of this curve.

9. Verify that the gradient of kx^2 is $2kx$, where k is any constant. Sketch the graphs of $y = x^2$, $y = 2x^2$, $y = 5x^2$, $y = -x^2$, $y = 2x^2 + 1$.

10. (i) Verify that $(x + \delta x)^3 = x^3 + 3x^2(\delta x) + 3x(\delta x)^2 + (\delta x)^3$.
    (ii) Sketch the graph of $y = x^3$, and find the gradient at $(3, 27)$ and at $(-3, -27)$.

11. (i) Verify that $(x + \delta x)^4 = x^4 + 4x^3(\delta x) + 6x^2(\delta x)^2 + 4x(\delta x)^3 + (\delta x)^4$.
    (ii) Find the gradient of x^4.
    (iii) Guess the gradient of x^n, where n is any positive integer.

12. Find the gradient of $\dfrac{1}{x}$, and sketch the shape of the curve $y = \dfrac{1}{x}$ (cf. Example 1.7).

13. Using the same pair of axes, sketch the curves $y = \dfrac{1}{x^n}$ for $n = 1, 2, 3, 4, 5$.

14. If a small number of bacteria is put into a closed vessel, they will first breed slowly (as there are few of them), then more rapidly (as there become more available to breed) and finally more slowly again (as they become cramped in

the full vessel).* Re-word this statement in terms of the gradient of the graph of number of bacteria present plotted against time, and give a rough sketch of this graph.

The remainder of this exercise is suitable for a 'refresher'.

15. (D) What sort of problems can the calculus help to solve?

16. Give the gradients of the following straight lines: (a) $y = 7x - 5$, (b) $y - 4 = \frac{1}{2}(x + 1)$, (c) the line joining (8, 3) to (11, 7), (d) the join of $(-3, 5)$ and $(12, -8)$, (e) $3y + 4x = 7$.

17. P is the point (2, 9) on the curve $y = x^3 + 1$ (Fig. 1.15). Find the gradient of the chord PQ, when the x-coordinate of Q has the following values: (i) 5; (ii) 4; (iii) 3; (iv) 2·5; (v) 2·1; (vi) 2·01. What is the gradient of the curve at P?

18. P is the point (3, 6) on the curve $y = x^2 - x$. Find the gradient of the chord PQ when the x-coordinate of Q has the values: (i) 4, (ii) 3·5, (iii) 3·1, (iv) 3·01, (v) $3 + h$. What is the gradient of the curve at P?

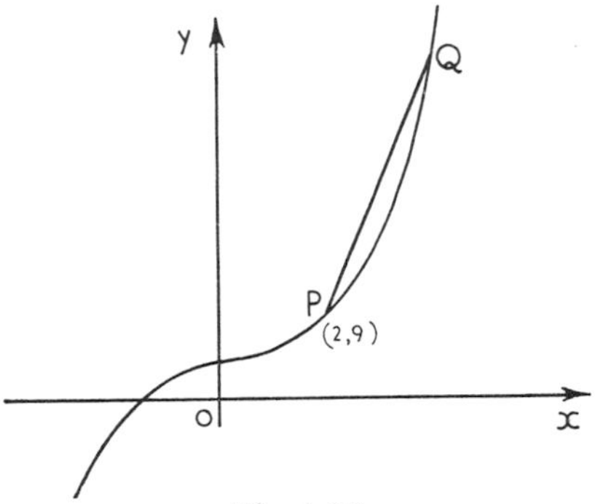

Fig. 1.15

19. Expand the following expressions involving x and δx. (i) $(x + \delta x)^2$, (ii) $(3x + 2\delta x)^2$, (iii) $\delta x(x + \delta x)$, (iv) $x(x + \delta x)$, (v) $4x(2x + 3\delta x)^2$, (vi) $(x + 3\delta x)(2x - 5\delta x)$, (vii) $(x + \delta x)^3$.

20. Simplify:

   (i) $3(x + \delta x)^2 - 3x^2$

   (ii) $\dfrac{3(x + \delta x)^2 - 3x^2}{\delta x}$

   (iii) $\lim\limits_{\delta x \to 0} \dfrac{3(x + \delta x)^2 - 3x^2}{\delta x}$

   Sketch the graph of $y = 3x^2$, and interpret the meaning of each of the above expressions with reference to it. What is the gradient of $3x^2$?

21. Using the same pair of axes, sketch the curves $y = x^n$, for $n = 1, 2, 3, 4, 5, 6$.

22. (i) Verify that
   $$(x + \delta x)^5 = x^5 + 5x^4(\delta x) + 10x^3(\delta x)^2 + 10x^2(\delta x)^3 + 5x(\delta x)^4 + (\delta x)^5.$$
   (ii) Find the gradient of x^5.

23. Find the gradient of (a) $3x^2 + 4x + 1$; (b) $5x^2 - 3x$.

24. Sketch the graph of $y = \dfrac{1}{x + 1}$, and find $\dfrac{dy}{dx}$.

25. (D) Criticize the statement: '*As $x \to \infty$, $y \to 0$' means that as x gets larger and larger, y gets smaller and smaller.*

---

* Cf. Steinhaus, *Mathematical Snapshots* (O.U.P.), p. 249.

[Consider the following examples:

(i) Suppose $y = 1 + \dfrac{1}{x}$. When x gets larger and larger, y gets 'smaller and smaller', but nevertheless remains greater than 1.

(ii) Let y be defined as follows. Suppose $y = \dfrac{1}{x^2}$ if x is not an integral power of 10, but $y = \dfrac{1}{x}$ if x is an integral power of 10. In this case, y does not get 'smaller and smaller' as x gets larger and larger; for example, when $x = 90$, $y = \dfrac{1}{8100}$, but when $x = 100$, $y = \dfrac{1}{100}$. Nevertheless, as $x \to \infty$, $y \to 0$.]

2
Differentiation

EXAMPLE 2.1 *A sketch of the graph of y = |x|, and an investigation of its
gradient.*

'$|x|$', or 'mod x' (short for 'the modulus of x'), means the numerical value
of x, discarding the minus sign if x is negative; e.g. $|2·5| = 2·5, |-7·1| = 7·1$.
The graph of $y = |x|$ is thus the same as that of $y = x$ if $x \geqslant 0$ and the same
as $y = -x$ if $x \leqslant 0$, and so forms a 'V' shape (Fig. 2.1).

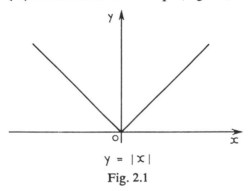

$$y = |x|$$

Fig. 2.1

If $x > 0$, the gradient at any point is $+1$; if $x < 0$, it is -1. But at $x = 0$
there is no obvious value for the gradient; in fact, we shall see that it doesn't
exist. This example suggests that we must look more closely still at the
meaning of *gradient*.

Again, consider the graph of $y^2 = x$. When $x = 4, y = \pm 2$ and the
'gradient' (if it existed) might be considered as either positive or negative.

Such ambiguities must be ruled out, and to clarify our ideas we must
express them in more general terms. So far we have worked out separately
the gradient of x^2, $3x + 1$, $\frac{1}{x}$ and so on. Each of these is a *function* of x, that
is, it has a *unique* value corresponding to certain values of x. (For $\frac{1}{x}$, we
must exclude $x = 0$, since $\frac{1}{0}$ is meaningless.)

15

We shall write $f(x)$ to denote a function of x.* If we wish to consider a particular function, say that which associates $x^2 + 4$ with x, we write $f(x) = x^2 + 4$. We can then substitute particular values for x; in this example, $f(7) = 7^2 + 4 = 53$, $f(0) = 4$, $f(-1) = 5$, and so on.

The values of x for which a function is defined constitute its *domain*; the set of corresponding values of $f(x)$ is the *range*.

As an example, we shall take $\sqrt{x}$ to mean the *positive* square root of x, so that $\sqrt{x}$ is a function of x whose domain consists of all numbers greater than or equal to zero.† Its graph is *only the part of the curve above the x-axis* in Fig. 2.2.

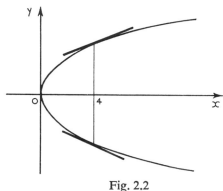

Fig. 2.2

EXERCISE IV

1. If $f(x) = x^2 - 4x + 1$, write down the values of (i) $f(3)$, (ii) $f(0)$, (iii) $f(-2)$, (iv) $f(a)$, (v) $f(a + h)$.

2. If $f(x) = x^2 - 4$, $g(x) = x - 2$, write down the values of (i) $f(0)$, (ii) $\dfrac{f(5)}{g(4)}$.

3. If $f(x) = x - \dfrac{2}{x}$, give the values of (i) $f(2)$, (ii) $f(y)$, (iii) $f(x + 1)$, (iv) $f(x + h)$, (v) $f(3)$, (vi) $f(-3)$. Show also that $f(-x) = -f(x)$.

4. Give an example of a function for which $f(x) = f(-x)$.

5. If $f(x) = \frac{1}{2}x(x + 1)$, give the values of $f(4)$, $f(8)$, and write down expressions for $f(x - 1)$, $f(-x)$ and $f(2x)$.

6. Give the domains over which the following functions can be defined and their corresponding ranges:

   (a) $2x - 5$, (b) $x^2 + 1$, (c) $\dfrac{1}{x - 2}$, (d) $\sqrt{(3 - x)}$, (e) $\sqrt{\{(x - 2)(x - 5)\}}$,

   (f) $\dfrac{1}{x} + \dfrac{1}{x - 3}$.

---

* Some authors prefer to write 'f is the function that maps x onto $f(x)$' or $f\colon x \rightarrow f(x)$. For them, $f(x)$ is the image of x.

† If, instead, $\sqrt{x}$ were defined as 'plus or minus the square root of x', it would not be a function of x for $x > 0$, since to any such x there would correspond two values, e.g. $x = 9$ would yield '$\sqrt{9} = \pm 3$' and our definition demands that there should be only one.

EXAMPLE 2.2 *To find the gradient of the curve $y = 2x^2 - 7$ at the point* $P(2, 1)$.

This is Example 1.4. We work it again to familiarize the new notation.

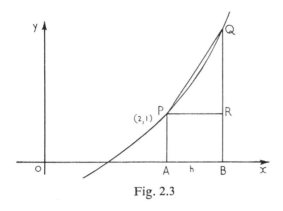

Fig. 2.3

We start by noting that $2x^2 - 7$ is indeed a function of x. We may therefore consider $f(x) = 2x^2 - 7$. Let Q be a point on the curve whose x-coordinate is $2 + h$ (Fig. 2.3). Then the gradient at P is

$$\lim_{h \to 0} \frac{RQ}{PR} = \lim_{h \to 0} \frac{BQ - BR}{PR}$$

$$= \lim_{h \to 0} \frac{f(2 + h) - f(2)}{h}$$

$$= \lim_{h \to 0} \frac{\{2(2 + h)^2 - 7\} - (2 \cdot 2^2 - 7)}{h}$$

$$= \lim_{h \to 0} \frac{8h + 2h^2}{h}$$

$$= \lim_{h \to 0} (8 + 2h), = 8, \text{ as before.}$$

Notice that we did not actually put $h = 0$ during the working, otherwise $\frac{8h + 2h^2}{h}$ would be $\frac{0}{0}$ which is meaningless, instead of equal to $8 + 2h$. But as h gets as small as we please, without actually reaching zero, the gradient of the chord, $\frac{8h + 2h^2}{h}$, is certainly equal to $8 + 2h$, and so tends to the limiting value 8.

More generally, if $y = f(x)$, the gradient at $P(x, y)$ is the limit as $h \to 0$ of the slope of the chord PQ, the coordinates of Q being $(x + h, f(x + h))$. $RQ = BQ - BR = BQ - AP = f(x + h) - f(x)$, and so the gradient is

$$\lim_{h \to 0} \frac{f(x + h) - f(x)}{h}$$

This expression for the gradient of $f(x)$ can be evaluated for a particular function without reference to its graph; we shall use it from now on as our final definition of the gradient of $f(x)$, and when a gradient is to be worked

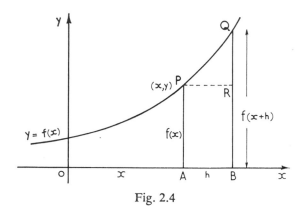

Fig. 2.4

out 'from first principles', this will mean by starting from the above expression.

EXAMPLE 2.3   *Find the gradient of* $6x^2 - 5x$.

Let $f(x) = 6x^2 - 5x$. Then the required gradient is given by

$$\lim_{h \to 0} \frac{f(x + h) - f(x)}{h} = \lim_{h \to 0} \frac{\{6(x + h)^2 - 5(x + h)\} - (6x^2 - 5x)}{h}$$

$$= \lim_{h \to 0} \left(\frac{12xh + 6h^2 - 5h}{h} \right)$$

$$= \lim_{h \to 0} (12x + 6h - 5)$$

$$= 12x - 5.$$

EXAMPLE 2.4   *To find the gradient of* $x + \dfrac{1}{x - 2}$.

Let $f(x) = x + \dfrac{1}{x - 2}$. (We must assume that $x \neq 2$, for $f(2)$ is meaningless.)

Then $\lim\limits_{h\to 0}\dfrac{f(x+h)-f(x)}{h} = \lim\limits_{h\to 0}\dfrac{\left(x+h+\dfrac{1}{x+h-2}\right)-\left(x+\dfrac{1}{x-2}\right)}{h}$

$$= \lim_{h\to 0}\left\{\frac{h+\dfrac{(x-2)-(x+h-2)}{(x+h-2)(x-2)}}{h}\right\}$$

$$= \lim_{h\to 0}\left\{1-\frac{1}{(x+h-2)(x-2)}\right\}$$

$$= 1-\frac{1}{(x-2)^2}.$$

Exercise V

1. Find the gradient of $4x^2 - 7x + 3$, using the '$f(x)$' notation, as in Example 2.3 above (cf. Example 1.6).
2. Find, from first principles, the gradient of $f(x)$ (i) when $f(x) = x^3$; (ii) when $f(x) = 3x^2 - 2$; (iii) when $f(x) = \dfrac{1}{x+2}$; (iv) when $f(x) = 2x - \dfrac{1}{x-4}$.

Limits and Continuity

There is a danger that

$$\lim_{h\to 0}\frac{f(x+h)-f(x)}{h}$$

will become a meaningless slogan, to be brought out whenever the word 'gradient' is mentioned. We now therefore digress briefly to re-examine some of the underlying ideas. This will give a bare preview of the rigorous treatment given in more advanced courses, but at least perhaps show the need for it.*

First, the idea '$h \to 0$' here implies that, as h changes from one value to a numerically smaller one, it takes up 'all' values in between. Just what do we mean by 'all'?

We consider, as a specimen interval, the set of numbers x such that $1 \leqslant x \leqslant 2$. By dividing the interval up successively into halves, thirds, and so on (Fig. 2.5), we obtain all the numbers between 1 and 2 of the form $\dfrac{p}{q}$, where p, q are integers. Such numbers are called *rational*. (For example, $1\cdot3 = 1\cdot333\ldots$ is rational, as it can be expressed as $\frac{4}{3}$.) Between any pair of rational numbers, however closely chosen, there is another rational

---

* Learning to speak, small children first pick up a number of words, and only later stop to analyse their precise meaning. Similarly here it is time to stop and at least reflect a little on what we are really talking about.

number (e.g. their average), so in any interval, however small, there is an infinity of rationals. All the same, the rationals do not account for all the numbers in any interval; the others, numbers which cannot be expressed in the form $\frac{p}{q}$, where p, q are integers, are called *irrational*. The rational and irrational numbers together form the set of *real* numbers.

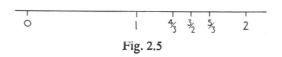

Fig. 2.5

EXAMPLE 2.5   *To prove that $\sqrt{2}$ is irrational.**

We give Euclid's famous proof, as a classical example of 'reductio ad absurdum'. Assume, then, that $\sqrt{2} = \frac{a}{b}$, where a, b are certain integers which are 'mutually prime' (i.e. the fraction has been 'cancelled' as far as possible, so that the H.C.F. of a, b is 1). Then, squaring, we have $\frac{a^2}{b^2} = 2$, so $a^2 = 2b^2$. Thus a^2 is even, and it follows that a is even. Let $a = 2c$, c thus also being an integer. Then $2b^2 = a^2 = 4c^2$, so $b^2 = 2c^2$. Thus b^2 is even, and so b is even.

As a, b are both even, we have our required contradiction, since we started with the understanding that they were mutually prime. It follows that the original assumption $\left(`\sqrt{2} = \frac{a}{b} \ldots ` \right)$ is false, i.e. $\sqrt{2}$ is irrational.

EXAMPLE 2.6 (*Harder*)   *If x is a real number such that $x^p = q$, where p, q are any positive integers, then x is either an integer or irrational.*

Suppose that $x = \frac{a}{b}$, where a, b are mutually prime integers and $b > 0$. Then $\left(\frac{a}{b} \right)^p = q$, so $a^p = qb^p = b(qb^{p-1})$. If $b > 1$, any prime factor k of b is a factor of a^p, and so of a.† But this contradicts the assumption that a, b are mutually prime. It follows that, if $x = \frac{a}{b}$, then $b = 1$; in other words, if x is rational, then it is an integer. The result follows.

---

* This proof may well be familiar. If so, the reader might tackle Example 2.6; this is a first generalization, from which the irrationality of numbers like $\sqrt{3}$, $\sqrt{7}$ can immediately be deduced.

† We are assuming here the 'Fundamental Theorem of Arithmetic', that any positive integer except 1 can be written as a product of prime factors in one way only apart from the order of the factors.

When we write '$f(x)$', we shall be considering a function of a real variable, x. Unless otherwise stated, we shall assume that x can take 'all' values (i.e. rational and irrational) in a certain interval, that is, it is a *continuous* variable.

We next look a little more closely at the meaning of 'limit'.

EXAMPLE 2.7   *To find* $\lim_{x \to 2} (x^2 + 3)$.

The answer is 'clearly' 7, since if x is given values closer and closer to 2, then $x^2 + 3$ gets nearer and nearer to 7. This, however, is careless talk. (How close is 'closer and closer'? Do we mean for *all* values of x close enough to 2? How near to 7 is near enough?)

If A writes, '$\lim_{x \to 2} (x^2 + 3) = 7$', this implies that if x is *any* number within a close enough range of 2, then $x^2 + 3$ is as near to 7 as any challenger, B, pleases. The 'close enough' range for x centred on 2 *does not necessarily apply to 2 itself*, as 'tending to 2' does not involve actually reaching it. The argument might go as follows.

B. You reckon $\lim_{x \to 2} (x^2 + 3) = 7$. Very well, I name 7 ± 0.01 as the boundaries for $x^2 + 3$. You find an interval centred on 2 such that, for all x in that interval (except possibly 2 itself), $x^2 + 3$ lies between $7 + 0.01$ and $7 - 0.01$.

A. Certainly. If x is between 2 ± 0.001, your condition is satisfied.

B. All right. Now I name 7 ± 0.0001 as the boundaries for $x^2 + 3$ . . . . . . (and so on).

If A is sure he can answer *any* such challenge, and come up with a corresponding interval for x, only then can he correctly assert that

$$\text{'}\lim_{x \to 2} (x^2 + 3) = 7\text{'}.$$

EXAMPLE 2.8   *To examine* $\lim_{x \to 2} \left(\dfrac{x^2 - 4}{x - 2} \right)$.

This is not quite so 'obvious' as the last example, for $\dfrac{x^2 - 4}{x - 2}$ is meaningless when $x = 2$. All the same, if $x \neq 2$, $\dfrac{x^2 - 4}{x - 2} = x + 2$, and, as $x \to 2$, this expression tends to 4. A could stand up to any challenge by B, the conversation running similarly to that of the previous example. The reader should verify this, remembering that as x is *tending* to 2 what happens, or fails to happen, if $x = 2$ doesn't affect the argument. Thus $\lim_{x \to 2} \dfrac{x^2 - 4}{x - 2} = 4$.

The graph of $y = \dfrac{x^2 - 4}{x - 2}$ is a straight line 'with a hole in it', that is, the line whose equation is $y = x + 2$ with a point missing at $x = 2$.

[If we had started instead with the equation $y(x - 2) = x^2 - 4$, this would have been equivalent to 'either $x = 2$ or $y = x + 2$', and the graph would then have consisted of the whole of the two straight lines $x = 2$ and $y = x + 2$.]

EXAMPLE 2.9   *To investigate* $\lim\limits_{h \to 0} \left(\dfrac{|h|}{h} \right)$.

This limit does not exist, for if $h > 0$, $\dfrac{|h|}{h} = 1$, while if $h < 0$, $\dfrac{|h|}{h} = -1$. It would therefore be impossible to find an interval for h centred on 0 such that $\dfrac{|h|}{h}$ was 'as near as you please' to any given number.

This last example shows that $|x|$ does not have a gradient when $x = 0$ (cf. Example 2.1), for the following problems are equivalent:

   (i) Investigate the gradient of $|x|$ at $x = 0$,

   (ii) Investigate $\lim\limits_{h \to 0} \dfrac{f(x + h) - f(x)}{h}$ if $f(x) = |x|$, when $x = 0$,

   (iii) Investigate $\lim\limits_{h \to 0} \dfrac{f(h) - f(0)}{h}$ if $f(x) = |x|$,

   (iv) Investigate $\lim\limits_{h \to 0} \left(\dfrac{|h|}{h} \right)$;

and we have just seen that the last limit doesn't exist.

We can now return with more conviction to our definition of the gradient of a function $f(x)$ for a particular value of x, say $x = a$. To avoid having to write $y = f(x)$ and then refer to $\dfrac{dy}{dx}$ $\left(\text{or to write clumsily '}\dfrac{d}{dx} \{f(x)\}' \right)$, we introduce the notation $f'(x)$ for the gradient of $f(x)$, and we have

$$f'(a) = \lim_{h \to 0} \frac{f(a + h) - f(a)}{h},$$

provided the limit on the right-hand side exists. (If it doesn't, then $f(x)$ has no gradient when $x = a$.)

The existence of the right-hand side implies that x can take all values within a certain range centred on a, and further that $f(a + h) - f(a) \to 0$ as $h \to 0$. This last condition means that the function is *continuous* at $x = a$[*]; that is, $f(a)$ exists, and by taking h small enough, $f(x) - f(a)$ is as small as you please for *all* x between $a \pm h$. But if a function is continuous at a point, this doesn't ensure that it has a gradient there (cf. $|x|$ at $x = 0$); it is, in fact, a necessary condition, but not a sufficient one.

---

[*] This is roughly equivalent to saying that the graph of $y = f(x)$ could be drawn in the neighbourhood of $x = a$ without taking the pencil off the paper.

EXAMPLE 2.10   *To find the value of x for which* $2x - \dfrac{1}{x-4}$ *is discontinuous and*

to sketch the graph of this function $\left(i.e.\ of\ y = 2x - \dfrac{1}{x-4} \right).$

The function is continuous everywhere except at $x = 4$, where it is undefined. For the graph, Fig. 2.6:

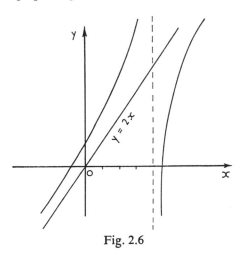

Fig. 2.6

(a) If x is just greater than 4, y is large and negative; if x is just less than 4, y is large and positive.

(b) If $x = 0$, $y = \tfrac{1}{4}$.

(c) As $x \to \infty$ or $-\infty$ (cf. p. 10), $\dfrac{1}{x-4} \to 0$, so the graph is very near to that of the straight line $y = 2x$ (below if $x > 4$, above if $x < 4$).

(d) If $y = 0$, $2x = \dfrac{1}{x-4}$, whence $x = 2 \pm \tfrac{3}{2}\sqrt{2}$, $\approx 4\cdot1$ or $-0\cdot1$.

EXERCISE VI

1. (D) Which of the following numbers are rational?
   (a) $0\cdot123$,   (b) $0\cdot\dot{1}2\dot{3}$ (i.e. $0\cdot123123123\ldots$),   (c) $0\cdot101001000100001\ldots$.

2. Evaluate the following limits: (a) $\lim_{x\to4} (3x^2 - 2)$; (b) $\lim_{x\to3} \dfrac{x^2 - 9}{x - 3}$; (c) $\lim_{x\to3} \dfrac{x^3 - 27}{x - 3}$
   [Hint: $x^3 - a^3 = (x - a)(x^2 + ax + a^2)$.]

3. (a) For what values of x does $\dfrac{x^2 - 16}{x - 4}$ differ from 8 by less than $0\cdot01$?

(b) What is the limit of $\dfrac{x^2 - 16}{x - 4}$ as $x \to 4$?

(c) Sketch the graph of $y = \dfrac{x^2 - 16}{x - 4}$.

4. Investigate whether the following functions have any discontinuities and sketch their graphs:

(a) $(x - 1)(x - 2)$; (b) $\dfrac{1}{x + 1}$; (c) $\dfrac{1}{(x + 1)^2}$; (d) $(x - 1)(x - 2)(x - 3)$;

(e) $(x - 1)(x - 1{\cdot}01)(x - 3)$; (f) $(x - 1)^2(x - 3)$; (g) $\dfrac{x - 1}{x + 1}$;

(h) $\dfrac{x - 1}{(x - 2)(x - 4)}$; (i) $\dfrac{(x - 1)(x - 3)}{(x - 2)(x - 4)}$; (j) $x + \dfrac{1}{3 - x}$.

(*The remainder of this exercise may be deferred for a second reading.*)

5. Prove that $\sqrt{3}$ is irrational.

6. Prove that if n is a positive integer which is not a multiple of 10, then $\log_{10} n$ is irrational. [Hint: suppose $\log_{10} n = \dfrac{a}{b}$, and compare Examples 2.5 and 2.6, p. 20; can the last digit of n^b be zero?]

7. Let $u_{n+1} = \dfrac{1}{2}\left(u_n + \dfrac{2}{u_n}\right)$, $(n = 1, 2, \ldots)$. Then, putting $u_1 = 1$, we have

$$u_2 = \frac{1}{2}\left(u_1 + \frac{2}{u_1}\right) = \frac{1}{2}\left(1 + \frac{2}{1}\right) = 1{\cdot}5, \quad u_3 = \frac{1}{2}\left(u_2 + \frac{2}{u_2}\right) = \ldots.$$

Calculate u_4, u_5 and u_6. [It can be shown that $u_n \to \sqrt{2}$ as $n \to \infty$; this demonstrates the important idea of an irrational number as the limit of a sequence of rational ones.]

8. (P) Find out what you can about the 'snowflake curve'* (see, for example, *Mathematical Models* by Cundy and Rollett, *Mathematics and the Imagination* by Kasner and Newman, or *Riddles in Mathematics* by Northrop).

Vocabulary

If $f(x)$ has a gradient, $f'(x)$, (cf. p. 22) the latter is another function of x, and to emphasize this it is often called the *gradient function*. Unfortunately, there are various words in use which mean more or less the same thing. This is simply because numbers of people worked independently to develop the calculus, and is not so confusing as might appear. Thus $f'(x)$ is sometimes called the *derivative*, *derived function*, or *differential coefficient* of $f(x)$†, and the process of finding the gradient function $f'(x)$ is also known as *differentiating* $f(x)$ with respect to x. If $f'(x)$ exists, $f(x)$ is said to be *differentiable*.

---

* This curve, admittedly a queer one, has the delightful property that it is continuous at every point, but has a gradient at none.

† 'Derivative' and 'derived function' now seem to be preferred.

EXERCISE VII (Oral)

We assume the following results, which have been worked out as examples in Chapter 1: (a) the gradient of $y = 2x^2 - 7$ is $4x$; (b) the gradient of $4x^2 - 7x + 3$ is $8x - 7$; (c) if $y = \dfrac{1}{x^2}$, $\dfrac{dy}{dx} = -\dfrac{2}{x^3}$.

1. If $f(x) = 2x^2 - 7$, give the values of: (i) $f(7)$; (ii) $f(-3)$; (iii) $f(a)$; (iv) $f'(x)$; (v) $f'(7)$; (vi) $f'(-3)$; (vii) $f'(a)$.
2. Differentiate $4x^2 - 7x + 3$ with respect to x.
3. Find the derived function of $\dfrac{1}{x^2}$.
4. If $g(x) = 4x^2 - 7x + 3$, evaluate: (i) $g'(1)$; (ii) $g'(0)$; (iii) $g'(\tfrac{7}{8})$.
5. Find the differential coefficient of $2x^2 - 7$ at $x = -2$.
6. What is the gradient function of $4x^2 - 7x + 3$?
7. If $F(x) = \dfrac{1}{x^2}$, evaluate: $F(a)$; $F(-a)$; $F(3 + h)$; $F'(\tfrac{1}{2})$.
8. Criticize the following statements:
    (a) 'Differentiate $4x^2 - 7x + 3$. Answer: $4x^2 - 7x + 3 = 8x - 7$';
    (b) '$\dfrac{dy}{dx}$ of $2x^2 - 7$ is $4x$.'

Differentiation of x^n

We next prove that, if n is a positive integer, the gradient of x^n is nx^{n-1}.*
The result in fact holds for all real values of n; we shall extend our proof to cover fractional and negative values in Chapter 6, but the extension to irrationals is beyond our present scope. We shall need the result of the Binomial theorem, that

$$(x + h)^n = x^n + nx^{n-1}h + c_2 x^{n-2} h^2 + c_3 x^{n-3} h^3 + \ldots + c_{n-1}x\, h^{n-1} + h^n.$$

This is proved in Appendix I, where the values of $c_2, \ldots, c_{n-1}$ are also given. All we need here is that they are numbers which are independent of x and h.
    Let $f(x) = x^n$; then

$$f'(x) = \lim_{h \to 0} \frac{f(x + h) - f(x)}{h}$$
$$= \lim_{h \to 0} \frac{(x + h)^n - x^n}{h}$$
$$= \lim_{h \to 0} \frac{(x^n + nx^{n-1}h + \ldots + c_{n-1}xh^{n-1} + h^n) - x^n}{h}$$

* For alternative proofs, cf. Exercise VIII, question 24, and Appendix I.

so
$$f'(x) = \lim_{h \to 0} (nx^{n-1} + c_2 x^{n-2} h + \ldots + h^{n-1})$$

$$= nx^{n-1} + \lim_{h \to 0} \{h(c_2 x^{n-2} + \ldots + h^{n-2})\}*$$

$$= nx^{n-1}, \text{ as required.}$$

Differentiation of $cf(x)$, $f(x) + g(x)$, **and** c

We have previously seen that $\dfrac{d}{dx}(4x^2 - 7x + 3) = 8x - 7$ (p. 10), so that

$$\frac{d}{dx}(4x^2 - 7x + 3) = 8x - 7$$

$$= \frac{d}{dx}(4x^2) - \frac{d}{dx}(7x) + \frac{d}{dx}(3)$$

$$= 4\frac{d}{dx}(x^2) - 7\frac{d}{dx}(x) + \frac{d}{dx}(3).$$

Again,

$$\frac{d}{dx}(6x^2 - 5x) = 12x - 5 \text{ (p. 18)}$$

$$= \frac{d}{dx}(6x^2) - \frac{d}{dx}(5x)$$

$$= 6\frac{d}{dx}(x^2) - 5\frac{d}{dx}(x).$$

These results suggest that in general

(i)
$$\frac{d}{dx}\{cf(x)\} = cf'(x) \qquad (c \text{ any constant}),$$

(ii)
$$\frac{d}{dx}\{f(x) + g(x)\} = f'(x) + g'(x).$$

To prove (i), we have

$$\frac{d}{dx}\{cf(x)\} = \lim_{h \to 0} \frac{cf(x + h) - cf(x)}{h} = \lim_{h \to 0} c\left\{\frac{f(x + h) - f(x)}{h}\right\}$$

$$= c \lim_{h \to 0} \frac{f(x + h) - f(x)}{h} = cf'(x).$$

The reader should verify (ii) similarly.

---

* We shall assume the truth of the following 'obvious' statements:
(i) If $\lim_{x \to a} f(x) = l$, then $\lim_{x \to a} cf(x) = cl$, where c is any constant.
(ii) If $\lim_{x \to a} f(x) = l$ and $\lim_{x \to a} g(x) = m$, then $\lim_{x \to a} \{f(x) + g(x)\} = l + m$.

If c is any constant, then $\dfrac{dc}{dx} = \lim\limits_{h \to 0} \dfrac{c - c}{h} = 0$. Thus *the gradient of any constant is zero.*

We are now in a position to differentiate on sight any *polynomial* in x, i.e. an expression of the form

$$a_n x^n + a_{n-1} x^{n-1} + \ldots + a_1 x + a_0,$$

where $a_0, a_1, \ldots, a_n$ are constants.

EXAMPLE 2.11 *To give the gradient functions of*

$$\text{(i) } 6x^2 - \tfrac{5}{2}x + 1, \quad \text{(ii) } (3x + 1)^2.$$

(i) The gradient of x^2 is $2x$, so the gradient of $6x^2$ is $12x$; the gradient of x is 1, so the gradient of $-\tfrac{5}{2}x$ is $-\tfrac{5}{2}$. The gradient of 1, being a constant, is zero. Thus

$$\frac{d}{dx}(6x^2 - \tfrac{5}{2}x + 1) = 12x - \tfrac{5}{2}.$$

(ii) $(3x + 1)^2 = 9x^2 + 6x + 1$, so the gradient is $18x + 6$.

[We must stick to the rules we have established, and resist the temptation to guess the answer, e.g. here a plausible guess would be $2(3x + 1)$, but this turns out to be wrong.]

EXAMPLE 2.12 *To differentiate $\sqrt{x}$ from first principles.*

Let $f(x) = \sqrt{x}$. Then

$$\begin{aligned}
f'(x) &= \lim_{h \to 0} \frac{\sqrt{(x + h)} - \sqrt{x}}{h} \\[2mm]
&= \lim_{h \to 0} \frac{\{\sqrt{(x + h)} - \sqrt{x}\}\{\sqrt{(x + h)} + \sqrt{x}\}}{h\{\sqrt{(x + h)} + \sqrt{x}\}} \\[2mm]
&= \lim_{h \to 0} \frac{h}{h\{\sqrt{(x + h)} + \sqrt{x}\}} \\[2mm]
&= \lim_{h \to 0} \frac{1}{\sqrt{(x + h)} + \sqrt{x}} \\[2mm]
&= \frac{1}{2\sqrt{x}}.
\end{aligned}$$

Writing $\sqrt{x} = x^{\frac{1}{2}}$, our result can be written $\dfrac{d}{dx}(x^{\frac{1}{2}}) = \tfrac{1}{2}x^{-\frac{1}{2}}$.

EXAMPLE 2.13   *To sketch the graph of* $y = \frac{1}{6}x^3 - 2x^2 + 6x + 2$.

(i) When $x = 0$, $y = 2$, but when $y = 0$ we have a cubic for x, which we shall not attempt to solve.

(ii) When x is very large (positive or negative), the graph looks like that of $y = \frac{1}{6}x^3$ (for example, if $x = 1000$, $2x^2$ at a mere two million is dominated by $\frac{1}{6}x^3$, which is over 80 times larger).

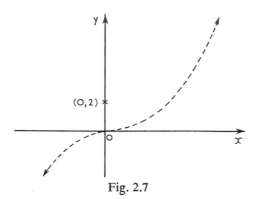

Fig. 2.7

So far, we have one point on the graph (Fig. 2.7) and its ultimate directions (the dotted line represents $y = \frac{1}{6}x^3$).

Now $\dfrac{dy}{dx} = \frac{1}{2}x^2 - 4x + 6$, and if there are values of x which make this zero, we can find where the graph is 'flat', i.e. the tangent is parallel to the x-axis. This will give us the decisive clue to enable us to make our sketch.

Solving the quadratic $\frac{1}{2}x^2 - 4x + 6 = 0$, we have $x = 2$ or 6. The corresponding values of y, computed from the equation

$$y = \frac{1}{6}x^3 - 2x^2 + 6x + 2,$$

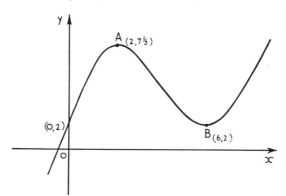

Fig. 2.8

are respectively $7\frac{1}{3}$ and 2. The points $A(2, 7\frac{1}{3})$ and $B(6, 2)$ are thus the only ones at which the tangent is parallel to the x-axis, and we can now make our sketch (Fig. 2.8). As the graph cuts the x-axis in one point only, we can deduce incidentally that there is only one value of x which satisfies the equation $\frac{1}{6}x^3 - 2x^2 + 6x + 2 = 0$, that is, the equation has only one real root.

EXAMPLE 2.14 *To investigate the function* $\dfrac{1}{x^2}$.

Cf. Example 1.7, p. 10, but the reader should go as far as he can by himself, before referring back as a check.

EXERCISE VIII

In questions 1 *to* 15, *the given functions are to be differentiated with respect to* x.

1. x^8.
2. $8x^7$.
3. $8x^7 + 4$.
4. x^{99}.
5. $x^2 - 6x$.
6. $8x^3 + 6x^2 - 37x + 4$.
7. $(x + 2)^2$.
8. $(7x - 3)^3$.
9. $\frac{1}{2}x - \sqrt{x}$.
10. $(2x^3)^2$.
11. $(2x^3 - 1)^2$.
12. $16 - 4x$.
13. $(x - 4)(2x + 3)$.
14. $x(x - 1)(x - 2)$.
15. $(6 - 7\sqrt{x})^2$.

In questions 16 *to* 22, *find the values of* x *for which* $\dfrac{dy}{dx} = 0$, *and sketch the graphs of* y *against* x.

16. $y = x^2 - x$.
17. $y = x^2 - 8x + 7$.
18. $y = (4 - x)^2$.
19. $y = x^3 - 6x^2 + 9x + 1$.
20. $y = 2x^3 - 3x^2 - 18x$.
21. $y = x(x - 1)(x - 3)$.
22. $y = -x^2 + 7x - 10$.

23. If $f(x) = x^3 - 5x^2 + 3x + 2$: (i) calculate $f(0)$, $f(\frac{1}{3})$, $f(3)$, $f'(x)$, $f'(1)$; (ii) find the values of x for which $f'(x) = 0$; (iii) sketch the graph of $y = f(x)$. How many real roots has the equation $x^3 - 5x^2 + 3x + 2 = 0$?

24. (i) Verify that $(x - a)(x^{n-1} + x^{n-2}a + x^{n-3}a^2 + \ldots + a^{n-1}) = x^n - a^n$.
    (ii) Evaluate $\lim\limits_{x \to a} \dfrac{x^n - a^n}{x - a}$.
    (iii) Writing $x = a + h$, deduce $\lim\limits_{h \to 0} \dfrac{(a + h)^n - a^n}{h}$.
    (iv) Interpret the last result as a gradient.

Miscellaneous questions (*in sets of* 4)

25. (i) Give the equation of the straight line through $(5, 7)$ whose gradient is 6.
    (ii) If $y = x^2 - 4x + 2$, evaluate y and $\dfrac{dy}{dx}$ when $x = 5$.

26. If $f(x) = x^3 - 7x + 1$, evaluate: (i) $f(2)$; (ii) $f(-4)$; (iii) $f'(0)$; (iv) $f'(\frac{3}{2})$; (v) $f(a) + \frac{1}{2}f'(a)$.

27. Give the gradient of: (i) $4x^3 - 6x + 2$; (ii) $(x^2 - x)^2$; (iii) $\frac{1}{2}x^{\frac{1}{2}}$.

28. Differentiate from first principles: (a) $2x^2$; (b) $\dfrac{1}{x - 2}$.

29. Draw accurately, on graph paper, the straight lines whose equations are $2x - 3y = 4$, $y = -\frac{3}{2}x + 1$. What are their gradients?

30. (i) '$[x]$' means the largest integer not larger than x: e.g. $[2 \cdot 5] = 2$, $[2] = 2$, $[2 \cdot 9] = 2$, $[3] = 3$, $[-1 \cdot 6] = -2$. Sketch the graph of $y = [x]$. At what points is it discontinuous?
    (ii) Sketch the graph of $y = x - [x]$.

31. Evaluate (i) $\lim\limits_{x \to -3} \dfrac{x^2 - 9}{x + 3}$; (ii) $\lim\limits_{x \to -a} \dfrac{x^2 - a^2}{x + a}$.

32. Sketch the curve $y = \dfrac{x}{(x - 1)(x - 2)(x - 5)}$. Do not attempt to work out $\dfrac{dy}{dx}$, but deduce from your sketch how many values of x make $\dfrac{dy}{dx}$ zero.

33. Give rough sketches of the graphs of distance against time and velocity against time for a boy running a 100 metre hurdle race.

34. Give the derivatives of: (a) $8x^3 - 7x + 3$; (b) $(2x - 1)^2$; (c) $3 + 4\sqrt{x} - 5x^5$.

35. Sketch the graph of $y = 2x^3 + 30x^2 - 9x + 1$, marking in the coordinates of the points for which $\dfrac{dy}{dx} = 0$.

36. Differentiate $7x + \dfrac{1}{3 - 2x}$ from first principles.

3
Applications

The theory of the previous chapter can now be applied to coordinate geometry and mechanics, problems on maxima and minima and approximations.

Tangent and Normal

We recall that the tangent at a point P on a curve is the straight line through P whose gradient is that of the curve at P. We can now calculate the gradient of various curves; then with the help of the formula

$$y - y_1 = m(x - x_1)$$

for the equation of the line through (x_1, y_1) whose gradient is m, we can write down the equation of the tangent at a given point.

EXAMPLE 3.1   *To find the tangent at* $(1, 9)$ *to* $y = 7x^3 - 3x^2 + 4x + 1$.

The general formula for the gradient at any point (x, y) on the curve is $\dfrac{dy}{dx} = 21x^2 - 6x + 4$. What we need is the gradient *at the point* $(1, 9)$, not the general value; substituting $x = 1$, we obtain $21 - 6 + 4 = 19$. The tangent is thus the line through $(1, 9)$ whose gradient is 19, so its equation is $y - 9 = 19(x - 1)$, which simplifies to $19x - y = 10$.

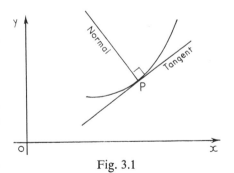

Fig. 3.1

The line, through a point on a curve, perpendicular to the tangent there is called the *normal* (Fig. 3.1). To determine the equation of a normal, we first seek the connexion between the gradients of two perpendicular lines.

Refer to Fig. 3.2. Let P be the point (a, b), so that the gradient of OP is given by $\tan \theta = \dfrac{b}{a}$. We now rotate OP through a right angle, so that P has moved to Q. Since $\triangle OPM \equiv \triangle QON$, the coordinates of Q are $(-b, a)$,

31

and the gradient of OQ is $\tan(90° + \theta) = -\dfrac{a}{b}$. The product of the gradients

of OP and OQ is therefore $\left(\dfrac{b}{a}\right) \cdot \left(-\dfrac{a}{b}\right) = -1$, and it can be seen that *if*

two straight lines are at right angles, the product of their gradients is -1.
Thus if the gradient of the tangent at a point P is m, that of the normal

is $-\dfrac{1}{m}$.

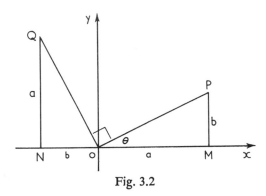

Fig. 3.2

EXAMPLE 3.2   *To find the tangent and normal to the curve* $y = 3x^2 - 5x + 1$
at $(2, 3)$.

$\dfrac{dy}{dx} = 6x - 5$, which equals 7 when $x = 2$; the tangent is therefore given by

$y - 3 = 7(x - 2)$, or $y = 7x - 11$. The gradient of the normal is $-\frac{1}{7}$, so
its equation is $y - 3 = -\frac{1}{7}(x - 2)$, i.e. $7y + x = 23$.

EXAMPLE 3.3   *To prove that the line* $y = 6x - 2$ *is a tangent to the curve*
$y = 2x^3 - 15x^2 + 42x - 30$, *and to find the equation of the other tangent*
to the curve which is parallel to this line.

PRELIMINARY REMARKS.   It would be possible to solve the equations
$y = 6x - 2$ and $y = 2x^3 - 15x^2 + 42x - 30$ simultaneously, to find where
the line meets the curve. Eliminating y gives $(x - 2)^2(2x - 7) = 0$; the line
therefore meets the curve 'twice' at $(2, 10)$ and (with the aid of a rough sketch)
it could be deduced that it is a tangent at this point. This method is sometimes
useful, but here it would be heavy-handed, and bring us no further towards
the equation of the parallel tangent. We therefore shall proceed instead as
follows. The gradient of $y = 6x - 2$ is 6; we shall accordingly seek the
equations of any tangents to the curve with gradient 6, and hope that one
of them will turn out to be this line.

Solution. The gradient of the curve is given by $\dfrac{dy}{dx} = 6x^2 - 30x + 42$, which equals 6 if $6x^2 - 30x + 36 = 0$, i.e. if $x = 2$ or 3. The points at which the gradient is 6 are therefore $(2, 10)$ and $(3, 15)$. The tangent at $(2, 10)$ is $y - 10 = 6(x - 2)$, i.e. $y = 6x - 2$, which establishes the first part of the question; and further there is just one parallel tangent, that at $(3, 15)$, whose equation is $y - 15 = 6(x - 3)$, or $y = 6x - 3$.

Exercise IX

1. Find the equations of the tangent and normal to the following curves at the given points:
 (a) $y = x^2$ at $(3, 9)$; (b) $y = x^3 - 3x + 2$ at $(2, 4)$;
 (c) $y = x^4 - 5x^3 + 6x + 1$ at $(0, 1)$; (d) $y = x^2 - 6x + 8$ at $(3, -1)$;
 (e) $y = 7x^7 - 8x^6 + x^3 - 4$ at $(1, -4)$.

2. Prove in two ways that the line $y = 8x - 16$ touches the parabola $y = x^2$ at the point $(4, 16)$.

3. Find the equation of the tangent to the curve $y = 7x^2 - 5x + 1$ whose gradient is 23.

4. Show that the line $y = 24x - 47$ is a tangent to the curve $y = x^3 - 3x + 7$, and find the equation of the other tangent parallel to this line.

5. Prove that the normal to $y = x^2 - x$ at $(3, 6)$ touches the parabola $x^2 + 660y = 0$.

6. (i) Criticize the following:
 '$y = x^2, \dfrac{dy}{dx} = 2x$. Therefore the tangent at $(2, 4)$ is $y - 4 = 2x(x - 2)$.'
 (ii) Find the equation of the tangent at $(2, 4)$ to the curve $y - 4 = 2x(x - 2)$.

Increasing and Decreasing Functions

If $f'(x)$ is positive over a certain domain of values for x, the function $f(x)$ is an *increasing* function over that domain (Fig. 3.3). For example, if $f(x) = x^2$, $f'(x) = 2x$, and so, if $x > 0$, x^2 is increasing. Similarly, if $f'(x)$ is negative, the function is *decreasing*; in particular, x^2 is decreasing if $x < 0$. The '$+$' and '$-$' refer to the sign of the gradient.

Example 3.4 *To find the values of x for which*

$$f(x) = \tfrac{3}{10}x^4 - \tfrac{4}{5}x^3 - 3x^2 + \tfrac{36}{5}x + 11$$

is (a) increasing, (b) decreasing; and to sketch the graph of $y = f(x)$.

$$f'(x) = \tfrac{6}{5}x^3 - \tfrac{12}{5}x^2 - 6x + \tfrac{36}{5}$$
$$= \tfrac{6}{5}(x^3 - 2x^2 - 5x + 6)$$
$$= \tfrac{6}{5}(x - 1)(x + 2)(x - 3).$$

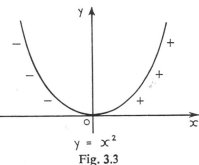

$y = x^2$

Fig. 3.3

Thus if

(a) $x > 3$, $f'(x) > 0$, and so $f(x)$ is increasing;
(b) $1 < x < 3$, $f'(x) < 0$, so $f(x)$ is decreasing;
(c) $-2 < x < 1$, $f'(x) > 0$, $f(x)$ is increasing;
(d) $x < -2$, $f'(x) < 0$, $f(x)$ is decreasing.

When $x = 1, 3$ or -2, $f'(x) = 0$, so the tangent to $y = f(x)$ is parallel to the x-axis. The corresponding values of y are $14 \cdot 7$, $8 \cdot 3$ and -11.

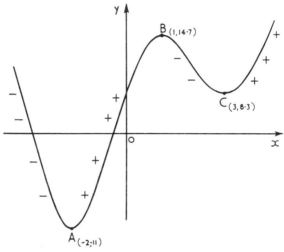

Fig. 3.4

Maxima and Minima; Stationary Values

A point such as A or C in the graph, Fig. 3.4, is called a *minimum* point, and a point such as B is a *maximum* point. More precisely, $f(x)$ has a maximum value at $x = a$ if: (i) $f'(x) > 0$ for values of x just less than a; (ii) $f'(a) = 0$; and (iii) $f'(x) < 0$ for values of x just greater than a. Minimum values of a function are defined similarly.

The values of $f(x)$ at maxima and minima are called *turning* values. It is, however, possible to have $f'(a) = 0$ without $f(a)$ being a turning value; as a simple example, if $f(x) = x^3$ (Fig. 3.5), $f'(0) = 0$, but $f'(x) > 0$ for all x other than zero. If $f'(a) = 0$ but $f'(x)$ has the same sign for values of x within a certain neighbourhood of a, then there is an S-bend and $x = a$ gives a *point of inflexion* on the graph of $y = f(x)$. All values of $f(x)$ for which $f'(x) = 0$ (whether maxima, minima or points of inflexion) are called *stationary* values.

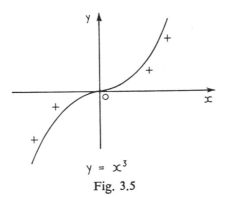

Fig. 3.5

EXAMPLE 3.5   *To find and classify the stationary values of*

$$y = 3x^4 - 28x^3 + 96x^2 - 144x + 80,$$

and sketch the graph of this function.

Let $f(x) = 3x^4 - 28x^3 + 96x^2 - 144x + 80.$

Then $f'(x) = 12x^3 - 84x^2 + 192x - 144,$

which is 0 if $12(x^3 - 7x^2 + 16x - 12) = 0$, i.e. $(x - 2)^2(x - 3) = 0$, so $x = 2$ or $x = 3$. By calculation, $f(2) = 0$ and $f(3) = -1$; these are the required stationary values.

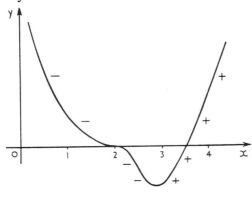

$$y = 3x^4 - 28x^3 + 96x^2 - 144x + 80$$

Fig. 3.6

If x is either just greater or just less than 2, $f'(x) < 0$, so $(2, 0)$ is a *point of inflexion*. But if x is just greater than 3, $f'(x) > 0$ while if x is just less than 3, $f'(x) < 0$, so $(3, -1)$ is a *minimum* point.

With the help of the usual other clues (when $x = 0$, $y = 80$; when x is large, the graph looks like that of $y = 3x^4$), we can now give the required sketch (Fig. 3.6).

EXERCISE X

1. For what values of x are the following functions positive?
   (a) $4x + 1$;   (b) $(x + 2)(x - 1)$;   (c) $(3 - x)(x + 2)$;   (d) $(x - 1)^2$.

2. Find the stationary values (distinguishing between them) of the following functions and the corresponding values of x. (The answers to the previous question may be helpful.) Also, in each case, sketch the graph of the function and indicate where it is increasing or decreasing.
   (a) $2x^2 + x + 3$;   (b) $2x^3 + 3x^2 - 12x + 4$;   (c) $-\frac{1}{3}x^3 + \frac{1}{2}x^2 + 6x$;
   (d) $x^3 - 3x^2 + 3x - 1$.

3. Verify that $x^3 - 3x^2 - 6x + 8 = (x + 2)(x - 1)(x - 4)$. For what values of x is $x^3 - 3x^2 - 6x + 8$ positive?

4. Investigate the stationary values of $f(x) = \frac{1}{4}x^4 - x^3 - 3x^2 + 8x + 7$, and sketch the graph of this function, indicating for what values of x it is increasing or decreasing. (Cf. the previous question.)

5. Investigate any stationary values of the following, illustrating in each case with a rough sketch.
   (a) $x^2 - 5x + 6$;   (b) $x^3 - 5x^2 + 3x - 2$;   (c) $x^3 - 5x^2 + x + 1$;
   (d) $x^3 - x^2 + 7x + 3$;   (e) $(x - 1)^2(x - 2)$;   (f) $(x - 1)^3$;   (g) $x^4 - x^3$;
   (h) $x^3(1 - x^2)$.

6. (D) What is the maximum value of y if $y^3 + x^2 = 0$? Sketch the graph, and explain why the gradient is not zero at the point concerned.

Problems on Maxima and Minima

EXAMPLE 3.6   *A rectangular piece of land is to be staked out such that the perimeter is 720 m. To find the dimensions which will give a maximum area.*

Method 1. Let the length be x m, then the breadth is $(360 - x)$ m, and the area A m² where $A = x(360 - x) = 360x - x^2$. Then $\frac{dA}{dx} = 360 - 2x$, which is zero when $x = 180$. This value of x gives the only stationary value of A, and by considering the graph of A against x (or the sign of $\frac{dA}{dx}$ when x is just above or below 180) this is seen to be a *maximum*. The required maximum area is therefore a square; length and breadth both being 180 m.

Method 2. As before, $A = 360x - x^2 = 180^2 - (x - 180)^2$, by completing the square. Thus $A \leqslant 180^2$, equality occurring when $x = 180$, i.e. the maximum area is a square of side 180 m.

Method 3. If l, b are positive, then $\frac{l + b}{2} \geqslant \sqrt{lb}$, equality occurring only if $l = b$.* Now if the length of the rectangle is l m and the breadth b m,

---

* We borrow this result ('Arithmetic Mean $\geqslant$ Geometric Mean') from algebra. It follows from $(\sqrt{l} - \sqrt{b})^2 \geqslant 0$.

we have $2(l + b) = 720$, so $\sqrt{lb} \leqslant \dfrac{l + b}{2} = 180$. The maximum value of A ($= lb$) is accordingly 180^2, this being attained when $l = b$, i.e. when the rectangle is a square.

EXAMPLE 3.7   *An open tank is to be designed in the shape of a cuboid with a square base, the total surface area of the base and four walls together being 100 m^2. To find the length of the side of the base which makes the volume of the tank a maximum.*

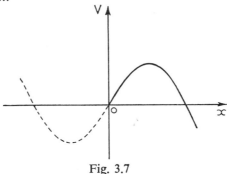

Fig. 3.7

Let the side of the square base be x m and the depth of the tank be y m. Then, considering the surface area, we have $x^2 + 4xy = 100$, whence $y = \dfrac{100 - x^2}{4x}$. Now the volume, V m^3, is given by

$$V = x^2 y = \frac{x^2(100 - x^2)}{4x} = 25x - \tfrac{1}{4}x^3.$$

$\dfrac{dV}{dx} = 25 - \tfrac{3}{4}x^2$, which is zero if $x = \pm \dfrac{10}{\sqrt{3}}$. We are only interested in positive values of x, and a rough sketch of the graph of V against x (Fig. 3.7) shows that $x = +\dfrac{10}{\sqrt{3}}$ gives the required maximum. Thus the required side of the square is $\dfrac{10}{\sqrt{3}} = \dfrac{10\sqrt{3}}{3} \approx 5 \cdot 8$ m.

EXERCISE XI

Do question 1 in at least two different ways: thereafter, use whatever method seems to give the neatest solution.

1. A variable rectangle has a constant perimeter of 28 cm. Find the lengths of the sides when the area is a maximum.

2. A man wishes to fence a rectangular enclosure of area 128 m^2. One side of the enclosure is formed by part of a brick wall, already in position. What is the least

possible length of fencing required for the other three sides? (Prove that your result gives a *minimum*.) $\left[\text{Hint: } \dfrac{d}{dx}\left(\dfrac{1}{x}\right) = -\dfrac{1}{x^2}.\right]$

3. A sealed cylindrical jam tin is of height h cm and radius r cm. The area of its *total* outer surface is A cm² and its volume is V cm³. Find an expression for A in terms of r and h.

   Taking $A = 24\pi$, find (i) an expression for h in terms of r and hence an expression for V in terms of r; (ii) the value of r which will make V a maximum.

4. A farmer has a certain length of fencing and uses it all to fence two square sheep-folds. Prove that the sum of the areas of the two folds is least when their sides are equal.

5. A right circular cone has radius $\frac{1}{2}h$ m and height h m. Find the radius of the right circular cylinder of greatest volume which can be inscribed in the given cone.

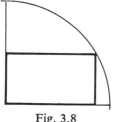

6. Prove that, if the sum of the radii of two circles remains constant, the sum of the areas of the circles is least when the circles are equal.

7. A piece of wire of length l m is cut into two portions, the length of one being x m. Each portion is then bent to form the perimeter of a rectangle whose length is twice its breadth. Find an expression for the sum of the areas of these rectangles.

   For what value of x is this area a minimum?

Fig. 3.8

8. (*Harder*) A rectangle is inscribed in a quadrant of a circle of radius 5 cm, two of the sides being along the bounding radii, as shown in Fig. 3.8. Find the maximum value of the area of the rectangle. [Hint: consider an expression for the square of the area. When this is a maximum, so also is the area itself.]

Kinematics

It is by now familiar that if y is a function of x, then $\dfrac{dy}{dx}$ is the *gradient* of y.

Another fair description would be 'the rate of change of y with respect to x', since it is the limit as $\delta x \to 0$ of the ratio of δy (the small increase of y) to δx (the corresponding small increment of x). There is, of course, no magic in calling the variables x and y; we have already considered, for example, $\dfrac{dA}{dx}$ where A was the area of a rectangle and x the length of one of its sides.

We next return to the notation at the beginning of Chapter 1, and the first two of the 'fair guesses' on p. 4 (which by now will look fairly tame). We consider, then, motion in a straight line of a particle, which after time t is a distance s from a fixed origin on the line, so that $s = f(t)$. If the particle moves a distance δs in time δt, the average velocity for this journey is $\dfrac{\delta s}{\delta t}$, and we define the velocity at an instant as $v = \lim\limits_{\delta t \to 0} \dfrac{\delta s}{\delta t}$. We thus have immediately that $v = \dfrac{ds}{dt}$, i.e. the gradient of the $s(t)$ graph (see Fig. 3.9). Similarly, the acceleration a is defined as $\lim\limits_{\delta t \to 0} \dfrac{\delta v}{\delta t}, = \dfrac{dv}{dt}$. When $s = f(t)$ has a stationary value, $\dfrac{ds}{dt} = 0$, and when v is stationary, $\dfrac{dv}{dt} = 0$.

It must be emphasized that s is a measure of the distance from O at a given instant and *not* of the total physical distance travelled, some of which may be negative (i.e. towards O).

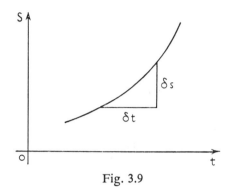

Fig. 3.9

EXAMPLE 3.8 *A particle moves in a straight line so that its distance s m from a point O on the line after t seconds is given by* $s = \frac{1}{5}t^3 - \frac{3}{2}t^2 + \frac{18}{5}t + 1$. *To investigate the motion.*

We have $v = \dfrac{ds}{dt} = \frac{3}{5}t^2 - 3t + \frac{18}{5}, a = \dfrac{dv}{dt} = \frac{6}{5}t - 3$. Thus

$$v = \tfrac{3}{5}(t-2)(t-3),$$

which is zero when $t = 2$ or $t = 3$. If $t < 2$ or $t > 3$, $v > 0$, but if $2 < t < 3$, then $v < 0$. Also $a = 0$ when $t = 2\cdot5$; before this, $a < 0$.

t	0	2	2·5	3	4
s	1	3·8	3·75	3·7	4·2
v	3·6	0	−0·15	0	1·2
a	−3	−0·6	0	0·6	1·8

Thus s has a maximum at $t = 2$ and a minimum at $t = 3$ (Fig. 3.10), and between these times the particle is moving backwards towards O. v has a minimum value of $-0\cdot15$, occurring when $a = 0$, $t = 2\cdot5$; when $t > 3$, s and v both increase.

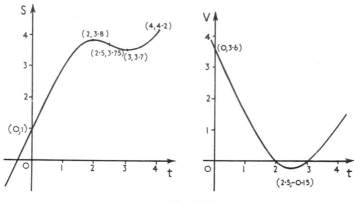

Fig. 3.10

EXERCISE XII

1. (D) If, for motion in a straight line, $s = ut + \frac{1}{2}lt^2$, where u, l are constants, show that the acceleration a is constant.

2. If a particle moves in a straight line subject to the relationship: $s = 4t^3 - 3t^2 - 18t + 4$, find at what times the velocity is zero, and interpret the negative answer. Also find the value of the acceleration: (i) when $t = 1$; (ii) when $t = \frac{1}{4}$.

3. If $s = 32t - 4t^2$, find: (i) the maximum value of s; (ii) the velocity when $t = 4$; (iii) the acceleration; (iv) the time taken for the velocity to decrease from 30 to 20 and the distance travelled in this time.

4. A point moves along a straight line OX so that its distance x m from the point O after t seconds is given by the formula $x = t^3 - 6t^2 + 9t$. Find at what times and in what positions the point will have zero velocity. Find also its acceleration at those instants, and its velocity when its acceleration is zero.

Approximations

So far our interest has been in finding $\dfrac{dy}{dx}$, considered as $\displaystyle\lim_{\delta x \to 0} \dfrac{\delta y}{\delta x}$. We can, however, look at things in reverse. Suppose y is a function of x for which $\dfrac{dy}{dx}$ is known; from $\displaystyle\lim_{\delta x \to 0} \dfrac{\delta y}{\delta x} = \dfrac{dy}{dx}$, it follows that if δy, δx are small, $\dfrac{\delta y}{\delta x} \approx \dfrac{dy}{dx}$. Thus we can compute approximately the small change in y arising from a small change in x.

EXAMPLE 3.9 *To find the approximate value of* $2x^3 - 5x^2 + 3x + 4$ *when* $x = 1 \cdot 003$.

Let $2x^3 - 5x^2 + 3x + 4 = y$. It is easily seen that when $x = 1$, $y = 4$; we shall therefore find an approximation for δy when $x = 1$, $y = 4$, $\delta x = 0 \cdot 003$.

We have $\dfrac{\delta y}{\delta x} \approx \dfrac{dy}{dx} = 6x^2 - 10x + 3$. Thus $\delta y \approx (6x^2 - 10x + 3)\delta x$.

Putting $x = 1$,
$$\delta y \approx (6 - 10 + 3)\ \delta x$$
$$= -1 \times 0\!\cdot\!003$$
$$= -0\!\cdot\!003,$$

so that if $x = 1\!\cdot\!003$, $y \approx 4 - 0\!\cdot\!003 = 3\!\cdot\!997$.

[The exact value of y when $x = 1\!\cdot\!003$ is $3\!\cdot\!997009054$; cf. the discussion after the next example.]

EXAMPLE 3.10   *The side of a cube is measured as 6 cm $\pm$ 0·05 cm. To find the range of possible error in calculating the volume from this measurement.*

Let the side of the cube be x cm, so that the volume, V cm^3, is given by $V = x^3$. Then

$$\frac{\delta V}{\delta x} \approx \frac{dV}{dx} = 3x^2, \quad \text{so } \delta V \approx 3x^2 \cdot \delta x.$$

Putting $x = 6$, $\delta x = 0\!\cdot\!05$, we have $\delta V \approx 3 \times 36 \times 0\!\cdot\!05 = 5\!\cdot\!4$. If δx is within the range $\pm 0\!\cdot\!05$, the corresponding error in the volume is thus approximately $\pm 5\!\cdot\!4$, and so $V = 216 \pm 5\!\cdot\!4$.

The question arises, how many figures in '$\delta V \approx 5\!\cdot\!4$' are reliable? $(6\!\cdot\!05)^3 - 6^3 = 5\!\cdot\!445125$, and it was a matter of luck rather than management that the second digit in $5\!\cdot\!4$ was correct. Looking more closely,

$$\delta V = (x + \delta x)^3 - x^3 = 3x^2 \cdot \delta x + 3x(\delta x)^2 + (\delta x)^3,$$

and by setting $\dfrac{\delta V}{\delta x} \approx \dfrac{dV}{dx} = 3x^2$, the larger of the terms which we have neglected is $3x(\delta x)^2$, $= 18 \cdot 0\!\cdot\!0025 = 0\!\cdot\!045$. Such a rough estimate of the amount neglected is necessary to ensure that too many figures are not given as 'reliable'. Reverting to Example 3.9,

$$\delta y = \{2(x + \delta x)^3 - 5(x + \delta x)^2 + 3(x + \delta x) + 4\} - \{2x^3 - 5x^2 + 3x + 4\}$$
$$= (6x^2 - 10x + 3)\delta x + (6x - 5)(\delta x)^2 + 2(\delta x)^3.$$

In effect, by taking $\dfrac{\delta y}{\delta x} \approx \dfrac{dy}{dx}$, we neglected the terms involving $(\delta x)^2$ and $(\delta x)^3$. Now, since $\delta x = 0\!\cdot\!003$, $(\delta x)^2 = 0\!\cdot\!000009$ and $(\delta x)^3 = 0\!\cdot\!0000000027$, so our answer to three places of decimals was quite safe.

In the '$f(x)$' notation, the idea we have been using above amounts to saying that, since $f'(a) = \lim\limits_{h \to 0} \dfrac{f(a + h) - f(a)}{h}$, then if h is small enough, $f'(a) \approx \dfrac{f(a + h) - f(a)}{h}$; that is, the gradient at P (where $x = a$) is approximately the same as the gradient of the chord PQ (where Q's x-coordinate is $a + h$), Fig. 3.11(i). If the curve is 'bending rapidly', as in Fig. 3.11(ii), this assumption may be dangerous.

With this warning in mind, we now apply the idea to the approximate solution of equations known as *Newton's method*. Suppose, then, that $f(x)$ is differentiable, and we are to investigate the roots of the equation $f(x) = 0$.

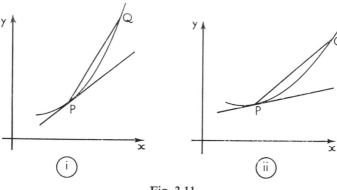

Fig. 3.11

Since $f(x)$ is differentiable, it is also continuous (cf. p. 22), and if, for example, $f(b)$ is negative and $f(c)$ is positive, there must accordingly be a *root*, α, of the equation (i.e. a number such that $f(\alpha) = 0$) between b and c (Fig. 3.12). By trial and error, it is possible to find a rough approximation for any such root α, that is, a neighbouring value, a, such that $f(a) \approx 0$, while $f(\alpha) = f(a + h)$, say, is zero. It is required, then, to find as accurate as possible an approximation for h.

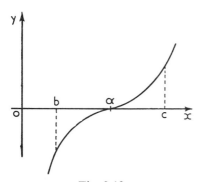

Fig. 3.12

Assuming $f'(a) \approx \dfrac{f(a + h) - f(a)}{h}$, it follows that $f'(a) \approx -\dfrac{f(a)}{h}$, since $f(a + h) = 0$, by hypothesis. Thus $h \approx -\dfrac{f(a)}{f'(a)}$ and so, subject to our assumption, $a - \dfrac{f(a)}{f'(a)}$ should be a better approximation to the root α than a itself. In Fig. 3.13(i), P is the point $(a, f(a))$, PT is the tangent at P and Q is the point $(\alpha, 0)$, so that $f(\alpha) = 0$ and α is the required root. The gradient of PT is $f'(a)$, so $TR = \dfrac{f(a)}{f'(a)}$, and $OT = a - \dfrac{f(a)}{f'(a)}$, which is nearer to α than a. It would be tempting to take it as an infallible rule that if a is an approximation to a root, then $a - \dfrac{f(a)}{f'(a)}$ is a better one; but Fig. 3.13(ii) shows that things are not always so conveniently placed.

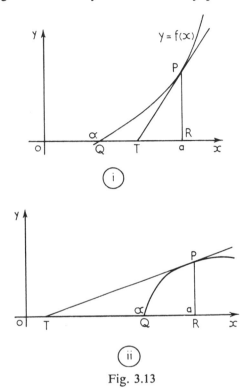

Fig. 3.13

As before, $OT = a - \dfrac{f(a)}{f'(a)}$, but there is no guarantee that this is nearer to α than a itself. It is, therefore, indeed necessary to proceed with care; the procedure is illustrated in the following example.

EXAMPLE 3.11 *Using Newton's method, to find, correct to two decimal places, the positive root of the equation* $12x^3 + 4x^2 - 15x - 4 = 0$.

Let $y = f(x) = 12x^3 + 4x^2 - 15x - 4$. Then $f(0) = -4$, $f(1) = -3$, $f(2) = 78$, and the root thus lies between 1 and 2, probably nearer 1. The next move is a rough sketch of the graph. We have $f'(x) = 36x^2 + 8x - 15$, which is zero if $x = \dfrac{-2 \pm \sqrt{139}}{18}$. Now $\dfrac{\sqrt{139} - 2}{18} < \dfrac{\sqrt{144} - 2}{18} = \dfrac{5}{9}$, which is rough and ready but sufficiently accurate for our sketch, Fig. 3.14. (We are not really interested in the negative values of x, but from the wording of the question there are probably two real negative roots as well as the positive one.) Now, since $f(1) < 0$, the root is rather larger than 1, and if we start

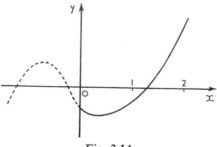

Fig. 3.14

from 1, we shall have the same sort of awkwardness as in Fig. 3.13(ii) above. (The reader should verify this with a rough sketch.) We therefore try to find an approximation on the 'correct' side; by computation, $f(1·2) = (14·4 + 4)1·44 - 22 = 4·496$. We shall accordingly take $a = 1·2$,

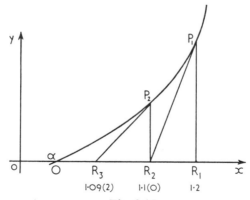

Fig. 3.15

whence $a - \dfrac{f(a)}{f'(a)} = 1·2 - \dfrac{4·496}{46·44} \approx 1·1(03)$. This is certainly a better approximation to the root than 1·2, but even the second decimal place is suspect, so we now reiterate the process, starting with $a = 1·10$. By repeating the process (Fig. 3.15), with the necessary patience or a calculator, we could go on to any degree of accuracy that might be required.

Taking $a = 1·10$, then, we have $a - \dfrac{f(a)}{f'(a)} = 1·10 - \dfrac{0·31}{37·36} \approx 1·09(2)$. The third decimal place here is suspect, but we are now reasonably safe in giving our answer as 1·09. To make sure, taking $a = 1·09$, we have $f(a) \approx -0·07$; thus $1·09 < \alpha < 1·092$. It could be argued that putting in this last step was being unnecessarily cautious, but the important thing about approximations is that they should be reliable so far as they go; when in doubt, re-insure.

It must by now be clear that the statement $f'(a) \approx \dfrac{f(a + h) - f(a)}{h}$ has to be treated with caution. We are on safer ground* with the remark that, if $f(x)$

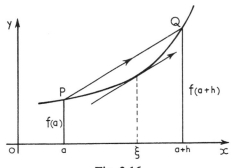

Fig. 3.16

is differentiable over the range $a \leqslant x \leqslant a + h$, then $\dfrac{f(a + h) - f(a)}{h} = f'(\xi)$,

where ξ is some value of x between a and $a + h$. In terms of the graph (Fig. 3.16), this simply expresses the fact that the gradient of the chord PQ is equal to the gradient of the curve somewhere in between P and Q. This ('*first mean-value theorem*') is not so trivial as may appear; '$f(a + h) = f(a) + hf'(\xi)$' will later be generalized into one of the key results of the differential calculus, Taylor's theorem.

EXERCISE XIII

1. Find, correct to two decimal places, the approximate value of
$$6x^3 - 7x^2 + x - 3, \quad \text{when} \quad x = 2·01.$$

2. Using $\dfrac{\delta y}{\delta x} \approx \dfrac{dy}{dx}$, find the approximate value of $x^4 - 5x^2 + 3$ when $x = 1·03$, giving your answer to as many decimal places as you feel justified.

3. The side of a square of side 6 m is incorrectly measured as 6·11 m. Find, to one decimal place, the resulting error in the calculation of the area of the square. (Cf. Example 3.10.)

4. The radius of a sphere is measured as 5 cm. Find the consequent range of error in the volume if there is a possible error of ±0·04 cm in the measurement of the radius.

---

* Though the formal proof is not, in fact, easy.

5. The vertex and the circumference of the base of a right circular cone lie on the surface of a sphere of radius R. The centre of the sphere lies inside the cone. It is a known result that, if the height of the cone is h, its volume is

$$\frac{\pi}{3} h^2(2R - h).$$

[This is to be assumed and should not be proved.]

   Prove that, if R is unaltered but h is increased by a small quantity x, the volume of the cone is increased by

$$\frac{\pi}{3} h(4R - 3h)x \quad \text{approximately.}$$

6. Factorize $12x^3 + 4x^2 - 15x - 5$, and then use Newton's method to find the positive root of the equation $12x^3 + 4x^2 - 15x - 5 = 0$.

7. Show that the equation $x^4 - 4x = 3$ has a root between 1 and 2, and find it correct to one decimal place.

8. Show that there is only one real root of the equation $x^3 + x = 3$, and find it correct to one decimal place.

9. Show that the equation $23x^3 + 6x^2 + 420x - 323 = 0$ has only one real root, and find its value correct to one decimal place.

Miscellaneous questions (in sets of 4)

10. Find the equations of the tangents to the curve

$$y = x^3 - 6x^2 + 12x + 2$$

which are parallel to the line $y = 3x$.

11. Find, correct to two decimal places, the value of $8x^4 - 4x^2 + 3x + 1$ when $x = 1·02$.

12. A particle is moving in a straight line and its distance (s m) from a fixed point in the line after t seconds is given by the equation $s = 12t - 15t^2 + 4t^3$. Find
   (i) the velocity and acceleration of the particle after 3 seconds;
   (ii) the distance travelled between the two times when the velocity is instantaneously zero.

13. The sum of two numbers is 24. Find the numbers if the sum of their squares is to be a minimum. (Try to find at least two good solutions, and illustrate graphically by considering the line $x + y = 24$ and the system of circles $x^2 + y^2 = r^2$.)

14. Prove that the point (5, 6) lies on the curve

$$y = 2x^2 - 7x - 9,$$

and find the equation of the tangent to the curve at this point.
   Find also the equation of the tangent which is parallel to the line $y = x$.

15. The formula $v = 8\sqrt{x}$ gives the velocity, v m s^{-1}, of a particle when it has travelled a distance x m from rest. Prove that, if δv and δx denote corresponding small increases in v and x, $\delta v \approx \dfrac{32\delta x}{v}$.

   Hence find the approximate increase in the velocity of the particle when x increases from 36 to 37.

16. Find the values of x for which $2x^3 - 9x^2 - 60x + 7$ is increasing; sketch the graph of this function.

17. A manufacturer of tinned foods uses tins with approximately rectangular bases, the sides being in the ratio 2:1, and holding 243 cm³ each. What height should the tins be to use the minimum amount of metal?

18. Find the equation of the tangent to the curve
$$y = x^3 - 9x^2 + 20x - 8$$
at the point (1, 4).
    At what points of the curve is the tangent parallel to the line $y + 4x = 3$?

19. A particle moves along the x-axis in such a way that its distance x m from the origin after t s is given by the formula $x = 27t - 2t^2$. What are its velocity and acceleration after $6\frac{3}{4}$ s?
    How long does it take for the velocity to be reduced from 15 m s⁻¹ to 9 m s⁻¹, and how far does the particle travel meanwhile?

20. Find the root of $x^3 - 9x^2 + 24x - 19 = 0$ which lies between 2 and 4, correct to one decimal place.

21. A thin rectangular piece of metal is 24 cm long and 6 cm wide. Equal squares of side x cm are cut from each corner and the sides are then turned up to form an open box. Find the value of x which makes the volume of the box a maximum.

22. Find the equation of the tangent at the point $P(6, 9)$ on the curve $x^2 = 4y$.
    Find the equation of the line joining (6, 9) and (0, 1) and determine the coordinates of the point Q at which this line cuts the curve again.
    Prove that the tangents at P and Q intersect on the line $y = -1$.

23. A cubical block of metal is heated and expands slightly, remaining cubical. If its volume increases by x per cent, show that the length of each edge increases by $\frac{x}{3}$ per cent approximately.
    By what approximate percentage is the surface area increased?

24. Find and classify the stationary values of
$$f(x) = -x^4 + 8x^3 - 18x^2 + 16x - 5; \quad \text{sketch the graph.}$$

25. The breadth of a rectangular box is x cm, its length is $2x$ cm, and its depth is $(18 - 3x)$ cm. Find the value of x which makes the volume of the box a maximum; and find also how many such boxes could be packed in a crate whose breadth, length, and depth are each 48 cm.

26. The point (h, k) lies on the curve $y = 2x^2 + 18$. Find the gradient at this point and the equation of the tangent there.
    Hence find the equations of the two tangents to the curve which pass through the origin.

27. Find one of the negative roots of $4x^4 - 20x^3 - 59x^2 + 15x + 42 = 0$ exactly, and the other correct to one decimal place.

28. While a train is travelling from its start at A to its next stop at B, its distance x miles from A is given by
$$x = 90t^2 - 45t^3,$$
where t hours is the time taken. Find in terms of t its velocity and its acceleration after time t.

Hence find (i) the time taken by the journey from A to B, (ii) the distance AB, (iii) the greatest speed attained.

29. A cylinder of radius x is inscribed in a cone of height h and base radius a, the two figures having a common axis of symmetry. Find an expression for the height of the cylinder in terms of x, a and h.

Prove that the cylinder of greatest volume that can be so inscribed has a volume $\frac{4}{9}$ of the volume of the cone.

4
Integration

Talk to him of Jacob's ladder, and he would ask the number of the steps.   D. W. JERROLD

We return to the three 'fair guesses' about motion in a straight line, on p. 4:

(i) The gradient of an $s(t)$ curve gives the velocity.

(ii) The gradient of a $v(t)$ curve gives the acceleration.

(iii) The area 'under' a section of a $v(t)$ graph gives the distance travelled in the time concerned.

The first two of these have been justified (p. 38) and can now be re-stated mathematically as (i) $\dfrac{ds}{dt} = v$, (ii) $\dfrac{dv}{dt} = a$. Guesses (i), (iii) form a pair:

| Gradient of $s(t)$ graph is v | Area under $v(t)$ graph is s |

This suggests looking for a connexion between gradients and areas as, roughly speaking, opposite processes. To test this idea out,

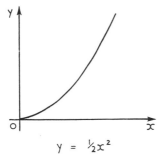 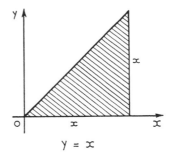

Fig. 4.1

| Gradient of $\frac{1}{2}x^2$ is x | Area 'under' $y = x$ is $\frac{1}{2}x^2$ |

The process towards which we have been groping above is *integration*, which will be investigated in this chapter. 'Fair guess (iii)' will fall into place as an application in Chapter 5.

[1] Integration can, in fact, be defined in three distinct ways, which we shall have to reconcile. The first of these is that *'integration is the opposite of differentiation'*. From this definition, it follows, for example, that since the gradient of x^2 is $2x$, an *integral* of $2x$ is x^2. However, the gradient of $x^2 + 3$ is also $2x$; indeed the gradient of $x^2 + c$ is $2x$, where c is an 'arbitrary constant', that is, it could be any number. We write

$$\int 2x \, dx = x^2 + c.$$

This reads 'the integral of $2x$ with respect to x is $x^2 + c$', and is simply another way of saying that the gradient of $x^2 + c$ is $2x$. The long S, $\int$, and the dx together form one symbol, meaning that we are writing the integral of whatever is between them. The reason for this notation will appear presently.

EXAMPLE 4.1 *Integrate $x^2 - 4x + 2$ with respect to x.*

This could have been phrased 'Evaluate $\int (x^2 - 4x + 2) \, dx$' and is equivalent to the problem of finding y such that $\dfrac{dy}{dx} = x^2 - 4x + 2$, this being the 'opposite of differentiation'. By trial and error (what function when differentiated becomes $x^2 - 4x + 2$?), we have

$$\frac{d}{dx} (\tfrac{1}{3}x^3 - 2x^2 + 2x + c) = x^2 - 4x + 2,$$

and so $\int (x^2 - 4x + 2) \, dx = \tfrac{1}{3}x^3 - 2x^2 + 2x + c$, c again being an arbitrary constant.

EXERCISE XIV (Oral)

1. Integrate the following with respect to x: (a) x^3; (b) $3x^2$; (c) $4x^2 + 2$; (d) 17; (e) $x^5 - 4x^3 + 3x + 2$; (f) x^4; (g) x^5; (h) x^6; (i) x^{73}.
2. Evaluate: (a) $\int (3x^4 - 7x^3 + 2x^2 + 29x - 3) \, dx$; (b) $\int (x - 1)^2 \, dx$; (c) $\int (x - 2)(x - 3) \, dx$.

Since $\dfrac{d}{dx} \left(\dfrac{x^{n+1}}{n + 1} \right) = x^n$, where n is a positive integer, we have

$$\int x^n \, dx = \frac{x^{n+1}}{n + 1} + c.$$

The reader should verify that this formula also holds when $n = 0$. In fact, it is true for all real values of n except -1, when the right-hand side becomes meaningless; we shall have to consider $\int x^{-1} \, dx$ separately, later.

The second definition of the integral of $f(x)$ is as an *area under the graph* of $y = f(x)$. Before being more precise about this, we shall work through an example, which will lead also later to the third definition, namely as the *limit of the sum of a number of small quantities.* *

EXAMPLE 4.2 *To find the area 'under' the graph of $y = x^2$, between $x = 0$ and $x = a$.*

The area required is that bounded by the curve, the x-axis and the line $x = a$ (Fig. 4.2).

We first divide the interval between 0 and a on the x-axis into n parts, each of length δx, so that $n \, \delta x = a$.

Thus $OA_1 = A_1A_2 = \ldots = A_{n-1}A_n = \delta x$. Since $OA_1 = \delta x$, $A_1P_1 = (\delta x)^2$; $OA_2 = 2 \, \delta x$, $A_2P_2 = (2 \, \delta x)^2$, and so on. $OA_{n-1} = (n-1) \, \delta x$, $A_{n-1}P_{n-1} = \{(n-1) \, \delta x\}^2$; $OA_n = n \, \delta x$, $A_nP_n = (n \, \delta x)^2$. We next construct two 'staircases' consisting of rectangles, as in the diagrams (Fig. 4.3). The area A which we seek is greater than the area under the first 'staircase', shaded vertically, and less than the area under the second, shaded horizontally. (The reader should superpose both staircases, preferably in

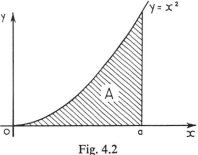

Fig. 4.2

different colours, on one large sketch of the parabola.)

Now the area under the first staircase is

$$A_1P_1 . \delta x + A_2P_2 . \delta x + A_3P_3 . \delta x + \ldots + A_{n-1}P_{n-1} . \delta x$$

$$= \delta x[(\delta x)^2 + (2 \, \delta x)^2 + (3 \, \delta x)^2 + \ldots + \{(n-1) \, \delta x\}^2]$$

$$= (\delta x)^3 \{1^2 + 2^2 + 3^2 + \ldots + (n-1)^2\}.$$

Similarly, the area under the second staircase is

$$(\delta x)^3 \{1^2 + 2^2 + 3^2 + \ldots + (n-1)^2 + n^2\}.$$

We thus have

$$(\delta x)^3 \{1^2 + 2^2 + \ldots + (n-1)^2\} < A < (\delta x)^3 \{1^2 + 2^2 + \ldots + n^2\}.$$

Now $1^2 + 2^2 + \ldots + n^2 = \sum_{r=1}^{n} r^2 = \frac{1}{6}n(n+1)(2n+1)$, (cf. Appendix 1 for the derivation of this formula, and the Σ notation), so

$$1^2 + 2^2 + \ldots + (n-1)^2 = \frac{1}{6}(n-1)n(2n-1),$$

and we have

$$(\delta x)^3 . \frac{1}{6}(n-1)n(2n-1) < A < (\delta x)^3 . \frac{1}{6}n(n+1)(2n+1).$$

---

* The reader should work through Example 4.2 very carefully, drawing his own diagrams and checking each step.

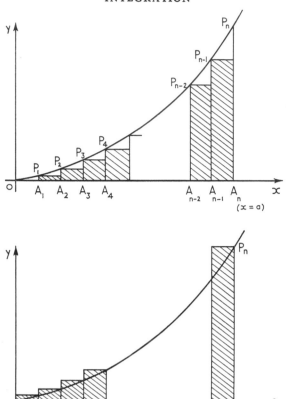

Fig. 4.3

This can be written

$$\tfrac{1}{6}(n\,\delta x - \delta x)(n\,\delta x)(2n\,\delta x - \delta x) < A < \tfrac{1}{6}(n\,\delta x)(n\,\delta x + \delta x)(2n\,\delta x + \delta x)$$

or, remembering that $n\,\delta x = a$,

$$\tfrac{1}{6}(a - \delta x)a(2a - \delta x) < A < \tfrac{1}{6}a(a + \delta x)(2a + \delta x).$$

We now let $\delta x \to 0$ and n become large in such a way that $n\,\delta x$ remains equal to a, that is, we increase the number of small sub-divisions of OA_n, correspondingly decreasing the length of each. Now, as $\delta x \to 0$,

$$\tfrac{1}{6}(a - \delta x)a(2a - \delta x) \quad \text{and} \quad \tfrac{1}{6}a(a + \delta x)(2a + \delta x)$$

both tend to $\tfrac{1}{6} . 2a^3 = \dfrac{a^3}{3}$.* As A lies between these amounts, however small we take δx to be, it follows that $A = \dfrac{a^3}{3}$. (This at least ties up roughly with the first definition of integration, by which $\int x^2\,dx = \tfrac{1}{3}x^3 + c$.)

---

* The difference between the areas under the two staircases is, in fact, $a^2\,\delta x$. The reader should interpret this in terms of his sketch.

1. Use the method of Example 4.2 to find the area under the graph between $x = 0$ and $x = a$ of (i) $y = x^3$, (ii) $y = x$. [Hint for (i): $1^3 + 2^3 + \ldots + n^3 = \frac{1}{4}n^2(n + 1)^2$, cf. Appendix 1.]

We shall now link the first two definitions of integration. Both are fruitful: the first, 'the opposite of differentiation', gives a method for evaluating a large class of integrals, while the second, 'area under graph', leads to the third, more general, definition and practical applications.

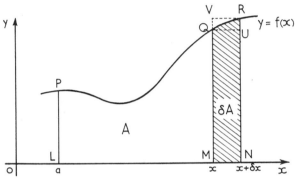

Fig. 4.4

Consider the graph (Fig. 4.4) of $y = f(x)$, a continuous function of x, and let the area $LMQP$ 'under' the arc PQ be A. Let the x-coordinate of L be a, and that of M be x (a will remain fixed but we shall give the area a small increment in such a way that x will vary). As x varies, not only is y a function of x, but so also is the area under the graph. If we make a small increase in x, say δx, then the corresponding small increase in A, δA, is the area $MNRQ$. Now we are considering two 'rival' definitions of integration, (i) as the opposite of differentiation, (ii) as the area under a graph. If the two are to be equivalent, when we differentiate the area we must come back to the function itself, i.e. we require that $\dfrac{dA}{dx} = y$.

We assume for the moment that $f(x)$ is increasing between x and $x + \delta x$; then from the diagram we have

$$MNRV > \delta A > MNUQ.$$

Now since $OM = x, ON = x + \delta x$, we have $MQ = f(x), NR = f(x + \delta x)$ so that $MNRV = f(x + \delta x) \cdot \delta x$, $MNUQ = f(x) \cdot \delta x$, and the inequality, can be written

$$f(x + \delta x) \cdot \delta x > A > f(x) \cdot \delta x.$$

Thus $f(x + \delta x) > \dfrac{\delta A}{\delta x} > f(x)$. Letting $\delta x \to 0$, and remembering that $f(x)$ is continuous, it follows that $\dfrac{dA}{dx} = \lim\limits_{\delta x \to 0} \dfrac{\delta A}{\delta x} = f(x)$, as required.

The reader should verify that the argument can be adapted if $f(x)$ is decreasing between x and $x + \delta x$, the inequality signs then being reversed. Assuming that we can always choose δx small enough so that $f(x)$ is either increasing or decreasing between x and $x + \delta x$, it follows that $\dfrac{dA}{dx} = f(x)$ for all x within a given interval for which $f(x)$ is continuous. Thus, *if A is the area under the graph of* $y = f(x)$, *we have* $\dfrac{dA}{dx} = y$, *and A is an integral of y with respect to x.*

We have seen, however, that if A is an integral of y, so also is $A + c$, where c is an arbitrary constant; $A + c$ is the *indefinite integral* of y with respect to x. This can be explained by the arbitrary choice of LP in Fig. 4.4; shifting LP to the right or left (by different choice of $a = OL$) will alter A by some definite amount but, once having fixed the 'start line' LP, the relationship $\dfrac{dA}{dx} = y$ will hold, independently of the choice of a. When computing an actual area under a curve between two definite limits, the arbitrary constant must, of course, disappear. This is illustrated in the next example.

EXAMPLE 4.3   *To find the area under the curve* $y = x^3 + 1$ *between* $x = 2$ *and* $x = 5$.

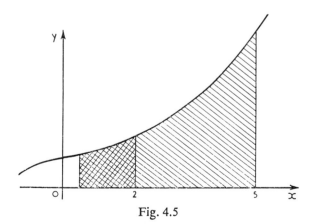

Fig. 4.5

The integral of $x^3 + 1$ is $\frac{1}{4}x^4 + x + c$, and this gives the area under the graph (Fig. 4.5) measured from some line parallel to the y-axis. Putting $x = 5$, we have $\dfrac{5^4}{4} + 5 + c$, while putting $x = 2$, we have $\dfrac{2^4}{4} + 2 + c$. The first expression is the area 'up to $x = 5$', the second the area 'up to $x = 2$'; the difference is the required area. This difference is written $[\frac{1}{4}x^4 + x + c]_2^5$, which is simply a compact way of writing 'the value of $\frac{1}{4}x^4 + x + c$ when $x = 5$ minus its value when $x = 2$'. Thus the required area is

$$[\tfrac{1}{4}x^4 + x + c]_2^5 = \left(\frac{5^4}{4} + 5 + \cancel{c}\right) - \left(\frac{2^4}{4} + 2 + \cancel{c}\right) = 155\tfrac{1}{4}.$$

To emphasize that the integral is between the values $x = 2$ and $x = 5$ (the 'limits of integration'), we insert these values at the ends of the $\int$ sign, the 'integral from $x = 2$ to $x = 5$' being written as $\displaystyle\int_2^5 (x^3 + 1)\,dx$. As the arbitrary constant c is 'cancelled' and so irrelevant, it is in practice left out altogether when such a *definite integral* (i.e. one between two fixed limits) is being considered.

EXAMPLE 4.4   *To find the area under the curve* $y = 3x^2 + 2x + 4$ *between* $x = -1$ *and* $x = 3$.

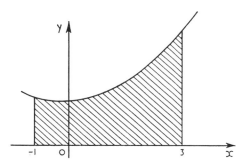

Fig. 4.6

The required area (Fig. 4.6) is given by

$$\int_{-1}^3 (3x^2 + 2x + 4)\,dx = [x^3 + x^2 + 4x]_{-1}^3$$

$$= (3^3 + 3^2 + 4 \times 3) - \{(-1)^3 + (-1)^2 + 4(-1)\}$$

$$= 48 - (-1 + 1 - 4) = 52.$$

EXERCISE XVI

1. Evaluate: (i) $\int x^4 \, dx$; (ii) $\int (x^3 - 3x + 1) \, dx$.

2. Evaluate: (i) $\int_3^5 x^4 \, dx$; (ii) $\int_1^4 (x^2 - 3x + 1) \, dx$;

 (iii) $\int_4^5 (x - 2)(x - 3) \, dx$; (iv) $\int_3^4 x(x - 1)(x - 2) \, dx$.

3. Find the area under the curve $y = 3x^2 + x + 1$ between $x = 0$ and $x = 2$.

4. Find the area under the curve $y = 4x^3 + 3x^2 + 2x + 1$ between $x = 1$ and $x = 3$.

5. Find the area under the curve $y = 2x^2 - x + 13$ between $x = -3$ and $x = -1$.

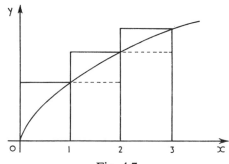

Fig. 4.7

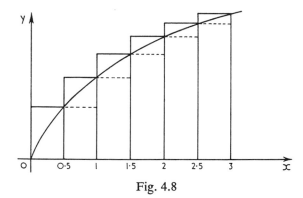

Fig. 4.8

6. (i) Figure 4.7 represents the graph of $y = x^{\frac{1}{2}}$; also indicated are two 'staircases' as in Example 4.2 above. By considering the areas under the two

staircases, prove that $1 + \sqrt{2} < \int_0^3 x^{\frac{1}{2}} \, dx < 1 + \sqrt{2} + \sqrt{3}$.

(ii) **By taking the intervals for the staircases along the x-axis as $\frac{1}{2}$ instead of 1
(Fig. 4.8), show that $\int_0^3 x^{\frac{1}{2}} \, dx$ lies between 2·96 and 3·83.**

7. Show that $0{\cdot}72 < \int_0^1 \frac{1}{1 + x^2} \, dx < 0{\cdot}85$. (See Fig. 4.9).

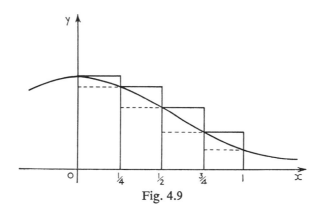

Fig. 4.9

8. Show that $\frac{1}{2} + \frac{1}{3} + \frac{1}{4} + \frac{1}{5} < \int_1^5 \frac{1}{x} \, dx < 1 + \frac{1}{2} + \frac{1}{3} + \frac{1}{4}$, and give a similar inequality for $\int_1^n \frac{1}{x} \, dx$, where n is any positive integer (see Fig. 4.10).

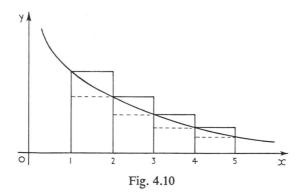

Fig. 4.10

We have seen that indefinite integrals include an arbitrary constant. It is instructive to look at this arbitrary constant from another point of view.

EXAMPLE 4.5   *To sketch the curves* $y = x^2$, $y = x^2 - 1$, $y = x^2 + 2$, $y = x^2 + 5$, *and give the gradient of each when* $x = 3$.

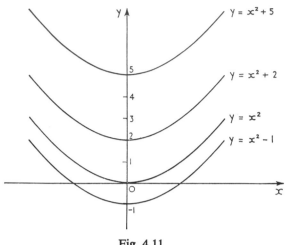

Fig. 4.11

The curves (Fig. 4.11) have the same gradient when $x = 3$, namely 6, and indeed the gradient of each curve is $2x$ for any given x.

Starting from the relationship 'gradient is $2x$', i.e. $\dfrac{dy}{dx} = 2x$, we can find the whole family of curves for which this relationship holds by integrating, viz. $y = x^2 + c$. To fix the arbitrary constant c, it is necessary to have a further piece of information, namely one point through which the curve passes.

EXAMPLE 4.6   *To find the equation of the curve whose gradient is $2x$ and which passes through the point* (7, 13).

From above, the equation of the curve must be of the form $y = x^2 + c$. The condition for it to pass through (7, 13) is that $13 = 7^2 + c$, so that $c = -36$, and the required equation is thus $y = x^2 - 36$.

Integral as the Limit of a Sum

We now return to the idea of the 'staircases' introduced in Example 4.2 and generalize as follows. Let $f(x)$ be a continuous function of x for $a \leqslant x \leqslant b$, and suppose we divide the interval from a to b into a number, n, of subintervals, not necessarily equal, so that $a = x_0 < x_1 < x_2 < x_3 \ldots < x_{n-1} < x_n = b$. (See Fig. 4.12.)

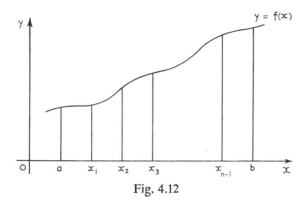

Fig. 4.12

Let $\xi_1, \xi_2, \ldots, \xi_n$ be any values of x such that $a < \xi_1 < x_1$, $x_1 < \xi_2 < x_2, \ldots, x_{n-1} < \xi_n < b$, and consider the sum

$$f(\xi_1)(x_1 - x_0) + f(\xi_2)(x_2 - x_1) + \ldots + f(\xi_n)(x_n - x_{n-1}).$$

This can be written more shortly as $\Sigma_{r=1}^{n} f(\xi_r)(x_r - x_{r-1})$, and represents the sum of the areas of the set of rectangles shown in Fig. 4.13, with base $x_r - x_{r-1}$ and height $f(\xi_r)$, $(r = 1 \text{ to } n)$.

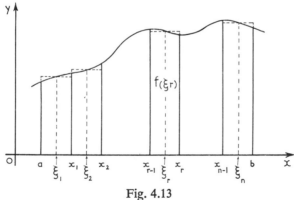

Fig. 4.13

We now increase n so that the intervals $(x_r - x_{r-1})$ all tend to zero, and assume that, *no matter how this is done, the sum $\Sigma_{r=1}^{n} f(\xi_r)(x_r - x_{r-1})$ tends to a definite limit.* This limit we take as our third definition of the integral of $f(x)$ with respect to x.* Thus

$$\lim_{n \to \infty} \Sigma_{r=1}^{n} f(\xi_r)(x_r - x_{r-1}) = \int_a^b f(x)\, dx.$$

---

* We have assumed that $f(x)$ is continuous for $a \leqslant x \leqslant b$. It can be proved that this is sufficient for the integral, defined in this way, to exist, but it is not necessary. Cf. Exercise XVII, question 12.

We can take the left-hand side as a formal definition of the area under the graph of $f(x)$, but this definition of the integral as the limit of a sum is *independent of the idea of area*, and indeed will be applied in the next chapter to other problems, such as the determination of volumes and centres of mass.

The mystery of the notation '$\int \ldots dx$' can now be cleared up. Our final definition of the integral is as the limit of the sum $\Sigma_{r=1}^{n} f(\xi_r) \delta x$ or, more shortly still, $\Sigma y \, \delta x$, where $\delta x = x_r - x_{r-1}$ when y takes the value of $f(x)$ for some x (namely ξ_r) between x_{r-1} and x_r. Writing $\lim_{\delta x \to 0} \Sigma_{x=a}^{b} y \, \delta x$ as

$\int_a^b y \, dx$, i.e. $\int_a^b f(x) \, dx$, the $\int$ is merely a long S for sum and the whole notation $\int \ldots dx$' is a reminder of the derivation from '$\lim \Sigma \ldots \delta x$'.

From the definition as limit of a sum, we have immediately that

$$\int_a^b f(x) \, dx = \int_a^c f(x) \, dx + \int_c^b f(x) \, dx.$$

Also $\int_b^a f(x) \, dx = -\int_a^b f(x) \, dx$, since the small increments $(x_r - x_{r-1})$ are replaced by $(x_{r-1} - x_r)$ if the range of integration is reversed as 'b to a' instead of 'a to b'. We also have

$$\int_a^b cf(x) \, dx = c\int_a^b f(x) \, dx \qquad (c \text{ a constant}),$$

and

$$\int_a^b \{f(x) + g(x)\} \, dx = \int_a^b f(x) \, dx + \int_a^b g(x) \, dx.$$

The following propositions are also fairly 'obvious' intuitively. We do not attempt here to give formal proofs, but the reader should draw his own sketch-graphs to illustrate them in terms of areas.

1. If $f(x)$, $g(x)$ are continuous and $f(x) > g(x)$ for all x such that $a \leqslant x \leqslant b$, then $\int_a^b f(x) \, dx > \int_a^b g(x) \, dx$.

2. If M, m are the greatest and least values attained by the continuous function $f(x)$ in the interval $a \leqslant x \leqslant b$, then $\int_a^b m \, dx \leqslant \int_a^b f(x) \, dx \leqslant \int_a^b M \, dx$, i.e. $m(b - a) \leqslant \int_a^b f(x) \, dx \leqslant M(b - a)$.

3. If $f(x)$ is continuous in the interval $a \leqslant x \leqslant b$, then there is a number ξ between a and b such that

$$\int_a^b f(x) \, dx = (b - a)f(\xi).$$

$f(\xi)$ as defined by this last equation is the *average* or *mean value* of $f(x)$ over the interval $a \leqslant x \leqslant b$.

Negative Areas

The areas 'under' graphs which we have considered so far have been conveniently above the x-axis. If, for a certain range of x, the graph is below the x-axis, the values of y in $\Sigma\, y\, \delta x$ will be negative and so the area $\int y\, dx$, being the limit of this sum, will also be reckoned as negative.

EXAMPLE 4.7   *To evaluate* $\displaystyle\int_{-1}^{3} (x^2 - 2x - 3)\, dx.$

$$\int_{-1}^{3} (x^2 - 2x - 3)\, dx = [\tfrac{1}{3}x^3 - x^2 - 3x]_{-1}^{3}$$
$$= -9 - (-\tfrac{1}{3} - 1 + 3) = -10\tfrac{2}{3}.$$

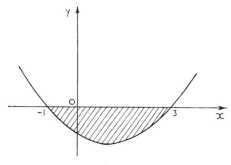

Fig. 4.14

Care must be taken if part of the graph is above the x-axis and part below; if the ·actual amount of space enclosed is required (instead of the mathematical 'area'), it is essential to draw a sketch-graph and compute the parts above and below separately.

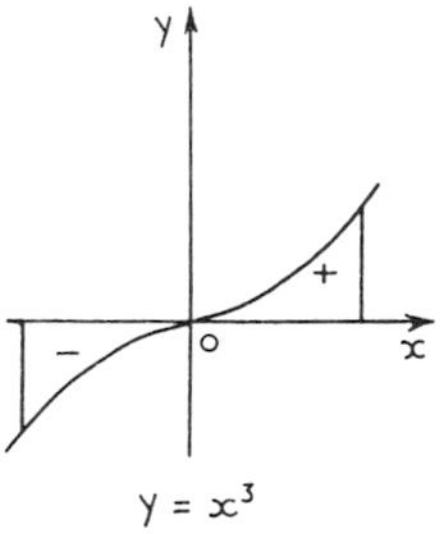

EXAMPLE 4.8   *To evaluate* $\displaystyle\int_{-1}^{1} x^3\, dx.$

$$\int_{-1}^{1} x^3\, dx = [\tfrac{1}{4}x^4]_{-1}^{1} = \tfrac{1}{4} - \tfrac{1}{4} = 0.$$

$y = x^3$

Fig. 4.15

The graphical interpretation of this is that the area between $x = -1$ and $x = 0$ is negative (Fig. 4.15). In fact,

$$\int_{-1}^{1} x^3\, dx = \int_{-1}^{0} x^3\, dx + \int_{0}^{1} x^3\, dx,$$

and

$$\int_{-1}^{0} x^3 \, dx = \left[\frac{x^4}{4}\right]_{-1}^{0} = 0 - \tfrac{1}{4} = -\tfrac{1}{4},$$

while

$$\int_{0}^{1} x^3 \, dx = \left[\frac{x^4}{4}\right]_{0}^{1} = \tfrac{1}{4} - 0 = +\tfrac{1}{4}.$$

More generally, if $f(-x) = -f(x)$, then $f(x)$ is an *odd* function of x, and for such a function $\int_{-a}^{a} f(x) \, dx = 0$. On the other hand, if $g(-x) = g(x)$, $g(x)$ is an *even* function, the graph is symmetrical about the y-axis, and $\int_{-a}^{a} g(x) \, dx = 2 \int_{0}^{a} g(x) \, dx$ (cf. Exercise XVII, question 10, below).

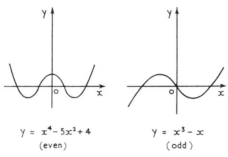

$$y = x^4 - 5x^2 + 4 \qquad\qquad y = x^3 - x$$
$$\text{(even)} \qquad\qquad\qquad \text{(odd)}$$

Fig. 4.16

Exercise XVII

In questions 1 to 7, evaluate the integrals, illustrating each in terms of areas by means of a rough sketch.

1. $\displaystyle\int_{-3}^{2} (x + 2) \, dx.$

2. $\displaystyle\int_{-2}^{4} (3x - 1) \, dx.$

3. $\displaystyle\int_{-1}^{2} (x^2 - x + 5) \, dx.$

4. $\displaystyle\int_{-1}^{2} (x^2 - x - 2) \, dx.$

5. $\displaystyle\int_{-1}^{2} (x^2 - x - 1) \, dx.$

6. $\displaystyle\int_{-2}^{3} (x^3 + x^2) \, dx.$

7. $\displaystyle\int_{0}^{2} x(x - 1)(x - 2) \, dx.$

8. Find y in terms of x given that $\dfrac{dy}{dx} = x^2 + 4$ and that $y = 1$ when $x = 2$.

9. Find the equation of the curve whose gradient is $x^3 - 3x^2 + 1$ and which passes through the point $(5, 4)$.

10. Classify the following functions as either: (i) *even*, (ii) *odd*, or (iii) *neither*:
(a) x^4; (b) $x^2 + 1$; (c) x^3; (d) $x^5 + 3x^3$; (e) $x^3 + 2x^2$. Integrate each of these functions with respect to x, (i) from 0 to 1; (ii) from -1 to 0; (iii) from -1 to $+1$.

11. *Write down* the values of the following integrals.

(i) $\displaystyle\int_{-5}^{5} x^3\, dx$; (ii) $\displaystyle\int_{-7}^{7} (5x^9 - 17x^7 + 4x^5 + 7x + 1)\, dx$.

12. Evaluate $\displaystyle\int_{0}^{3} [x]\, dx$. (For definition of $[x]$, see Exercise VIII, question 30, p. 30.

Satisfy yourself that $\displaystyle\lim_{\delta x \to 0} \Sigma_{x=0}^{3} f(x)\, \delta x$ exists when $f(x) = [x]$.)

13. Find the average value of x^2 over the range $0 \leqslant x \leqslant 1$.

Miscellaneous questions (*in sets of* 4).

14. The curve $y = 6 - x - x^2$ cuts the x-axis in two points A and B. By integration find the area enclosed by the x-axis and that portion of the curve which lies between A and B.

15. (i) Find $\dfrac{dy}{dx}$ when $y = (1 + 3x)^2$.

(ii) Deduce the equation of the tangent at $(2, 49)$ to the curve whose equation is $y = (1 + 3x)^2$.

16. Evaluate: (i) $\dfrac{d}{dx}(7x^3 - 4x^2 + 13x)$; (ii) $\int (21x^2 - 8x + 13)\, dx$.

17. Calculate

$$\int_{-1}^{1} x(x^2 - 1)\, dx.$$

Find the area bounded by the curve $y = x(x^2 - 1)$ and the x-axis (i) between $x = -1$ and $x = 0$ and (ii) between $x = 0$ and $x = 1$. Explain with the aid of a rough figure the connexion between your results and the value of the integral found in the first part.

18. On squared paper, taking 2 cm as unit on both axes, sketch the curve $10y = (x + 2)^3$ between $x = -4$ and $x = 2$.

By integration, find the area between the curve, the x-axis, and the line $x = 2$.

19. Find the equation of the curve which passes through the point $(0, 1)$ and whose gradient at the point (x, y) is given by

$$\frac{dy}{dx} = 4x^3 - 4x.$$

Find the values of x for which y is a minimum and draw a rough sketch of the curve.

20. Evaluate: (i) $\displaystyle\int_{0}^{3} (2x^4 - 7x^2 + 3)\, dx$; (ii) $\displaystyle\int_{-3}^{3} (2x^4 - 7x^2 + 3)\, dx$;

(iii) $\displaystyle\int_{-3}^{3} (x^5 + 2x^4 - 17x^3 - 7x^2 + 41x + 3)\, dx$.

21. Make a rough sketch of the curve $y = x(x - 1)(x - 3)$ from $x = -1$ to $x = 4$. Find
    (i) the equation of the tangent to the curve at the origin;
    (ii) the area enclosed by the axis of x and the portion of the curve lying between $x = 1$ and $x = 3$.

22. Sketch the curve $y = (x - 2)^2$ between the values $x = 0$ and $x = 5$. Find by integration the area enclosed by the x-axis, the line $x = 1$, and the curve.

23. (i) Use 'staircases' to show that

$$\frac{1}{11^2} + \frac{1}{12^2} < \int_{10}^{12} x^{-2}\, dx < \frac{1}{10^2} + \frac{1}{11^2}.$$

(ii) Show that $0{\cdot}174 < \displaystyle\int_{10}^{12} \frac{1}{x}\, dx < 0{\cdot}191$.

24. Integrate $6x^2 - 4x + 5$ with respect to x. If the result is equal to 30 when $x = 3$, find its value when $x = 4$.

25. Find the equation of the tangent at $(3, 6)$ to the curve $y = x^3 - 6x^2 + 11x$, and find the coordinates of the point where this tangent meets the curve again.
    Sketch the curve for values of x from 0 to 3, and find the area cut off between the curve, the line $x = 3$, and the x-axis.

5
Applications of Integration

Many a quaint and curious volume.

E. A. POE

We have seen that the area 'under' the graph of $y = f(x)$ between $x = a$ and $x = b$ is $\int_a^b y \, dx$ (Fig. 5.1). It is, however, essential not to learn '$\int y \, dx$' as a formula, but to go back, mentally at least, to the idea of this integral

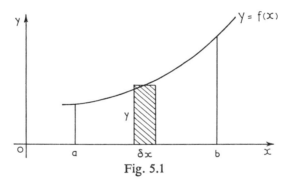

Fig. 5.1

as the limit as $\delta x \to 0$ of $\Sigma_{x=a}^b y \, \delta x$. We recall also that the continuity of $f(x)$ ensures that $\lim_{\delta x \to 0} \Sigma_{x=a}^b f(x) \, \delta x = \int_a^b f(x) \, dx$, and that this relationship does not necessarily refer to areas under graphs at all. In this chapter, we shall give a number of other applications of integration, in addition.

EXAMPLE 5.1 *To find the area bounded by the curve $y^2 = x$, the y-axis and the line $y = 3$.*

We divide the y-axis between 0 and 3 into a number of small intervals such as AB, of length δy. The required area is the limit of the sum of rectangles such as $ABCD$ (Fig. 5.2), i.e.

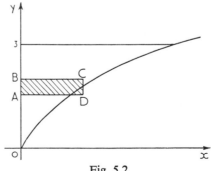

Fig. 5.2

$\lim\limits_{\delta y \to 0} \Sigma_{y=0}^{3} x \, \delta y$. Now as x is a continuous function of y in the interval $0 \leqslant y \leqslant 3$, it follows that this limit

$$= \int_{0}^{3} x \, dy$$

$$= \int_{0}^{3} y^2 \, dy$$

$$= [\tfrac{1}{3} y^3]_{0}^{3}$$

$$= 9.$$

EXAMPLE 5.2 *To find the area enclosed by the curves $y = x^2$ and $y = x^3$.*

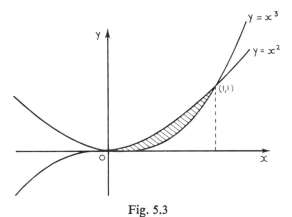

Fig. 5.3

Method 1. The curves meet (Fig. 5.3) at the origin and at $(1, 1)$. The required area is the difference between the areas 'under' the two curves, from $x = 0$ to $x = 1$, i.e.

$$\int_{0}^{1} x^2 \, dx - \int_{0}^{1} x^3 \, dx = \int_{0}^{1} (x^2 - x^3) \, dx$$

$$= [\tfrac{1}{3} x^3 - \tfrac{1}{4} x^4]_{0}^{1}$$

$$= \tfrac{1}{3} - \tfrac{1}{4} = \tfrac{1}{12}.$$

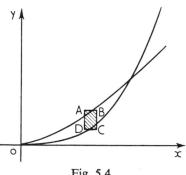

Fig. 5.4

Method 2. Alternatively, the area may be considered as the limit of the sum of small rectangles such as $ABCD$ in Fig. 5.4 (the sides of $ABCD$ being parallel to the axes and a point on $y = x^2$ lying between A and B, a point on $y = x^3$ between C and D). If y_2, y_1 are the respective heights of AB, CD above the x-axis, the required area is

$$\lim_{\delta x \to 0} \sum_{x=0}^{1} (y_2 - y_1)\, \delta x = \int_0^1 (x^2 - x^3)\, dx = \tfrac{1}{12},$$

as above.

EXERCISE XVIII

Solutions of questions 2, 5 and 6 should be preserved, for use with Exercise XX.

1. Find the area bounded by the curve $y^3 = x$, the y-axis and the line $y = 4$.
2. Find the area bounded by the curve $y^2 = 4x$, the y-axis and the lines $y = 1$ and $y = 3$.
3. Sketch the curve $x = y(y - 2)$ and find the area enclosed between this curve and the y-axis.
4. Find the area enclosed between the curve $y = x^4$ and the straight line $y = 8x$. [Hint: Where the line meets the curve, we have $x^4 = 8x$, i.e. $x = 0$ or 2.]
5. Find the area enclosed between the parabola $y = x^2$ and the straight line $y = 5x - 6$.
6. Find the area bounded by the curve $y = x^5 + 1$ and the lines $x = 3$ and $y = 1$.
7. Find the area, enclosed between the curves $y = x^3 + 1$ and $y = 4x^2 - 5x + 3$.

Volume of Solid of Revolution

If a curve is rotated about an axis which it does not cut, the area between the curve and the axis will generate a *solid of revolution*. For example, if the

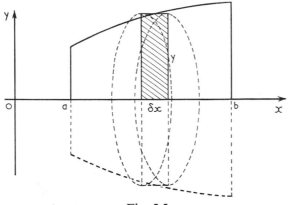

Fig. 5.5

portion of the curve $y = f(x)$ in Fig. 5.5 ($a \leqslant x \leqslant b$) is rotated about the x-axis, the area under the curve will generate a solid in the shape of a pudding-basin. Just as the area under the curve can be considered as the limit of the sum of rectangles $y \cdot \delta x$, so the volume generated is the limit of the sum of the volumes of cylinders formed by the rotation of these rectangles.

When the shaded area $y \cdot \delta x$ is rotated, a penny-shaped disc is formed, namely a cylinder with radius y and height δx whose volume, $\pi r^2 h$, is $\pi y^2 \, \delta x$.

Thus the volume of the solid of revolution is $\lim\limits_{\delta x \to 0} \Sigma_{x=a}^{b} \pi y^2 \, \delta x$, i.e. $\displaystyle\int_a^b \pi y^2 \, dx$.

EXAMPLE 5.3   *The part of the circle $x^2 + y^2 = a^2$ which lies above the x-axis is rotated about this axis through four right angles. To find the volume of the sphere thus generated.*

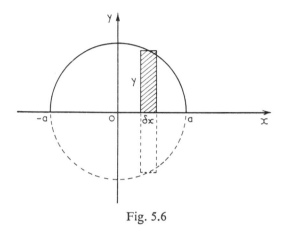

Fig. 5.6

Refer to Fig. 5.6. The volume required is

$$\lim_{\delta x \to 0} \Sigma_{x=-a}^{a} \pi y^2 \, \delta x = \int_{-a}^{a} \pi y^2 \, dx$$

$$= \pi \int_{-a}^{a} (a^2 - x^2) \, dx$$

$$= \pi \left[a^2 x - \tfrac{1}{3} x^3 \right]_{-a}^{a}$$

$$= \pi \{ (a^3 - \tfrac{1}{3} a^3) - (-a^3 + \tfrac{1}{3} a^3) \}$$

$$= \tfrac{4}{3} \pi a^3 .$$

(Note that we have used the cylinder formula $\pi r^2 h$ in the course of establishing the formula $\tfrac{4}{3} \pi a^3$ for the volume of a sphere.)

EXAMPLE 5.4   *The portion of the curve $y = 3x^2$ between $y = 1$ and $y = 3$ is rotated through 4 right angles about the y-axis. To find the volume of the resulting solid.*

Dividing the area in the first quadrant into strips as shown in Fig. 5.7, the required volume is

$$\lim_{\delta y \to 0} \sum_{y=1}^{3} \pi x^2 \, \delta y = \int_1^3 \pi x^2 \, dy$$

$$= \int_1^3 \tfrac{1}{3}\pi y \, dy$$

$$= \tfrac{1}{3}\pi \left[\tfrac{1}{2}y^2 \right]_1^3$$

$$= \tfrac{4}{3}\pi.$$

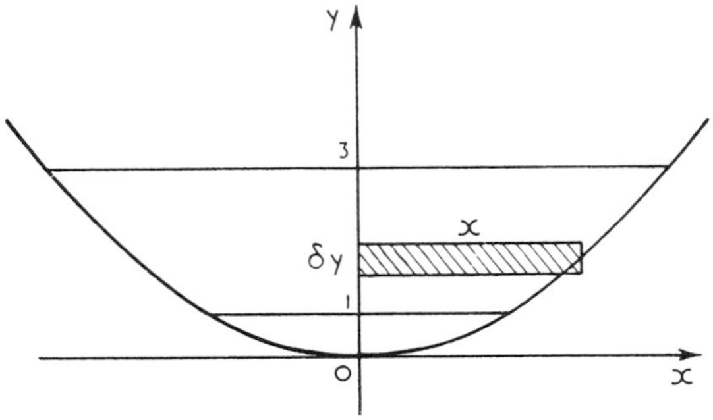

Fig. 5.7

EXAMPLE 5.5 *The area enclosed by the curves $y^2 = 8x$ and $y = \tfrac{1}{2}x^3$ is rotated through 4 right angles about the axis of x. To find the volume of the solid formed.*

Where the curves meet, $8x = y^2 = \tfrac{1}{4}x^6$, yielding the origin and the point (2, 4). The required volume is the difference between the volumes formed by rotating the areas under the curves between $x = 0$ and $x = 2$ about the x-axis, i.e.

$$\int_0^2 \pi(8x) \, dx - \int_0^2 \pi(\tfrac{1}{2}x^3)^2 \, dx$$

$$= \pi \int_0^2 (8x - \tfrac{1}{4}x^6) \, dx = \pi \left[4x^2 - \frac{x^7}{28} \right]_0^2 = \frac{80\pi}{7}.$$

EXERCISE XIX

Solutions of questions 3, 5 and 8 should be kept, for use with Exercise XX.

1. (D) The segment of the straight line $y = mx$ between the origin and the point (a, ma) is rotated about the x-axis. (i) What is the shape of the solid formed? (ii) Use integration to find the volume generated, and reconcile your answer with the formula $\tfrac{1}{3}\pi r^2 h$.

In questions 2 to 7, find the volumes of the solids formed when the areas whose boundaries are described are rotated through 4 right angles about the given axes.

2. The curve $y^2 = 4x$, the x-axis and the line $x = 3$; about the x-axis.

3. The curve $y = x^2 + 3$, the x-axis and the lines $x = 1$ and $x = 2$; about the x-axis.

4. The curve $y = x^2 + 2$ and the lines $x = 1$, $x = 3$ and $y = 1$; about the line $y = 1$.

5. The curve $y = x^3$, the y-axis and the line $y = 1$; about the y-axis.

6. The curve $y = x^2$, the x-axis and the line $x = 2$; (i) about the x-axis, (ii) about the y-axis.

7. The circle $x^2 + y^2 = 25$, the x-axis and the lines $x = 2$ and $x = 3$; about the x-axis.

8. Find the volume generated when the area enclosed by the curve $y = x(1 - x)$ and the x-axis is rotated about the x-axis.

9. The area cut off between the curve $6y = x^2 + x$ and the straight line $y = x - 1$ is rotated through 4 right angles about the x-axis. Find the volume of the resulting solid.

10. The area between the curves $y^2 = 128x$ and $y = x^4$ is rotated about the x-axis. Find the volume of the solid thus generated.

Centre of Mass

The x-coordinate of the *centre of mass* of a set of particles of mass m_i at points (x_i, y_i) $(i = 1, 2, \ldots, n)$ is denoted by $\bar{x}$, and is defined by the equation

$$m_1 x_1 + m_2 x_2 + \ldots + m_n x_n = (m_1 + m_2 + \ldots + m_n)\bar{x},$$

or, more briefly, $\Sigma(mx) = (\Sigma m)\bar{x}$. Similarly the y-coordinate, $\bar{y}$, is given by $\Sigma(my) = (\Sigma m)\bar{y}$.

It is shown in books on mechanics that if the acceleration due to gravity is constant in magnitude and direction, then the *centre of gravity* of a body (made up of a system of particles) is the same as its centre of mass. The word *centroid* is often used indiscriminately to denote either, though strictly it should be confined to the centre of mass of a lamina.

We shall abbreviate 'centre of mass' to C.M.

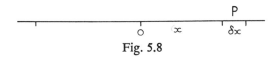

Fig. 5.8

The C.M. of a rigid body may be defined by considering it as made up of a number of small elements. As a simple example, we consider a uniform*

---

* I.e. the *density*, or mass per unit length (rod), area (lamina) or volume (solid) is constant. We assume such uniformity throughout this chapter.

rod of mass M and length $2a$. Let the point P in Fig. 5.8 be within a small element of length δx and at a distance x from the mid-point of the rod, which we take as origin. The mass of the element is $\rho\,\delta x$, where ρ is the density of the rod, and by analogy with $\Sigma mx = (\Sigma m)\bar{x}$ we define the C.M. of the rod by means of the equation

$$\lim_{\delta x \to 0}\sum_{x=-a}^{a} x \cdot \rho\,\delta x = M\bar{x},$$

so that $M\bar{x} = \displaystyle\int_{-a}^{a} x\rho\,dx = 0$, whence $\bar{x} = 0$ and the C.M. is (reasonably) at the mid-point.

Similarly, it follows from considerations of symmetry that the C.M. of a rectangle is also at its geometrical centre.

If two bodies of masses m_1, m_2 have $\bar{x}_1$, $\bar{x}_2$ as the respective x-coordinates of their C.M.'s (referred to given axes) then $\bar{x}$, the x-coordinate of the C.M. of the two combined, is given by

$$(m_1 + m_2)\bar{x} = m_1\bar{x}_1 + m_2\bar{x}_2.$$

This can be seen by considering each term in this equation as the limit of a sum; and the result can be generalized to cover any number of bodies.

We next consider the C.M. of the area under a curve. The procedure for other areas and for volumes is also illustrated in the examples which follow.

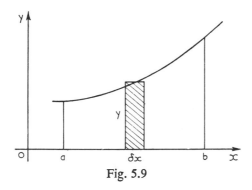

Fig. 5.9

Fig. 5.9 represents the area of a lamina bounded by the curve $y = f(x)$, the x-axis and the lines $x = a$ and $x = b$. As usual we consider the area as the limit of the sum of rectangles $y \cdot \delta x$, and we define the C.M. of the lamina by the equations

$$\lim_{\delta x \to 0}\sum_{x=a}^{b} x \cdot \rho y\,\delta x = \bar{x} \cdot \lim_{\delta x \to 0}\sum_{x=a}^{b} \rho y\,\delta x;$$

$$\lim_{\delta x \to 0}\sum_{x=a}^{b} \frac{y}{2} \cdot \rho y\,\delta x = \bar{v} \cdot \lim_{\delta x \to 0}\sum_{x=a}^{b} \rho y\,\delta x.$$

Given uniformity, the factor ρ for the density may be omitted. Note that the C.M. of the shaded rectangle is a distance x from the y-axis but $\dfrac{y}{2}$ from the x-axis; this of course accounts for the factor of $\frac{1}{2}$ in the equation for $\bar{y}$. Denoting the area by A, the above equations reduce to

$$\int_a^b xy\, dx = A\bar{x}; \qquad \int_a^b \tfrac{1}{2}y^2\, dx = A\bar{y}.$$

These equations should *not* be memorized, but each example should be worked out afresh, using limits of sums.

EXAMPLE 5.6 *To find the coordinates of the C.M. of the area under the curve* $y = x^2$ *from* $x = 0$ *to* $x = 3$.

Let the coordinates of the C.M. be $(\bar{x}, \bar{y})$. Then (at least mentally considering the area as the limit of the sum of rectangles), we have

$$\int_0^3 xy\, dx = \bar{x} \cdot \int_0^3 y\, dx,$$

i.e.

$$\int_0^3 x^3\, dx = \bar{x} \cdot \int_0^3 x^2\, dx,$$

whence

$$\bar{x} = \frac{\left[\dfrac{x^4}{4}\right]_0^3}{\left[\dfrac{x^3}{3}\right]_0^3} = \frac{3}{4} \times \frac{3^4}{3^3} = \frac{9}{4}.$$

Also

$$\int_0^3 \frac{y}{2} \cdot y\, dx = \bar{y} \cdot \int_0^3 y\, dx,$$

or

$$\tfrac{1}{2}\int_0^3 x^4\, dx = \bar{y} \cdot \int_0^3 x^2\, dx,$$

whence

$$\bar{y} = \frac{\dfrac{1}{2}\left[\dfrac{x^5}{5}\right]_0^3}{\left[\dfrac{x^3}{3}\right]_0^3} = \frac{1}{2} \times \frac{3}{5} \times \frac{3^5}{3^3} = \frac{27}{10}.$$

Thus the required coordinates of the C.M. are $(\frac{9}{4}, \frac{27}{10})$.

EXAMPLE 5.7 *To find the C.M. of the area enclosed by the curves* $y = x^2$ *and* $y = x^3$.

From Example 5.2 above, the area is $\frac{1}{12}$. Considering rectangles as in Method 2 of that example (p. 66), the x-coordinate of the C.M. is given by

$$\int_0^1 x(x^2 - x^3)\, dx = \tfrac{1}{12}\bar{x},$$

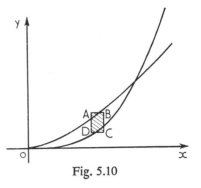

whence $\quad \bar{x} = 12\left[\dfrac{x^4}{4} - \dfrac{x^5}{5}\right]_0^1 = \dfrac{3}{5}.$

Now, using the notation of Example 5.2, Method 2, the height of the C.M. of the rectangle $ABCD$ above the x-axis is $\dfrac{y_1 + y_2}{2}$, so that the y-coordinate of the C.M. of the area, $\bar{y}$ is given by

Fig. 5.10

$$\lim \sum \frac{y_1 + y_2}{2} \cdot (y_2 - y_1)\, \delta x = \bar{y} \cdot \lim \sum (y_2 - y_1)\, \delta x$$

i.e.

$$\int_0^1 \frac{x^3 + x^2}{2}(x^2 - x^3)\, dx = \bar{y} \cdot \int_0^1 (x^2 - x^3)\, dx = \tfrac{1}{12}\bar{y};$$

whence

$$\bar{y} = 12\int_0^1 \tfrac{1}{2}(x^4 - x^6)\, dx = 6\left[\frac{x^5}{5} - \frac{x^7}{7}\right]_0^1 = \tfrac{12}{35}.$$

The required coordinates of the C.M. are therefore $(\tfrac{3}{5}, \tfrac{12}{35})$.

EXAMPLE 5.8 *The area bounded by the curve $y = x^2 + 1$, the axes of co-ordinates and the line $x = 3$ is rotated through 4 right angles about the x-axis. To find the C.M. of the volume generated.*

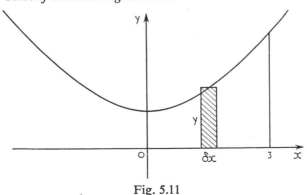

Fig. 5.11

Refer to Fig. 5.11. By symmetry, $\bar{y} = 0$. For $\bar{x}$, we consider the volume as the limit of the sum of penny-shaped cylinders of radius y and height δx.

We have, accordingly,

$$\int_0^3 x \cdot \pi y^2 \, dx = \bar{x} \int_0^3 \pi y^2 \, dx$$

i.e.

$$\pi \int_0^3 x(x^2 + 1)^2 \, dx = \pi \bar{x} \int_0^3 (x^2 + 1)^2 \, dx.$$

Thus

$$\int_0^3 (x^5 + 2x^3 + x) \, dx = \bar{x} \int_0^3 (x^4 + 2x^2 + 1) \, dx,$$

and

$$\bar{x} = \frac{\left[\dfrac{x^6}{6} + \dfrac{2x^4}{4} + \dfrac{x^2}{2}\right]_0^3}{\left[\dfrac{x^5}{5} + \dfrac{2x^3}{3} + x\right]_0^3} = \frac{\dfrac{3^6}{6} + \dfrac{3^4}{2} + \dfrac{3^2}{2}}{\dfrac{3^5}{5} + \dfrac{2 \times 3^3}{3} + 3} = \frac{555}{232}.$$

The C.M. is therefore at $(\frac{555}{232}, 0)$.

EXERCISE XX

1. (D) The part of the circle $x^2 + y^2 = a^2$ in the first quadrant is rotated about the x-axis, thus generating a hemisphere.
   (i) Interpret the equation

$$\int_0^a x \cdot \pi y^2 \, dx = \tfrac{2}{3}\pi a^3 \cdot \bar{x}.$$

   (ii) Verify that the C.M. of a uniform hemisphere of radius a is at a distance of $\frac{3}{8}a$ from the centre of its plane face.

2. Find the coordinates of the C.M. of the area bounded by the curve $y = x^3$, the x-axis and the line $x = 1$.

3–5. Find the C.M. of each of the areas described in Exercise XVIII, questions 2, 5 and 6.

6–8. Find the coordinates of the C.M. of each of the volumes described in Exercise XIX, questions 3, 5 and 8.

Kinematics

Cf. p. 38. Since $\dfrac{dv}{dt} = a$, $\dfrac{ds}{dt} = v$, we have $v = \int a \, dt$, $s = \int v \, dt$.

The last relationship disposes of the third 'fair guess' (p. 4), the area under a velocity-time graph being $\int v \, dt$, between appropriate limits.

EXAMPLE 5.9   *A particle starts with a velocity of 2 m s^{-1} and moves along a straight line. Its acceleration after t seconds is $(t + 3)$ m s^{-2}. Find its velocity at the end of 2 seconds and the distance travelled in the next 2 seconds.*

Since $a = \dfrac{dv}{dt} = t + 3$, on integrating with respect to t we have

$$v = \tfrac{1}{2}t^2 + 3t + c,$$

where $c = 2$, since we are given that $v = 2$ when $t = 0$. Thus

$$v = \tfrac{1}{2}t^2 + 3t + 2,$$

so that after 2 s, $v = \tfrac{1}{2} \times 2^2 + 3 \times 2 + 2 = 10$.

The distance required is that between $t = 2$ and $t = 4$. Since

$$v = \dfrac{ds}{dt} = \tfrac{1}{2}t^2 + 3t + 2,$$

this distance is given by

$$\int_2^4 (\tfrac{1}{2}t^2 + 3t + 2)\,dt = \left[\tfrac{1}{6}t^3 + \tfrac{3}{2}t^2 + 2t \right]_2^4$$

$$= (\tfrac{32}{3} + 24 + 8) - (\tfrac{4}{3} + 6 + 4) = 31\tfrac{1}{3}.$$

Thus the required velocity is 10 m s⁻¹, and the distance travelled in the next 2 seconds is $31\tfrac{1}{3}$ m.

EXERCISE XXI

1. A particle starts from rest and moves in a straight line. Its velocity in m s⁻¹ is given to be $8t - t^2$, where t is the time in seconds from the commencement of motion. How far will the particle have moved in 3 seconds?

   Find also its greatest distance from the starting point, and the value of t when this distance is reached.

2. A particle, starting from rest, moves along a straight line with a velocity of $8t - t^2$ m s⁻¹ at the end of t seconds. Find its velocity when its acceleration vanishes and the distance travelled up to that time.

   What distance will have been travelled when the velocity vanishes instantaneously?

3. A particle P moves in a straight line with velocity $7t - t^2 - 6$ m s⁻¹ at the end of t seconds. What is its acceleration when $t = 2$ and when $t = 4$?

   When $t = 3$ the particle is at A; when $t = 5$ the particle is at B. Find the length of AB.

   For what values of t is the particle momentarily at rest?

4. A train starts from rest and its acceleration t seconds after the start is $\tfrac{1}{10}(20 - t)$ m s⁻². What is its speed after 20 seconds?

   Acceleration ceases at this instant and the train proceeds at this uniform speed. What is the total distance covered 30 seconds after the start from rest?

5. The point P moves in a straight line with an acceleration of $2t - 4$ m s⁻² after t seconds. When $t = 0$, P is at O and its velocity is 3 m s⁻¹. Find

   (i) the velocity of P after t seconds,

   (ii) the value of t when P starts to return to O, and

   (iii) the distance of P from O at this moment.

Numerical Integration

There are many integrations which are either difficult or even impossible to carry out exactly $\left(\text{e.g. } \int_0^1 \sqrt{(1 + x^3)}\, dx\right)$, and so methods have been devised to find approximate values over a given range. One such method is afforded by the 'staircases', as indicated in Exercise XVI, questions 6–8 (p. 56). Another method is to approximate by means of trapezia, as illustrated in the next example.

EXAMPLE 5.10 *To find an approximate value of* $\int_0^1 \dfrac{1}{1 + x^2}\, dx$, *using the 'trapezium method'*.

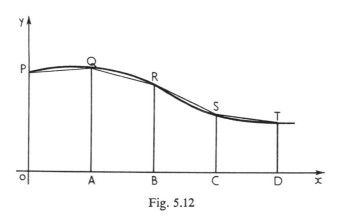

Fig. 5.12

Figure 5.12 represents the graph of $y = \dfrac{1}{1 + x^2}$, from $x = 0$ to 1. Successive intervals of length $\frac{1}{4}$ have been marked off along the x-axis, and trapezia constructed as shown. The points on the curve have coordinates as follows: $P(0, 1)$; $Q(\frac{1}{4}, 0\cdot941)$; $R(\frac{1}{2}, 0\cdot8)$; $S(\frac{3}{4}, 0\cdot64)$; $T(1, 0\cdot5)$. The required approximation is obtained by taking the sum of the areas of the trapezia

$$OAQP + ABRQ + BCSR + CDTS$$
$$= \frac{1}{4}\left(\frac{1 + 0\cdot941}{2}\right) + \frac{1}{4}\left(\frac{0\cdot941 + 0\cdot8}{2}\right) + \frac{1}{4}\left(\frac{0\cdot8 + 0\cdot64}{2}\right) + \frac{1}{4}\left(\frac{0\cdot64 + 0\cdot5}{2}\right)$$
$$= \tfrac{1}{8}(1 + 2 \times 0\cdot941 + 2 \times 0\cdot8 + 2 \times 0\cdot64 + 0\cdot5) \approx 0\cdot78$$

(cf. Exercise XVI, question 7, p. 56).

A more efficient method, which we now describe, is known as *Simpson's rule*.

Figure 5.13 represents the graph of a function $y = f(x)$ which passes through the points $P(-h, y_1)$, $Q(0, y_2)$ and $R(h, y_3)$. We obtain an approximation to $\int_{-h}^{h} f(x)\, dx$ by considering the area under the parabola through the points P, Q and R whose axis is parallel to the y-axis. The equation of such a parabola has the form $y = ax^2 + bx + c$, the area under it between $x = \pm h$ being

$$\int_{-h}^{h} (ax^2 + bx + c)\, dx = \left[\tfrac{1}{3}ax^3 + \tfrac{1}{2}bx^2 + cx \right]_{-h}^{h} = \tfrac{2}{3}h(ah^2 + 3c).$$

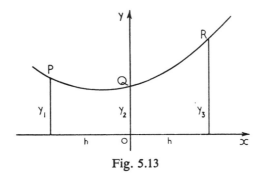

Fig. 5.13

To eliminate a, c and so give a formula in terms of y_1, y_2, y_3 and h only, we substitute the coordinates of P, Q, R in the equation $y = ax^2 + bx + c$, obtaining

$$y_1 = ah^2 - bh + c \tag{1}$$

$$y_2 = c \tag{2}$$

$$y_3 = ah^2 + bh + c. \tag{3}$$

Adding (1) and (3), we have $y_1 + y_3 = 2(ah^2 + c) = 2(ah^2 + y_2)$, from (2) whence $ah^2 = \tfrac{1}{2}(y_1 + y_3) - y_2$, $c = y_2$. The area under the parabola is accordingly equal to $\tfrac{2}{3}h\left(\dfrac{y_1 + y_3}{2} - y_2 + 3y_2 \right) = \dfrac{h}{3}(y_1 + 4y_2 + y_3)$, and this is Simpson's approximation for $\int_{-h}^{h} f(x)\, dx$.

This 'rule' is independent of the choice of origin, and the following (Fig. 5.14) is an immediate generalization.

To obtain an approximation for $\int_{a}^{b} f(x)\, dx$, the interval from a to b is

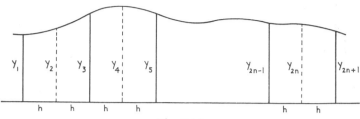

Fig. 5.14

divided into $2n$ sub-intervals, each of length h. If $y_1, y_2, \ldots, y_{2n+1}$ are the successive *ordinates*, i.e. values of $f(x)$, at the extremities of these sub-intervals, then

$$\int_a^b f(x)\, dx$$

$$\approx \frac{h}{3}\{(y_1 + 4y_2 + y_3) + (y_3 + 4y_4 + y_5) + \ldots$$

$$+ (y_{2n-1} + 4y_{2n} + y_{2n+1})\}$$

$$= \frac{h}{3}(y_1 + 4y_2 + 2y_3 + 4y_4 + 2y_5 + 4y_6 + \ldots$$

$$+ 4y_{2n-2} + 2y_{2n-1} + 4y_{2n} + y_{2n+1}).$$

There is no point in trying to remember this result, as it is so readily obtained from the rule for three ordinates.

EXAMPLE 5.11 *To find an approximate value for* $\displaystyle\int_0^1 \frac{1}{1+x^2}\, dx$, *using Simpson's rule with five ordinates.*

Refer to Fig. 5.15. Using five ordinates for the interval $0 \leqslant x \leqslant 1$ implies that $h = \frac{1}{4}$. The required approximation is accordingly

$$\frac{h}{3}(y_1 + 4y_2 + 2y_3 + 4y_4 + y_5)$$

$$= \tfrac{1}{12}(1 + 4 \times 0 \cdot 941 + 2 \times 0 \cdot 8 + 4 \times 0 \cdot 64 + 0 \cdot 5) = 0 \cdot 785.$$

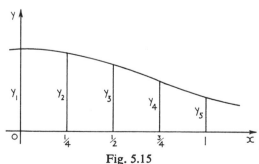

Fig. 5.15

(cf. Example 5.10, p. 76; we shall see later that

$$\int_0^1 \frac{1}{1 + x^2}\, dx = \frac{\pi}{4} = 0 \cdot 78539 \ldots)$$

The next example emphasizes that Simpson's rule refers essentially to a definite integral, which need not have arisen in connexion with an area.

EXAMPLE 5.12   *The cap cut off from the curve $y = x(2 - x)$ by the x-axis is revolved through 4 right angles about the x-axis. Find by exact integration the volume enclosed. What is the numerical value for the volume by Simpson's rule for five ordinates?*

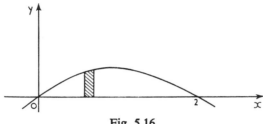

Fig. 5.16

The curve (Fig. 5.16) meets the x-axis at $x = 0$ and $x = 2$, and the volume is accordingly

$$\int_0^2 \pi y^2\, dx = \int_0^2 \pi x^2 (2 - x)^2\, dx$$

$$= \pi \int_0^2 (4x^2 - 4x^3 + x^4)\, dx$$

$$= \pi \left[\tfrac{4}{3}x^3 - x^4 + \tfrac{1}{5}x^5 \right]_0^2$$

$$= \frac{16\pi}{15}.$$

The values of $\pi x^2 (2 - x)^2$ when $x = 0, \tfrac{1}{2}, 1, \tfrac{3}{2}, 2$ are respectively $0, \dfrac{9\pi}{16}, \pi, \dfrac{9\pi}{16}$, 0, and the approximate value of $\int_0^2 \pi y^2\, dx$ as given by Simpson's rule is

$$\frac{1}{6}\left(0 + 4 \times \frac{9\pi}{16} + 2\pi + 4 \times \frac{9\pi}{16} + 0 \right) = \frac{13\pi}{12}.$$

Thus Simpson's rule for 5 ordinates gives the value $\dfrac{13\pi}{12}$ ($\approx 3 \cdot 40$), against the correct value of $\dfrac{16\pi}{15}$ ($\approx 3 \cdot 35$).

EXERCISE XXII

1. Evaluate approximately $\int_0^1 \sqrt{(1 + x^2)} \, dx$, using Simpson's rule with 3 ordinates.

2. Evaluate each of the integrals $\int_0^4 x^3 \, dx$ and $\int_0^1 x^{\frac{1}{2}} \, dx$,

   (i) approximately, using four trapezia, as in Example 5.10,
   (ii) using Simpson's rule with 5 ordinates. Verify that Simpson's rule gives the exact answer for the first integral.

3. Use Simpson's rule to find the approximate area under a curve in which successive ordinates at intervals of 0·5 are 1·3, 2·1, 2·8, 3·4, 2·7, 2·4, 2·1.

4. Evaluate approximately $\int_0^1 \dfrac{1}{\sqrt{(1 + x^4)}} \, dx$, using Simpson's rule with 5 ordinates.

5. Use Simpson's rule with 5 ordinates to evaluate approximately $\int_1^3 \dfrac{1}{x} \, dx$.

6. (P) Investigate the circumstances under which Simpson's rule with 3 ordinates gives the *exact* value for an integral.

Miscellaneous questions (in sets of 3).

7. Sketch the curve
$$y = x^2 - x - 2$$
   for values of x from $x = -2$ to $x = 3$.
   Use the calculus to find the area bounded by the curve and the x-axis.

8. At a certain instant a body is moving in a straight line at 20 cm s⁻¹; its acceleration during the first 5 s of the subsequent motion is $(30 - 6t)$ cm s⁻², where t is the time in seconds; after 5 s it travels with a constant speed. Find:
   (i) its velocity after 2 s;
   (ii) the greatest velocity attained;
   (iii) the total distance travelled in 10 s.

9. The arc of the curve $y = 3x - x^2$, between the points where it cuts the x-axis, is rotated through four right angles about that axis. Find the volume described. (Give your answer as a multiple of π.)

10. Taking 1 cm as unit along both axes, draw the graph of the curve $y = \frac{1}{4}x^2$ for values of x between $x = -3$ and $x = 6$.

    Show, by shading, the areas denoted by $\int_3^5 y \, dx$ and $\int_1^4 x \, dy$. Evaluate *one* of these areas.

11. Evaluate
$$\int_{-1}^{+3} (2x + 4) \, dx \quad \text{and} \quad \int_{-1}^{+3} (7 + 4x - x^2) \, dx.$$
    Find the area between the line $2x - y + 4 = 0$ and the curve
$$y = 7 + 4x - x^2.$$

12. The velocity of a particle travelling in a straight line is given by $v = t^2 - t - 6$, where v is measured in m s^{-1} and t is the time in seconds reckoned from a definite instant of the motion. Find the time that elapses before the particle is instantaneously at rest, and find the acceleration when $t = 5$.

Find also the distance between its position when $t = 3$ and its position when $t = 9$.

13. The coordinate x m of a moving point A at time t seconds is given by the formula

$$x = t^3 - 2t^2 + 3t - 4.$$

Find the velocity and the acceleration of A at the instant $t = 4$.

The acceleration a of a moving point B is given by the formula $a = 3t - 2$; B starts from rest at the origin O at time $t = 0$. Find its velocity and its distance from O at time $t = 4$.

14. The area enclosed between the line $x = 1$, the x-axis, the line $x = 3$, and the line $3x - y + 2 = 0$, is rotated through four right angles about the x-axis. Find the volume generated.

[You need not substitute for π in your answer.]

15. From the point P (2, 4) on the curve $y = x^2$, PN is drawn perpendicular to the axis of x. Find the area bounded by PN, the axis of x and the curve.

Find also the x and y coordinates of the centre of gravity of this area.

16. Show without calculation why

$$1 < \int_0^1 (1 + x)^{\frac{1}{3}} \, dx < \int_0^1 (1 + x)^{\frac{1}{2}} \, dx < \int_0^1 (1 + x)^{\frac{2}{3}} \, dx$$

and find the value of the first integral, as given by Simpson's rule with three ordinates.

17. The acceleration of a particle t seconds after starting from rest is $(2t - 1)$ m s^{-2}. Prove that the particle returns to the starting-point after $1\frac{1}{2}$ s, and find the distance of the particle from the starting-point after a further $1\frac{1}{2}$ s.

Find also at what time after starting the particle attains a velocity of 20 m s^{-1}.

18. Find the area bounded by the curve

$$y = (x + 1)(x - 2)^2$$

and the x-axis from $x = -1$ to $x = 2$.

Also find the x-coordinate of the centre of gravity of this area.

6

Exploitation of Ideas

Piling Ossa on Pelion.
OVID

The next two chapters are principally concerned with developing technique, so that a wider range of problems can be tackled.

Differentiation of Product and Quotient

Let $u(x)$, $v(x)$ be differentiable functions of x, which we shall write shortly as u, v. We shall show that

(i)
$$\frac{d}{dx}(uv) = u\frac{dv}{dx} + v\frac{du}{dx},$$

and

(ii)
$$\frac{d}{dx}\left(\frac{u}{v}\right) = \frac{v\dfrac{du}{dx} - u\dfrac{dv}{dx}}{v^2}.$$

If δu, δv are the small increases in u, v due to an increment δx of x, then from first principles we have

(i)
$$\frac{d}{dx}(uv) = \lim_{\delta x \to 0} \frac{(u + \delta u)(v + \delta v) - uv}{\delta x}$$

$$= \lim_{\delta x \to 0} \left(u\frac{\delta v}{\delta x} + v\frac{\delta u}{\delta x} + \delta u \cdot \frac{\delta v}{\delta x}\right)$$

$$= u\frac{dv}{dx} + v\frac{du}{dx}, \text{ as required.}*$$

Also,

(ii)
$$\frac{d}{dx}\left(\frac{u}{v}\right) = \lim_{\delta x \to 0} \frac{\dfrac{u + \delta u}{v + \delta v} - \dfrac{u}{v}}{\delta x}$$

$$= \lim_{\delta x \to 0} \frac{(u + \delta u)v - u(v + \delta v)}{v(v + \delta v)\,\delta x}$$

---

* We are assuming that $\lim_{\delta x \to 0} \delta u \cdot \dfrac{\delta v}{\delta x} = \left(\lim_{\delta x \to 0} \delta u\right)\left(\lim_{\delta x \to 0} \dfrac{\delta v}{\delta x}\right) = \left(\lim_{\delta x \to 0} \delta u\right)\dfrac{dv}{dx} = 0.$

so

$$\frac{d}{dx}\left(\frac{u}{v}\right) = \lim_{\delta x \to 0} \frac{v\,\dfrac{\delta u}{\delta x} - u\,\dfrac{\delta v}{\delta x}}{v(v + \delta v)}$$

$$= \frac{v\,\dfrac{du}{dx} - u\,\dfrac{dv}{dx}}{v^2}.$$

The last expression can be memorized inelegantly as 'the derivative of the top times the bottom minus the derivative of the bottom times the top, all over the bottom squared'.

The 'product' rule for $\dfrac{d}{dx}(uv)$ can be extended to any number of functions, say $f_1, f_2, \ldots, f_n$. For example,

$$\frac{d}{dx}(f_1 f_2 f_3) = \frac{d}{dx}\{f_1(f_2 f_3)\}$$

$$= f_1\frac{d}{dx}(f_2 f_3) + \frac{df_1}{dx}f_2 f_3$$

$$= f_1\left(f_2\frac{df_3}{dx} + \frac{df_2}{dx}f_3\right) + \frac{df_1}{dx}f_2 f_3$$

$$= \frac{df_1}{dx}f_2 f_3 + f_1\frac{df_2}{dx}f_3 + f_1 f_2\frac{df_3}{dx}.$$

The general result

$$\frac{d}{dx}(f_1 f_2 \ldots f_n) = \frac{df_1}{dx}f_2 f_3 \ldots f_n + f_1\frac{df_2}{dx}f_3 \ldots f_n$$

$$+ f_1 f_2\frac{df_3}{dx}\ldots f_n + \ldots + f_1 f_2 f_3 \ldots f_{n-1}\frac{df_n}{dx}$$

can be proved by induction (cf. Appendix 1).

EXERCISE XXIII

1. Differentiate with respect to x (i) as a product, (ii) after removing brackets:
   (a) $x(x + 1)$, (b) $(x + 1)(x - 2)$, (c) $(x - 4)(x^2 - 3x + 1)$.

In questions 2 to 7, differentiate the expressions with respect to x.

2. $(x + 1)(x^2 - 3)$.

3. $(x^2 - 5x + 6)(x^2 - 3x + 1)$.

4. $\dfrac{x}{x + 1}$.

5. $\dfrac{x - 3}{x + 2}$.

6. $\dfrac{1}{x^2 - 4x + 1}$.

7. $\dfrac{x^2 - 3x + 4}{x^3 + 2}$.

The Chain Rule ('Function of a Function')

It may happen that z is a differentiable function of y, which, in turn, is a differentiable function of x. (For example, if a spherical balloon is being blown up, the volume V is a function of r, namely $\frac{4}{3}\pi r^3$, while r, in turn, is a function of the time, t.) We then have

$$\frac{dz}{dx} = \frac{dz}{dy} \times \frac{dy}{dx}. \tag{1}$$

This is a shorthand for a clumsier expression; strictly, we should write $z = f(x) = g(y)$ and $y = h(x)$, and the rule would then read

$$\frac{d}{dx}\{f(x)\} = \frac{d}{dy}\{g(y)\} \times \frac{d}{dx}\{h(x)\}. \tag{2}$$

The validity of the rule can be seen from the fact that a change δx in the value of x gives a corresponding change δy in y, and since z is a function of y this, in turn, leads to a change, say δz, in z. Then as $\delta x \to 0$, $\dfrac{\delta y}{\delta x} \to \dfrac{dy}{dx}$ and since then δy also tends to zero, it follows as well that $\dfrac{\delta z}{\delta y} \to \dfrac{dz}{dy}$. Thus

$$\frac{dz}{dx} = \lim_{\delta x \to 0} \frac{\delta z}{\delta x} = \lim_{\delta x, \delta y \to 0} \frac{\delta z}{\delta y} \times \frac{\delta y}{\delta x} = \frac{dz}{dy} \times \frac{dy}{dx}.$$

$\left(\text{We assume that the limit of the product } \dfrac{\delta z}{\delta y} \times \dfrac{\delta y}{\delta x} \text{ is equal to the product of the separate limits } \dfrac{\delta z}{\delta y}, \dfrac{\delta y}{\delta x}, \text{ and that } \delta y \neq 0.\right)$ In eqn. (1), z is a 'function of a function', i.e. a function of y which, in turn, is a function of x. The rule is also known as the 'chain rule' as it can be generalized for a whole chain of variables $z, u, v, \ldots, y, x$ in the form

$$\frac{dz}{dx} = \frac{dz}{du} \cdot \frac{du}{dv} \ldots \frac{dy}{dx},$$

where z is a function of u, u a function of v, and so on. As a special case, if y is a function of x then

$$\frac{d}{dx}(y^n) = \frac{d}{dy}(y^n) \times \frac{dy}{dx} = ny^{n-1}\frac{dy}{dx}.$$

EXAMPLE 6.1   *To differentiate* $(3x^2 - 4x + 1)^3$.

Let $(3x^2 - 4x + 1)^3 = y$, $3x^2 - 4x + 1 = s$, so that $y = s^3$. Then

$$\frac{dy}{dx} = \frac{dy}{ds} \times \frac{ds}{dx} = 3s^2(6x - 4) = 3(3x^2 - 4x + 1)^2(6x - 4).$$

(After a little practice, some of these steps are omitted. The reader may even find himself whispering: 'First we differentiate with respect to the stuff inside [namely $3x^2 - 4x + 1$], getting $3(3x^2 - 4x + 1)^2$, and then multiply by d(stuff) by dx, i.e. $6x - 4$.')

Gradient of x^n

We have already shown that $\dfrac{d}{dx}(x^n) = nx^{n-1}$ when n is a positive integer (p. 25). We can now extend the proof to cover all rational n. First let $n = \dfrac{p}{q}$ where p and q are positive integers ($q \neq 1$). Writing $y = x^n = x^{p/q}$, we have $y^q = x^p$ and so

$$qy^{q-1}\frac{dy}{dx} = px^{p-1}.$$

Thus

$$\frac{dy}{dx} = \frac{px^{p-1}}{qy^{q-1}} = \frac{px^{p-1}}{qx^{p-(p/q)}} = \frac{p}{q}x^{(p/q)-1},$$

i.e. $\dfrac{d}{dx}(x^n) = nx^{n-1}$ when $n = \dfrac{p}{q}$ (>0).

Finally, if $n = -m$, where m is a positive rational number,

$$\frac{d}{dx}(x^n) = \frac{d}{dx}\left(\frac{1}{x^m}\right) = \frac{-mx^{m-1}}{x^{2m}},$$

by the 'quotient' rule, $= -mx^{-m-1} = nx^{n-1}$. This completes the proof that $\dfrac{d}{dx}(x^n) = nx^{n-1}$ *for all rational values of n.** *

It follows that $\int x^n\, dx = \dfrac{x^{n+1}}{n+1} + c$ for all rational n except -1.

EXAMPLE 6.2 *To differentiate* $f(x) = \dfrac{(6x + 1)^{\frac{1}{2}}(x + 2)}{(x - 1)^7}$.

By successive use of the foregoing rules, we have

$$f'(x) = \frac{\{\frac{1}{2}(6x + 1)^{-\frac{1}{2}} \times 6(x + 2) + (6x + 1)^{\frac{1}{2}}\}(x - 1)^7 - (6x + 1)^{\frac{1}{2}}(x + 2) \times 7(x - 1)^6}{(x - 1)^{14}}$$

$$= \frac{\{3(x + 2) + 6x + 1\}(x - 1) - 7(6x + 1)(x + 2)}{(x - 1)^8(6x + 1)^{\frac{1}{2}}}$$

$$= -\frac{33x^2 + 93x + 21}{(x - 1)^8(6x + 1)^{\frac{1}{2}}}.$$

---

* The above proof breaks down if $x = 0$ (the reader should discover where), but the result still holds *if the gradient exists*. As a counter-example, $x^{\frac{2}{3}}$ is not differentiable at $x = 0$; it is not even continuous there, being undefined when $x < 0$.

EXAMPLE 6.3   *To find* $\int (3x + 4)^7 \, dx$.

As integration is the 'opposite of differentiation', the answer may be guessed to be a multiple of $(3x + 4)^8$. Now

$$\frac{d}{dx}(3x + 4)^8 = 8(3x + 4)^7 \times 3 = 24(3x + 4)^7,$$

and so

$$\int (3x + 4)^7 \, dx = \tfrac{1}{24}(3x + 4)^8 + c.$$

EXERCISE XXIV

In questions 1 to 24, differentiate the given functions with respect to x.

1. $x^{\frac{1}{2}}$.
2. $3x^{-\frac{1}{2}}$.
3. $x + \dfrac{1}{x}$.
4. x^{-7}.
5. $x^{-\frac{3}{2}}$.
6. $(x + 1)^4$.
7. $(3x^2 - 4x + 2)^7$.
8. $\sqrt{(x^3 + 1)}$.
9. $\dfrac{1}{(x + 3)^2}$.
10. $(3x + 7)^{-3}$.
11. $(x - 5)^3$.
12. $(5 - x)^3$.
13. $(x - 2)^3(x + 1)^2$.
14. $(3x + 1)^2(4 - x)^3$.
15. $(3x)^{-\frac{1}{2}}$.
16. $(1 - 5x)^{-\frac{3}{2}}$.
17. $\dfrac{x - 4}{x^2 + 1}$.
18. $\dfrac{(x - 4)^3}{(x^2 + 1)^2}$.
19. $\dfrac{(3x - 4)^3(2 - x)}{5x^2 + 1}$.
20. $\sqrt{(x^2 + 1)}(3x + 2)$.
21. $\dfrac{\sqrt{(x^2 + 1)}}{3x + 2}$.
22. $\sqrt{\left(\dfrac{x + 1}{x - 1}\right)}$.
23. $\dfrac{1}{(x - 2)(x + 1)}$.
24. $\{1 + \sqrt{(7 - 3x)}\}\left(2 - \dfrac{1}{2x + 1}\right)$.

25. Find the gradient of the curve $y = x^{\frac{3}{2}} - \dfrac{1}{x^2}$ at the point $(1, 0)$.

In questions 26 to 32, integrate the functions with respect to x.

26. $x^{\frac{1}{2}}$.
27. x^{-2}.
28. $3x^{\frac{3}{2}}$.
29. $(x + 1)^{-\frac{1}{2}}$

30. $(5x - 1)^3$.

31. $(1 - 5x)^3$.

32. $\dfrac{4x^3 - 7x^2 + 1}{x^2}$.

33. Evaluate $\displaystyle\int_1^2 \left(x^{\frac{1}{2}} - \dfrac{1}{x^2}\right) dx$.

Trigonometric Functions

The 'circular' or trigonometric functions ($\sin x$, $\cos x$, etc.) crop up frequently, especially in the application of calculus to mechanics. If necessary, the reader should brush up his formulae such as (i) $\tan \theta = \dfrac{\sin \theta}{\cos \theta}$, (ii) $\sin X + \sin Y = 2 \sin \dfrac{X + Y}{2} \cos \dfrac{X - Y}{2}$; he will also probably have come across

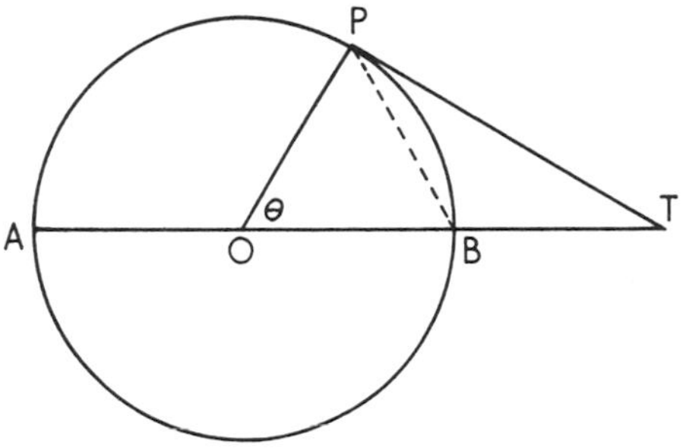

Fig. 6.1

'circular measure', i.e. using radians instead of degrees (2π radians $= 360$ degrees). This considerably simplifies our results, and we shall use radians throughout, mainly because the relationship

$$\lim_{\theta \to 0} \frac{\sin \theta}{\theta} = 1 \tag{1}$$

holds if θ is measured in radians but not if it is measured in degrees. We demonstrate (1) informally. Figure 6.1 represents a circle of radius r, centre O. The tangent at a point P meets the diameter AB produced at T. Let angle $POB = \theta$ radians; then the area of the sector OPB is a fraction $\dfrac{\theta}{2\pi}$ of the whole circle, and so is $\dfrac{\theta}{2\pi} \times \pi r^2 = \frac{1}{2}\theta r^2$. Also the area of triangle OPB is $\frac{1}{2}r^2 \sin \theta$ (using '$\frac{1}{2}bc \sin A$') and that of triangle OPT is $\frac{1}{2}r^2 \tan \theta$ (using '$\frac{1}{2}bh$', since $PT = r \tan \theta$).

Now area of triangle OPB < area of sector OPB < area of triangle OPT,

i.e. $\frac{1}{2}r^2 \sin \theta < \frac{1}{2}r^2 \theta < \frac{1}{2}r^2 \tan \theta$

or

$$1 < \frac{\theta}{\sin \theta} < \frac{1}{\cos \theta} .$$

As $\theta \to 0$, $\cos \theta \to 1$, and so $\dfrac{\theta}{\sin \theta} \to 1$, whence $\lim\limits_{\theta \to 0} \dfrac{\sin \theta}{\theta} = 1$.

We can now differentiate $\sin x$. From first principles, i.e.

$$\frac{d}{dx} f(x) = \lim_{h \to 0} \frac{f(x + h) - f(x)}{h},$$

we have

$$\frac{d}{dx} (\sin x) = \lim_{h \to 0} \frac{\sin (x + h) - \sin x}{h}$$

$$= \lim_{h \to 0} \frac{2 \cos \left(x + \dfrac{h}{2} \right) \sin \dfrac{h}{2}}{h}$$

$$= \lim_{h \to 0} \cos \left(x + \frac{h}{2} \right) . \frac{\sin \dfrac{h}{2}}{\dfrac{h}{2}}$$

$$= \cos x \text{ (using (1) above with } \theta = \tfrac{1}{2}h).*$$

Thus $\dfrac{d}{dx} (\sin x) = \cos x$. The reader should prove similarly that

$$\frac{d}{dx} (\cos x) = -\sin x.$$

EXAMPLE 6.4 *To differentiate* (i) $\tan x$, (ii) $\tan 3x$, (iii) $\tan^3 x$.

(i) $\dfrac{d}{dx} (\tan x) = \dfrac{d}{dx} \left(\dfrac{\sin x}{\cos x} \right) = \dfrac{+\cos^2 x + \sin^2 x}{\cos^2 x}$ (using the 'quotient'
rule) $= \sec^2 x.$

(ii) $\tan 3x$ is a function ('tangent') of a function ('3 times') of x. We differentiate first with respect to $3x$, obtaining $\sec^2 3x$ and then multiply by the derivative of $3x$ with respect to x, namely 3. Thus $\dfrac{d}{dx} (\tan 3x) = 3 \sec^2 3x.$

(iii) $\tan^3 x = (\tan x)^3$, so $\dfrac{d}{dx} (\tan^3 x) = 3(\tan x)^2 . \dfrac{d}{dx} (\tan x) = 3 \tan^2 x \sec^2 x.$

---

* We are assuming here that $\lim\limits_{h \to 0} \cos \left(x + \dfrac{h}{2} \right) . \dfrac{\sin \dfrac{h}{2}}{\dfrac{h}{2}} = \lim\limits_{h \to 0} \cos \left(x + \dfrac{h}{2} \right) . \lim\limits_{h \to 0} \dfrac{\sin \dfrac{h}{2}}{\dfrac{h}{2}}.$

EXERCISE XXV

1. (D) Sketch the graphs of sin x, cos x for values of x from 0 to 2π and verify roughly that $\frac{d}{dx}(\sin x) = \cos x$, $\frac{d}{dx}(\cos x) = -\sin x$, e.g. by checking that cos $x < 0$ when sin x is decreasing.

2. (D) Differentiate sin x + cos x (i) directly, (ii) by first writing it as $\sqrt{2}\sin\left(x + \frac{\pi}{4}\right)$, and reconcile your answers.

In questions 3 to 16, differentiate with respect to x.

3. sin $3x$.

4. sin $2x$ + cos $4x$.

5. $\sin\left(\dfrac{\pi}{4} - x\right)$.

6. cosec x. $\left[\text{take as } \dfrac{1}{\sin x}\right]$.

7. cot x.

8. sec x.

9. $\sin^2 x \cos x$.

10. $\sin^3 x$.

11. sec x + tan x.

12. sec x tan x.

13. sin $4x$ cos $3x$.

14. $(1 - 3\tan^2 x)^{\frac{1}{2}}$.

15. x sin x.

16. sin $(x°)$, i.e. sine of x degrees; write $x°$ as $\dfrac{\pi x}{180}$ radians.

In questions 17 to 26, integrate with respect to x.

17. cos $2x$.

18. sin $(5x + 1)$.

19. sin x + cos x.

20. $\sin^2 x$. $[1 - \cos 2x = 2\sin^2 x]$.

21. $\cos^2 x$.

22. sin $4x$ cos $3x$. $[\sin X + \sin Y = 2\ldots]$

23. cos $7x$ cos $5x$.

24. $\sec^2 x$ (cf. Example 6.4).

25. $\tan^2 x$.

26. $\text{cosec}^2 x$.

27. Evaluate $\displaystyle\int_0^{\frac{\pi}{2}} \cos x \, dx$.

28. Evaluate $\displaystyle\int_0^{\frac{\pi}{2}} \sin x \, dx$, and explain why $\displaystyle\int_{-\frac{\pi}{2}}^{\frac{\pi}{2}} \sin x \, dx = 0$.

Inverse Functions

Let $f(x)$ be a function defined over a certain domain which takes up its values once only; such a function is called *one-to-one* over this domain. For example, $f(x) = x^3$ is one-to-one over the domain of real numbers, Fig. 6.2(i), but $g(x) = x^2$ is not, Fig. 6.2(ii) (since $g(x) = a^2$ for two distinct values of x, namely $\pm a$).

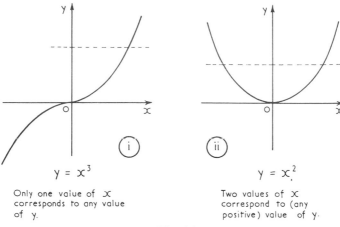

$y = x^3$

Only one value of x corresponds to any value of y.

$y = x^2$

Two values of x correspond to (any positive) value of y.

Fig. 6.2

If, however, $g(x) = x^2$ and x is restricted so that $x \geqslant 0$, then $g(x)$ is one-to-one over this domain (Fig. 6.3).

If a function is one-to-one over a domain, then it has an *inverse function*. If $y = f(x)$, the inverse function is obtained by changing over x and y, i.e. it is the new 'y' defined as a function of x by the equation $x = f(y)$). For example, if $y = x^3$, so that $x = \sqrt[3]{y}$, the inverse is given by $y = \sqrt[3]{x}$, i.e. the inverse function of x^3 is $\sqrt[3]{x}$ (Fig. 6.4).

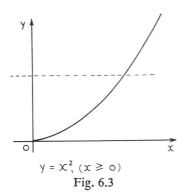

$y = x^2$, $(x \geqslant 0)$

Fig. 6.3

Given the graph of a function, we obtain that of the inverse function by re-labelling the axes, i.e. 'changing over x and y'; having done so, the axes can be put back in their normal positions by turning the paper through a right-angle and then looking at it from the back, holding it up to the light. The reader should experiment with this, e.g. with x^5, x^4 $(x \geqslant 0)$, $x + 1$.

If f is a differentiable one-to-one function, so that $y = f(x)$ can be re-written as $x = g(y)$, g also being differentiable, then

$$\frac{dy}{dx} = \lim_{\delta x, \delta y \to 0} \frac{\delta y}{\delta x} = \frac{1}{\displaystyle\lim_{\delta x, \delta y \to 0} \frac{\delta x}{\delta y}} = \frac{1}{\dfrac{dx}{dy}}$$

$\left(\text{provided that } \dfrac{dy}{dx}, \dfrac{dx}{dy} \neq 0\right)$.

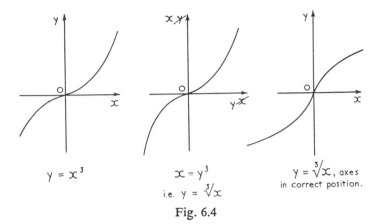

$y = x^3$

$x = y^3$
i.e. $y = \sqrt[3]{x}$

$y = \sqrt[3]{x}$, axes
in correct position.

Fig. 6.4

The trigonometric functions have inverses, Fig. 6.5, provided a suitable domain is chosen. Consider, for example, sin x. This is one-to-one if we restrict x to the interval $\left[-\dfrac{\pi}{2}, \dfrac{\pi}{2} \right]$.*

We accordingly define the inverse function $\sin^{-1} x$ ('the angle whose sine is x') as the angle y such that $\sin y = x$ and $-\dfrac{\pi}{2} \leqslant y \leqslant \dfrac{\pi}{2}$.

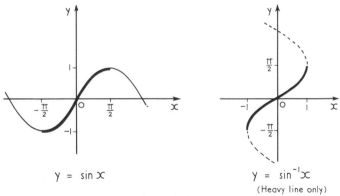

$y = \sin x$

$y = \sin^{-1} x$
(Heavy line only)

Fig. 6.5(a)

Let $y = \sin^{-1} x \left(-\dfrac{\pi}{2} < y < \dfrac{\pi}{2} \right)$; then $\sin y = x$ and, differentiating with

---

* I.e. $-\dfrac{\pi}{2} \leqslant x \leqslant \dfrac{\pi}{2}$. Square brackets imply that the end-values are included, round brackets that they are not. E.g., if x is in the interval (0, 1), then $0 < x < 1$; if in [0, 1] then $0 \leqslant x \leqslant 1$; if in [0, 1), then $0 \leqslant x < 1$.

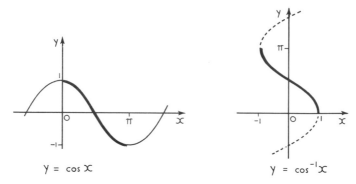

Fig. 6.5(b)

respect to x, we have $\qquad\qquad \cos y \dfrac{dy}{dx} = 1,$

so $\qquad \dfrac{dy}{dx} = \dfrac{1}{\cos y} = \dfrac{1}{\sqrt{(1 - \sin^2 y)}} = \dfrac{1}{\sqrt{(1 - x^2)}}.$ $\qquad\qquad$ (1)

(The positive square root is chosen since $\dfrac{dy}{dx}$ is clearly positive.)

Similarly, $y = \cos^{-1} x$ is defined so that $0 \leqslant y \leqslant \pi$ (Fig. 6.5b) and $y = \tan^{-1} x$ so that $-\dfrac{\pi}{2} < y < \dfrac{\pi}{2}$ (Fig. 6.6).

The result (1) above may be written

$$\frac{d}{dx} (\sin^{-1} x) = \frac{1}{\sqrt{(1 - x^2)}}.$$

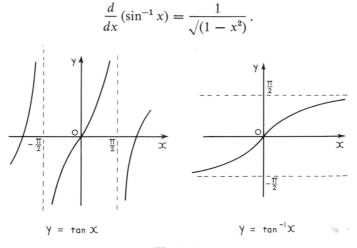

Fig. 6.6

The reader should now verify, in a similar manner, that

$$\frac{d}{dx}(\tan^{-1} x) = \frac{1}{1 + x^2}.$$

EXAMPLE 6.5 *To evaluate* $I = \int \frac{1}{4 + 9x^2} dx.$

Since $\frac{1}{4 + 9x^2} = \frac{\frac{1}{4}}{1 + \left(\frac{3x}{2}\right)^2}$ and $\int \frac{1}{1 + x^2} dx = \tan^{-1} x,$ it can be guessed

that I is a multiple of $\tan^{-1} \frac{3x}{2}$. Now $\frac{d}{dx}\left(\tan^{-1} \frac{3x}{2}\right) = \frac{\frac{3}{2}}{1 + \frac{9x^2}{4}} = \frac{6}{4 + 9x^2}.$

Thus $I = \frac{1}{6}\tan^{-1}\frac{3x}{2}.$

EXERCISE XXVI

1. (i) Evaluate $\frac{d}{dx}(\cot^{-1} x)$ by the method used for $\sin^{-1} x$ above; (ii) taking positive acute angles, explain why $\tan^{-1} x + \cot^{-1} x = \frac{\pi}{2}.$

2. Copy and complete the following table (these results are probably worth remembering; allied ones can easily be worked out when required):

$f(x)$	$f'(x)$
$\sin x$	
$\cos x$	
$\tan x$	
$\sec x$	
$\sin^{-1} x$	
$\cos^{-1} x$	
$\tan^{-1} x$	

(When differentiating $\cos^{-1}x$ use Fig. 6.5(b) to help determine the sign of $f'(x)$.)

3. Differentiate with respect to x (a) $\sin^{-1} 3x$, (b) $\cos^{-1} 4x$, (c) $\sin^{-1}\{(1 - x^2)^{\frac{1}{2}}\}.$

4. If $0 < x < 1$, explain why $\frac{d}{dx}(\sin^{-1} x + \cos^{-1} x) = 0.$

5. Evaluate (i) $\int_0^1 \frac{1}{1 + x^2} dx,$ (ii) $\int_{\frac{1}{2}}^{\sqrt{3}/2} \frac{1}{\sqrt{(1 - x^2)}} dx.$

6. Integrate with respect to x (i) $\dfrac{1}{9 + 16x^2}$; (ii) $\dfrac{1}{4 + 3x^2}$; (iii) $\dfrac{1}{(1 - 9x^2)^{\frac{1}{2}}}$;

(iv) $\dfrac{1}{(3 - 2x^2)^{\frac{1}{2}}}$.

7. Differentiate (a) $\cos^{-1}\left(\dfrac{1}{x}\right)$, (b) $\tan^{-1}\{(x^2 - 1)^{\frac{1}{2}}\}$. Why are the answers the same?

8. Differentiate, with respect to x, $\tan^{-1}\left(\dfrac{3x + 4}{x^2 + 1}\right)$.

9. Criticize (i) $\displaystyle\int_0^{\frac{1}{2}} \dfrac{1}{\sqrt{(1 - x^2)}}\, dx = [\sin^{-1} x]_0^{\frac{1}{2}} = 30°$;

(ii) $\displaystyle\int_1^{\sqrt{3}} \dfrac{1}{1 + x^2}\, dx = [\tan^{-1} x]_1^{\sqrt{3}} = \dfrac{4\pi}{3} - \dfrac{\pi}{4} = \dfrac{13\pi}{12}$.

Implicit Functions

Often y is not given explicitly in terms of x, e.g. $y = \sin^2 x + \dfrac{1}{x}$, but by means of an equation such as $x^2 + y^2 = 1$. In this particular example (the

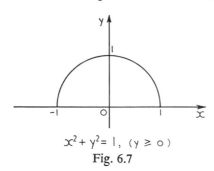

$x^2 + y^2 = 1$, $(y \geqslant 0)$

Fig. 6.7

equation representing a circle, centre the origin and radius 1) y represents a function of x (namely $\sqrt{(1 - x^2)}$) if we restrict ourselves to the upper 'half-plane' given by $y \geqslant 0$ (Fig. 6.7).

A function defined in such a way (y not being the subject) is an *implicit* function. It will be tacitly assumed that the ranges of values of x, y are restricted in such a way that y does represent a function of x. The gradient of such a function is generally best found directly, without even attempting to express y explicitly in terms of x.

EXAMPLE 6.6 *To find the gradient of the curve* $x^3 + x^2y + y^4 = 3y + 10$ *at the point* (2, 1).

Using the 'product' rule, $\dfrac{d}{dx}(x^2y) = 2xy + x^2\dfrac{dy}{dx}$. Also, using the chain rule, $\dfrac{d}{dx}(y^4) = \dfrac{d}{dy}(y^4)\cdot\dfrac{dy}{dx} = 4y^3\dfrac{dy}{dx}$. Thus, differentiating both sides of the given equation with respect to x, we have

$$3x^2 + 2xy + x^2\dfrac{dy}{dx} + 4y^3\dfrac{dy}{dx} = 3\dfrac{dy}{dx}.$$

This gives a general expression for the gradient at any point (x, y) on the curve. Substituting $x = 2, y = 1$, the gradient at $(2, 1)$ is thus given by

$$12 + 4 + 4\frac{dy}{dx} + 4\frac{dy}{dx} = 3\frac{dy}{dx}, \quad \text{i.e.} \quad \frac{dy}{dx} = -\frac{16}{5}.$$

Exercise XXVII

1. (D) If $x^2 + y^3 = 2$, find an expression for $\frac{dy}{dx}$ (i) directly, as in Example 6.6, (ii) by first writing $y = (2 - x^2)^{\frac{1}{3}}$. Reconcile the two answers.

2. Find, in two different ways, the gradient of the curve $y^2 = x^2 - 4x + 7$ at the point $(1, 2)$.

3. Find $\frac{dy}{dx}$ in terms of x, y if (i) $\sin x + \cos y = 3$, (ii) $x^3 - 3xy + y^2 = 4$.

4. Find $\frac{dy}{dx}$ in terms of x only if $x^4 + y^4 = 1$.

5. Find the gradient of the curve $x^3 + 3x^2y + xy^2 + 2x - 2y + 9 = 0$ at the point $(-1, 3)$.

6. Find the gradient at the origin of the curve $x^3 + 2xy^2 - y^4 + 3y - 4x = 0$.

Stationary Values: Repeated Differentiation

We recall that, if $f'(a) = 0$, $f(x)$ has a *stationary* value at $x = a$, which may be one of three types: (i) minimum, (ii) maximum, (iii) point of inflexion (cf. p. 34, see Fig. 6.8). For a minimum, there is a neighbourhood of a in which the gradient $f'(x)$ is increasing; for a maximum, there is one in which $f'(x)$ is decreasing.

For a point of inflexion, $f'(x)$ is increasing on one side of a and decreasing on the other, so that $f'(x)$ has either a maximum or minimum at such a point. Any point a satisfying this last condition is called a *point of inflexion*, $f'(a)$ not necessarily being zero (Fig. 6.9).

We are thus led to consider maximum and minimum values not of $f(x)$ itself but of $f'(x)$, and therefore we shall require to investigate the gradient of $f'(x)$. If $y = f(x)$, $\frac{dy}{dx} = f'(x)$, we write $\frac{d}{dx}\left(\frac{dy}{dx}\right) = \frac{d^2y}{dx^2} = f''(x)$. Later we shall require to differentiate a number of times, writing for example

$$\frac{d}{dx}\left(\frac{d^2y}{dx^2}\right) = \frac{d^3y}{dx^3} = f'''(x), \quad \frac{d}{dx}\left(\frac{d^3y}{dx^3}\right) = \frac{d^4y}{dx^4} = f^{iv}(x),$$

and so on.

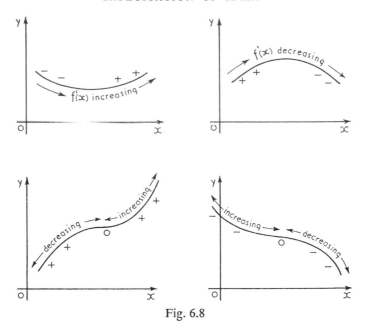

Fig. 6.8

In terms of this new notation, the remarks at the beginning of this section can be re-stated as follows.

(i) $f'(a) = 0, f''(a) > 0$ implies a minimum of $f(x)$ at $x = a$.

(ii) $f'(a) = 0, f''(a) < 0$ gives a maximum at $x = a$.

(iii) $f''(a) = 0, f''(x)$ *changes sign as x passes through a* (i.e. there are neighbourhoods $(a - h, a)$, $(a, a + h)$ such that $f''(x)$ is positive for all x in one and negative for all x in the other) gives a point of inflexion for $f(x)$ at $x = a$.

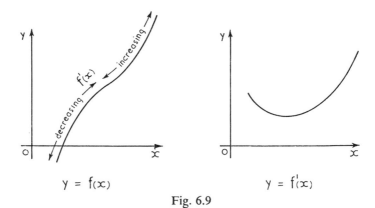

$y = f(x)$ $y = f'(x)$

Fig. 6.9

If $f'(a) = f''(a) = 0$ and $f''(x)$ does not change sign, then if $f''(x) > 0$ throughout some interval containing a (apart from a itself), $x = a$ gives a minimum; similarly, if $f''(x) < 0$ throughout such an interval, $x = a$ gives a maximum. In practice, the nature of stationary points can often be determined with the aid of a sketch-graph, without repeated differentiation.

EXAMPLE 6.7 *Prove that the function* $27 \sec \theta + 64 \operatorname{cosec} \theta$ *has stationary values whenever* $\tan \theta = \frac{4}{3}$. *If* θ *is acute, calculate the stationary value and verify by a rough sketch that it is a minimum.*

Let $f(\theta) = 27 \sec \theta + 64 \operatorname{cosec} \theta$, then

$f'(\theta) = 27 \sec \theta \tan \theta - 64 \operatorname{cosec} \theta \cot \theta$,

which is 0 if $27 \dfrac{\sin \theta}{\cos^2 \theta} = 64 \dfrac{\cos \theta}{\sin^2 \theta}$, i.e.

$\tan^3 \theta = \frac{64}{27}$, $\tan \theta = \frac{4}{3}$. Thus the only stationary values occur when $\tan \theta = \frac{4}{3}$, that is, at intervals of π for θ. Now when $\theta \searrow 0, f(\theta) \to \infty$ and when $\theta \nearrow \dfrac{\pi}{2}$,

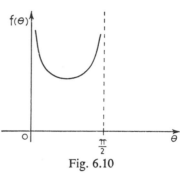

Fig. 6.10

$f(\theta) \to \infty$ (Fig. 6.10), and so the stationary value corresponding to the acute angle $\alpha = \tan^{-1} \frac{4}{3}$ is a minimum. This value is given by $27 \times \frac{5}{3} + 64 \times \frac{5}{4} = 125$.

EXERCISE XXVIII

1. (D) Evaluate $f'(0)$, $f''(0)$ for each of the following, and illustrate with sketch-graphs: (i) $f(x) = x^4$, (ii) $f(x) = -x^4$, (ii) $f(x) = x^3$.

2. Evaluate $\dfrac{dy}{dx}$, $\dfrac{d^2y}{dx^2}$ and $\dfrac{d^3y}{dx^3}$ if (i) $y = x^3 - 4x^2 + 3x + 2$, (ii) $y = x^5 + \dfrac{1}{x}$,

   (iii) $y = \dfrac{1}{1 + x}$, (iv) $y = \sin x + \cos 2x$.

3. If $y = 4 \sin 3x + 5 \cos 3x$, verify that $\dfrac{d^2y}{dx^2} + 9y = 0$.

4. Find the value of x for which $2x^3 - 9x^2 + 12x + 1$ has a point of inflexion.

5. Investigate the stationary values of $x^3 - x^2 - 4x + 10$, and illustrate graphically.

Parametric Equations*

When considering graphs, it is often convenient to consider x, y as functions of a third variable, say t, a 'parameter', rather than try to manipulate an

---

* This section may be regarded as a preview. Extensive use of parametric equations will be made in Chapters 11 and 12.

awkward equation of the form $F(x, y) = 0$. For example, all points on the curve $(y - 1)^3 = x^2$ have coordinates $x = t^3$, $y = 1 + t^2$ for different values of t. Equations of the form $x = f(t)$, $y = g(t)$, (elimination of t leading back to $F(x, y) = 0$) are called *parametric equations*.

EXAMPLE 6.8   *To investigate the curve $x^{\frac{2}{3}} + y^{\frac{2}{3}} = a^{\frac{2}{3}}$.*

The given equation is satisfied by values of x, y such that

$$x = a \cos^3 t, \qquad y = a \sin^3 t$$

and by taking all values of t in the interval $[0, 2\pi)$ these parametric equations give the complete curve. It can be seen that x, y are both restricted to the range $[-a, a]$.

For a particular value of t, the coordinates x, y give a point on the curve which for short is referred to as 'the point t'. Replacing t by $(\pi - t)$ leaves y unaltered but changes the sign of x; the curve thus has symmetry about the y-axis. The reader should verify that it also has symmetry about the x-axis (replace t by $-t$). There are now two likely shapes for the curve (Fig. 6.11).

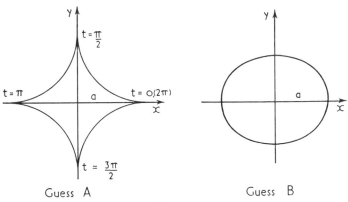

Fig. 6.11

To settle between them, we compute the gradient. We restrict the domain of t in turn to intervals such as $\left(0, \dfrac{\pi}{2}\right)$ so that for each of these intervals x and y are one-to-one functions of t; in particular, $\dfrac{dt}{dx} = 1 \Big/ \dfrac{dx}{dt}$. We then have

$$\frac{dy}{dx} = \frac{dy}{dt} \cdot \frac{dt}{dx} = \frac{\dfrac{dy}{dt}}{\dfrac{dx}{dt}} = \frac{3a \sin^2 t \cos t}{-3a \cos^2 t \sin t} = -\tan t.$$

Thus in the first quadrant the gradient is negative and decreases as t increases. If t is just greater than zero, the gradient is numerically small; t nearly $\frac{\pi}{2}$, the gradient is numerically large. Guess A is, in fact, the correct one.

Note that the upper half of the curve does not have a gradient where it meets the y-axis. We cannot, of course, define y as a one-to-one function of x near the points $(\pm a, 0)$.

't' is used in a double sense. While differentiating $\left(\text{e.g. } \dfrac{dy}{dt} \right)$, t is a variable of which x, y are functions, but later we consider a particular point t. For example, the tangent at 't' is given by

$$y - a \sin^3 t = -\tan t (x - a \cos^3 t)$$

which reduces to

$$x \sin t + y \cos t = a \cos t \sin t.$$

It is, however, clear from the context whether at any stage 't' is being considered as a variable or a constant.

Exercise XXIX

1. In each of the following, verify that elimination of t leads to the given 'cartesian' equation in terms of x, y; also find the equations of the tangent and normal at 't'.
   (a) The parabola $x = at^2$, $y = 2at$; $y^2 = 4ax$.
   (b) The ellipse $x = a \cos t$, $y = b \sin t$; $\dfrac{x^2}{a^2} + \dfrac{y^2}{b^2} = 1$.

   (c) The rectangular hyperbola $x = ct$, $y = \dfrac{c}{t}$; $xy = c^2$.

   (d) The hyperbola $x = a \sec t$, $y = b \tan t$; $\dfrac{x^2}{a^2} - \dfrac{y^2}{b^2} = 1$.

2. (i) Sketch the 'semi-cubical parabola' $y = at^3$, $x = at^2$,
   (ii) if $x^3 = ay^2$, investigate the existence of the gradient at $x = 0$.

3. (D) Figure 6.12 (p. 100) represents a circular hoop rolling along the x-axis. The point P was initially at 0, and angle PCT is measured in radians as θ. (i) Explain why $OT = \text{arc } PT = a\theta$. (ii) Show that the coordinates of P are given by $x = OT - PN$, $y = CT - CN$, i.e. $x = a(\theta - \sin \theta)$, $y = a(1 - \cos \theta)$. (iii) Sketch the locus of P as θ varies, a *cycloid*. (iv) Find the equation of the tangent to the cycloid at the point 'θ'.

4. P is the point 't' on the astroid $x = a \cos^3 t$, $y = a \sin^3 t$ (situated in the first quadrant). The tangent at P meets the x-axis at T and the y-axis at T'. Show that: (i) 't' may be interpreted as the magnitude of angle OTT'; (ii) $OT = a \cos t$; (iii) $OT' = a \sin t$, (iv) $TT' = a$.

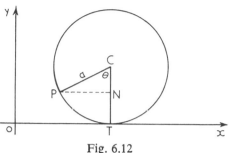

Fig. 6.12

Applications

The wider range of problems which can now be tackled is illustrated by the following examples.

EXAMPLE 6.9   *The radius of a sphere is r cm, the area of its surface is $4\pi r^2$ cm^2 and its volume is $\frac{4}{3}\pi r^3$ cm^3. If the area of the surface increases at a constant rate of 4 cm^2/sec, find the rate of increase of the volume when $r = 3$.*

Let the surface area be denoted by S and the volume by V. 'The rate of increase of surface area being 4 cm^2 s^{-1}' can be written mathematically as $\dfrac{dS}{dt} = 4$; we require $\dfrac{dV}{dt}$ when $r = 3$.

By the chain rule, we have $\dfrac{dV}{dt} = \dfrac{dV}{dr} \cdot \dfrac{dr}{dS} \cdot \dfrac{dS}{dt}$. Now $V = \frac{4}{3}\pi r^3$, so $\dfrac{dV}{dr} = 4\pi r^2$. Also $S = 4\pi r^2$, so $\dfrac{dS}{dr} = 8\pi r$ and thus $\dfrac{dr}{dS} = \dfrac{1}{8\pi r}$; finally, $\dfrac{dS}{dt} = 4$. Substituting, we have $\dfrac{dV}{dt} = 4\pi r^2 \cdot \dfrac{1}{8\pi r} \cdot 4 = 2r$, so that, when $r = 3$, $\dfrac{dV}{dt} = 6$, and the required rate is thus 6 cm^3 s^{-1}.

EXAMPLE 6.10   *The volume of water in a vessel is given by the formula $V = \dfrac{3\pi x^3}{16}$, where x is the depth of the water and V is its volume. When the depth is 1 m, the water is running into the vessel at the rate of 0·16 m^3 every minute. Find, by the method of 'small increases', the approximate increase in the depth of the water during the next 10 seconds. Give your answer correct to two significant figures.*

    [Take π to be 3·142.]

Working with metres and minutes as units, when $x = 1$, $\dfrac{dV}{dt} = 0\cdot16$. Thus $\dfrac{\delta V}{\delta t} \approx 0\cdot16$; we require δx when $\delta t = \frac{1}{6}$.

Now $\dfrac{dV}{dx} = \dfrac{9\pi x^2}{16}$, so $\dfrac{\delta x}{\delta V} \approx \dfrac{16}{9\pi x^2} = \dfrac{16}{9\pi}$ when $x = 1$. Thus

$$\delta x \approx \dfrac{16}{9\pi} \delta V = \dfrac{16}{9\pi} \dfrac{\delta V}{\delta t} \delta t \approx \dfrac{16}{9\pi} \times 0{\cdot}16 \times \dfrac{1}{6}.$$

The required approximate increase in centimetres is therefore

$$\dfrac{16}{9\pi} \times 0{\cdot}16 \times \dfrac{1}{6} \times 100 \approx 1{\cdot}5.$$

EXAMPLE 6.11   *Gas flows in a cylindrical pipe of variable cross-section. The speed of the gas is measured by the 'Mach' number M, which is defined as the ratio of the actual speed to the speed of sound. If S(x) is the area of the cross-section at a distance x along the pipe, the Mach number may be calculated from the formula*

$$MS(x) = C\{1 + AM^2\}^{(A+1)/2A},$$

where C and A are positive constants. Prove that at the smallest cross-section either the speed of the gas has a stationary value or the speed of sound is reached.

The Mach number M and the cross-section of the pipe S are both functions of x. For simplicity, we write $(A + 1)/2A = B$, thus $MS = C(1 + AM^2)^B$. Differentiating with respect to x, we have

$$M \frac{dS}{dx} + S \frac{dM}{dx} = CB(1 + AM^2)^{B-1} . 2AM \frac{dM}{dx}.$$

Now at the smallest cross-section $\dfrac{dS}{dx} = 0$, so either $\dfrac{dM}{dx} = 0$ (i.e. the speed has a stationary value) or $S = CB(1 + AM^2)^{B-1} . 2AM$, so that

$$C(1 + AM^2)^B = MS = 2AM^2 CB(1 + AM^2)^{B-1},$$

whence $1 + AM^2 = M^2(A + 1)$, and so $M^2 = 1$. Thus at the smallest cross-section either the speed is stationary or the speed of sound is reached ($M = 1$).

EXAMPLE 6.12   *Calculate the maximum and minimum values of the function* $\cos x + \cos 3x$ *in the range* $0 \leqslant x \leqslant \pi$. *Sketch the graph of the function in the given range.*

Let $y = f(x) = \cos x + \cos 3x$, then $\dfrac{dy}{dx} = f'(x) = -\sin x - 3 \sin 3x$, which is zero if

$$\sin x = -3 \sin 3x = -3(3 \sin x - 4 \sin^3 x) = -3 \sin x(3 - 4 \sin^2 x).$$

Thus stationary values are given by $\sin x = 0$ or $\sin x = \pm\sqrt{\tfrac{5}{6}}$, or (within the given range) $x = 0$, $1\cdot1(5)$, $2\cdot0$, π. To determine their nature, we differentiate again, obtaining $\dfrac{d^2y}{dx^2} = f''(x) = -\cos x - 9\cos 3x$.

x	0	$1\cdot1(5)$	$2\cdot0$	π
$f(x)$	2	$-0\cdot54$	$0\cdot54$	-2
$f''(x)$	$-^{\text{ve}}$	$+^{\text{ve}}$	$-^{\text{ve}}$	$+^{\text{ve}}$
	Max	Min	Max	Min

We can now make the required sketch (Fig. 6.13):

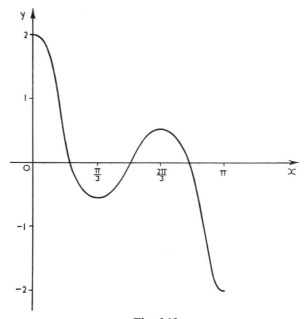

Fig. 6.13

The *period* of $f(x)$ is 2π, i.e. $f(x) = f(x + 2\pi)$. An alternative method for finding the shape of the graph is to sketch $y = \cos x$ and $y = \cos 3x$ separately (Fig. 6.14) and then add together the ordinates corresponding to the various values of x.

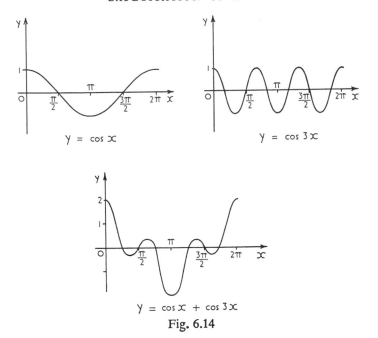

$y = \cos x$

$y = \cos 3x$

$y = \cos x + \cos 3x$

Fig. 6.14

EXAMPLE 6.13 *To sketch the curve* $y^2 = \dfrac{x(x^2 - 1)}{x - 2}$.

We shall first find the shape of $y = \dfrac{x(x^2 - 1)}{x - 2}$ and then replace each ordinate by its square root (taking both positive and negative values). We shall then obtain the graph of $y = \pm \sqrt{\left(\dfrac{x(x^2 - 1)}{x - 2} \right)}$, i.e. $y^2 = \dfrac{x(x^2 - 1)}{x - 2}$, as required.

For $y = \dfrac{x(x^2 - 1)}{x - 2}$,

(i) When $y = 0$, $x = 0$, 1 or -1.

(ii) If x is just less than 2, y is large and negative; as $x \nearrow 2$, $y \to -\infty$. Also as $x \searrow 2$, $y \to +\infty$. The line $x = 2$ is an *asymptote*, i.e. a line which the curve approaches without actually reaching.

(iii) As $x \to \pm\infty$, $\dfrac{y}{x^2} = \dfrac{x(x^2 - 1)}{x^2(x - 2)} = \dfrac{x^3 - x}{x^3 - 2x^2} = \dfrac{1 - \dfrac{1}{x^2}}{1 - \dfrac{2}{x}} \to 1.$

Thus as $x \to \pm\infty$, the curve resembles the parabola $y = x^2$. (This can be seen from the fact that when x is numerically large the terms of highest degree in x will dominate the numerator and denominator separately, so that $y \approx \dfrac{x^3}{x} = x^2$, but it is essential to check this observation, at least mentally, as above.)

We now have enough information to make a rough sketch (Fig. 6.15).

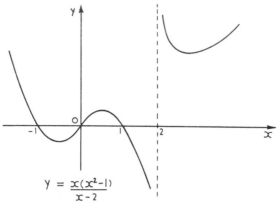

$$Y = \frac{x(x^2-1)}{x-2}$$

Fig. 6.15

Replacing y by $\pm\sqrt{y}$, we can now sketch the required graph (Fig. 6.16).

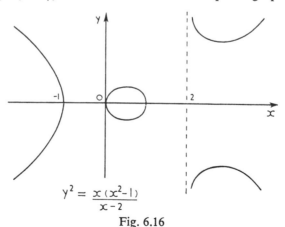

$$y^2 = \frac{x(x^2-1)}{x-2}$$

Fig. 6.16

There are gaps $-1 < x < 0$ and $1 < x < 2$ since $\dfrac{x(x^2-1)}{x-2}$ is negative for these values of x. To verify the shape near $x = 0$ and ± 1, if $y^2 = \dfrac{x(x^2-1)}{x-2} = \dfrac{x^3-x}{x-2}$, we have

$$2y \frac{dy}{dx} = \frac{(3x^2-1)(x-2) - (x^3-x)}{(x-2)^2},$$

i.e.

$$\frac{dy}{dx} = \frac{x^3 - 3x^2 + 1}{y(x - 2)^2}$$

When x is near 0, 1 or -1, so that $y \approx 0$, $\dfrac{dy}{dx}$ is in each case numerically large, in accordance with the sketch.

REVISION EXERCISE I

In questions 1 to 11, differentiate the given functions with respect to x.

1. $3x - 4 + x^{-2}$.

2. $x^{\frac{2}{3}} - 2x^{\frac{1}{2}}$.

3. $\dfrac{1}{1 + x}$.

4. $(3x - 2)^2$.

5. $\dfrac{3x - 4}{x^2 + 1}$.

6. $3 \sin x + \cos 2x$.

7. $\dfrac{x(x - 1)}{(x - 2)(x - 3)}$

8. $\sin 4x \cos^3 x$.

9. $\sin^{-1} 4x - \tan^{-1}(2x + 1)$.

10. $\cos^{-1}\left(\dfrac{1}{1 + x}\right)^{\frac{1}{2}}$.

11. $\sin^{-1}(m \sin x)$.

12. (i) Differentiate with respect to x:

$$3x^2 + \frac{5}{x^2}, \quad (3x^2 + 4x - 2)^2.$$

(ii) Integrate with respect to x:

$$3x^2 + 4x - 2.$$

In questions 13 to 16, differentiate from first principles.

13. $x^3 + 4x$.

14. $\dfrac{2}{x}$.

15. $\cos 2x$.

16. $\tan x$.

17. Differentiate and integrate with respect to x: $\sin 2x + (3x - 1)^2$.

18. If $y = 4 \cos 3x - 3 \sin 3x$, verify that $\dfrac{d^2y}{dx^2} = -9y$.

In questions 19 to 23, integrate the given functions with respect to x.

19. $3x^3 - x^{-2}$.

20. $\cos x - 4 \sin 3x$.

21. $\cos x \sin 7x$.

22. $\dfrac{1}{7 + x^2}$.

23. $\dfrac{1}{(4 - 25x^2)^{\frac{1}{2}}}$.

24. Evaluate (i) $\displaystyle\int_0^\pi \cos 4x \cos 5x \, dx$, (ii) $\displaystyle\int_0^\pi \cos^2 4x \, dx$.

25. Investigate $\displaystyle\int_0^\pi \cos px \cos qx\, dx$, where p and q are integers.

26. (i) Differentiate $(1 + x^2)^{\frac{1}{2}}$, (ii) integrate $\dfrac{x}{(1 + x^2)^{\frac{1}{2}}}$.

27. (D) Criticize the following argument.

$$\text{`} \frac{d}{dx}(\sin x) = \lim_{h\to 0} \frac{\sin(x + h) - \sin x}{h}$$

$$= \lim_{h\to 0} \frac{\sin x \cos h + \cos x \sin h - \sin x}{h}$$

$$= \lim_{h\to 0} \frac{\cos x \sin h}{h}$$

$$= \cos x.\text{'}$$

In questions 28 to 57, sketch the graphs.

28. With the same axes; x, $2x$, $3x + 4$, $-x$, $-\frac{1}{2}x$, $-2x$, $4 - x$.

29. x^2, $2x^2$, $-2x^2$, $x^2 + 1$, $1 - 2x^2$.

30. $y = x^n$ for (i) $n = 2, 3, 4, 5$; (ii) $n = -1, -2, -3, -4$; (iii) $n = \frac{1}{2}, \frac{1}{3}, \frac{1}{4}$.

31. $y^2 = x^n$ for $n = 1, 2, 3, -1$.

32. $x(x - 1)$.

33. $x(x - 1)(x - 2)$.

34. $x(x - 1)(x - 1\cdot01)$.

35. $x(x - 1)^2$.

36. $y^2 = x^2$.

37. $y^2 = x^2 - 1$.

38. $y^2 = x(x - 1)$.

39. $y^2 = x(x - 1)(x - 2)$.

40. $y^2 = x^2(x - 1)$.

41. $y^2 = x^2(1 - x)$.

42. $(x^2 - 1)(x^2 - 4)$.

43. $x^2(x^2 - 1)$.

44. $\dfrac{1}{1 + x}$.

45. $\dfrac{1}{1 + x^2}$.

46. $\dfrac{x}{1 + x^2}$.

47. $\dfrac{1}{1 - x^2}$.

48. $\dfrac{x}{1 - x^2}$.

49. $y^2 = \dfrac{x}{1 - x^2}$.

50. $\dfrac{x^2}{1 - x^2}$.

51. $y^2 = \dfrac{x^2}{1 - x^2}$.

52. $y^2 = \dfrac{x^2}{x^2 - 1}$.

53. $\dfrac{1}{(x - 1)(x - 2)(x - 3)}$.

54. $\dfrac{1}{(x - 1)(x - 3)^2}$.

55. $\dfrac{x}{(x - 1)(x - 2)(x - 3)}$.

56. $\dfrac{x - 1}{x(x - 2)(x - 3)}$.

57. $y^2 = \dfrac{x - 1}{x(x - 2)(x - 3)}$.

58. The curves $y = x^2 - x - 5$ and $y = \dfrac{3}{x}$ cut at the point $(3, 1)$. Find the equation of the tangent to each curve at this point, and also find the area of the triangle formed by these two tangents and the x-axis.

59. (i) Find $\dfrac{dy}{dx}$ if (a) $y = (x^2 - 3)^2$,

   (b) $x^2y - x^3 + 1 = 0$.

   (ii) Find the equation of the tangent at the point $(2, 2)$ on the curve $y = x^2 - 3x$ and the coordinates of the point at which this tangent meets the curve again.

60. Find the equation of the tangent to the curve $y = \dfrac{4}{x}$ at the point $(2, 2)$; and prove that the area of the triangle formed by the axes and this tangent is equal to 8.

   Prove that the area of the triangle formed by the axes and the tangent at the point $\left(a, \dfrac{4}{a}\right)$ is also equal to 8 for all values of a.

61. Find where the normal at $(1, \tfrac{1}{2})$ to the curve $y = \dfrac{1}{x + 1}$ meets the tangent at $(2, -7)$ to the curve $y = x^3 - 4x^2 + 1$.

62. Find the equations of the tangent and normal to $x^3 + y^3 = 9$ at the point $(1, 2)$.

63. Find the tangent and normal at the point 't' of the curve $x = \cos^5 t$, $y = \sin^5 t$.

64. Find the equation of the tangent at the point 't' of the curve $x = \dfrac{t}{1 + t^3}$, $y = \dfrac{t^2}{1 + t^3}$.

65. Find the equation of the tangent at the point (x_1, y_1) on the conic

$$ax^2 + 2hxy + by^2 + 2gx + 2fy + c = 0.$$

66. Evaluate $\lim_{x \to 1} (x^4 - 2x + 2)$.

67. Evaluate $\lim_{x \to 1} \dfrac{x^3 - 1}{x - 1}$.

68. Evaluate $\lim_{v \to 0} \dfrac{x^2 + x}{x^{\frac{1}{2}}}$.

69. (i) Use a diagram to illustrate why

$$1 < \frac{\theta}{\sin \theta} < \sec \theta \text{ if } 0 < \theta < \frac{\pi}{2}$$

(ii) Deduce that $\lim\limits_{\theta \to 0} \frac{\sin \theta}{\theta} = 1$.

70. By writing $\frac{\sin m\theta}{\theta} = m \frac{\sin m\theta}{m\theta}$, evaluate $\lim\limits_{\theta \to 0} \frac{\sin m\theta}{\theta}$.

71. Evaluate the limit as $\theta \to 0$ of (i) $\frac{\theta}{\sin 2\theta}$, (ii) $\frac{\theta \cos 2\theta}{\sin \theta}$, (iii) $\theta \cot \theta$, (iv) $\frac{\tan 5\theta}{\sin 3\theta}$.

72. A function $f(x)$ is defined as follows:

$$f(x) = x^2 + 1 \quad (x \leqslant 2);$$
$$f(x) = 5(x - 1) \quad (x > 2).$$

Investigate whether $f(x)$ is (i) continuous, (ii) differentiable, at $x = 2$.

73. $g(x) = 7 \cos x, \left(x \leqslant \frac{\pi}{2}\right); \ g(x) = \cos 3x + 2 \cos 2x, \left(x > \frac{\pi}{2}\right)$. Investigate the existence of $g'\left(\frac{\pi}{2}\right)$.

74. (i) For what values of x is the function $\frac{1 + \sin x}{\cos x}$ discontinuous?

(ii) Sketch the graph of $y = \sec x + \tan x$.

75. (i) If $f(x) = x^n \sin \frac{1}{x}$, $(x \neq 0)$, $f(0) = 0$, investigate the existence of $f'(0)$, i.e.

$$\lim\limits_{h \to 0} \frac{f(h) - f(0)}{h}, \quad (a) \text{ when } n = 2, \quad (b) \text{ when } n = 1.$$

(ii) Criticize the following argument for the case $n = 2$.

'$f(x) = x^2 \sin \frac{1}{x}$, therefore $f'(x) = 2x \sin \frac{1}{x} - \cos \frac{1}{x}$. As $x \to 0$, this oscillates between values near ± 1, and so $f'(0)$ does not exist.'

76. The velocity v m s^{-1} of a particle travelling in a straight line is given by $v = 32 - \frac{1}{2}t^2$, where t is the time in seconds. Find its initial velocity; and find at what instant the particle will reverse the direction of its motion. Use the Calculus to find its acceleration at this instant, and also the distance it has then travelled from the starting point.

77. A particle is moving in a straight line; its velocity v m s^{-1} at time t s is given by $v = t^2 - 6t + 11$. Find its acceleration when its velocity is 2 m s^{-1}.

Given that its distance from a fixed point A on the line at 6 s is 8 m, find its distance from A at 12 s.

78. The acceleration of a car t seconds after starting from rest is $\frac{75 + 10t - t^2}{20}$ m s^{-2} until the instant when this expression vanishes. After this instant the speed of the car remains constant. Find:

(i) the maximum acceleration;
(ii) the time taken to attain the greatest speed;
(iii) the greatest speed attained.

79. A particle moves in a straight line such that its distance s m from a fixed point O on the line after t s is given by $s = 12 \cos 3t + 5 \sin 3t$. Find (i) the velocity and acceleration when $t = \dfrac{\pi}{3}$, (ii) the maximum and minimum values of s.

80. (i) Differentiate

    (a) $(1 - x)(2 - 3x + x^2)$;   (b) $\dfrac{6x^3 - 4x}{x^2}$.

    (ii) Find the range of values of x for which the function $x^3 - 3x^2 - 9x + 11$ decreases in value as x increases.

81. (i) Differentiate with respect to x the expression
$$4x^5 + x - 5x^{-4}.$$

    (ii) Find a maximum value, and also a minimum value, of the expression
$$x^3 - 3x^2 - 9x + 27.$$

82. (i) Differentiate $(x + 2)(x^2 + 3)$ and $\left(x - \dfrac{3}{x}\right)^2$ with respect to x.

    (ii) Find the coordinates of points on the curve
$$y = 2x^3 + 3x^2 - 12x + 6$$
at which y has a stationary value and state in each case whether the value of y is a maximum or a minimum.

83. (a) Differentiate with respect to x the functions

    (i) $\dfrac{1}{x^2 - x + 1}$;   (ii) $x^2 \sin x + 2x \cos x - 2 \sin x$.

    (b) Find the maximum and minimum values of the function
$$x + \dfrac{1}{x + 1}.$$
Sketch the graph of the function.

84. Show that the maximum and minimum values of the function $(x + 3)^2/(x^2 + 1)$ are 10 and 0 respectively. Draw a rough graph of this function for the range $-5 \leqslant x \leqslant 3$, indicating clearly the positions of the turning points.

85. An open tank is to be constructed with a horizontal square base and four vertical rectangular sides. It is to have a capacity of 32 m³. Find the least area of sheet metal of which it can be made.

86. A variable rectangle has a constant perimeter of 20 cm. Find the lengths of the sides when the area is a maximum.

    A variable rectangle has a constant area 36 cm². Find the lengths of the sides when the perimeter is a minimum.

87. The cost £C and revenue £R of a freight plane per mile flown are given by the relations
$$C = V^2 + 16W^2, \qquad R = K + 8WV,$$
where $100V$ is the average speed in miles per hour, W is the pay load in tons, and £K is a constant subsidy. Show that the profit per mile cannot exceed £K.

    Find the values of V and W for this profit to be made when V and W are connected by the relation $WV^2 = 2$.

88. (i) Integrate with respect to x the expression

$$2x^3 + 3x^{\frac{1}{2}} - x^{-2}.$$

    (ii) After making a rough sketch, find by integration the area cut off between the x-axis and the curve

$$y = x(6 - x).$$

89. Draw a rough sketch of the curve $y^2 = 16x$ between $x = 0$ and $x = 4$. Calculate the area contained between the curve and the line $x = 4$, and the volume obtained by completely rotating that area about the axis of x.

90. Find the volume obtained by rotating the part of the curve $y = \dfrac{1}{x}$ between $x = 1$ and $x = 3$ about the axis of x.

91. The gradient at a point (x, y) on a curve which passes through the point $(3, 3)$ is $3x^2 - 8x + 4$. Find the equation of the curve, and indicate in a sketch the *form* of the curve, paying attention to the turning points.

    Find (i) the area between the curve and the part of the x-axis from $x = 0$ to $x = 2$, and (ii) the volume enclosed by the surface formed by revolving this area about the x-axis.

92. An area in the first quadrant is bounded by the ellipse $4x^2 + 9y^2 = 36$ and the axes of coordinates. This area is rotated through four right angles about the x-axis. Find (i) the volume generated and (ii) the x-coordinate of the centre of gravity of this volume.

93. Find the x-coordinates of the points on the curve

$$y = (x + 1)(x - 2)^2$$

at which the gradient is zero, and test whether y has a maximum or minimum value at each of the points that you have found.

    Also find the x-coordinate of the point at which the tangent is parallel to the tangent at the point $(3, 4)$.

    Draw a rough sketch of the curve.

94. The straight line $y = a^2$ meets the curve $y = x^2$ at the points P, Q. The co-ordinates of a point C are $(0, 12)$. Given that a^2 is less than 12, find an expression in terms of a for the area of the triangle CPQ, and find the value of a when this area is a maximum.

    When $a = 3$, find the area between the curve, the x-axis, and the perpendicular on the x-axis from P; and hence find the area between CP, CQ, and the curve.

95. Find the coordinates of the maximum and minimum points on the curve

$$y = x^3 - 9x^2 + 24x.$$

    Sketch the curve.

    Find the area of the region enclosed by the curve, its maximum and minimum ordinates and the x-axis.

96. (i) Find the equation of the tangent to the curve

$$3y^2 = 2x^3 + x^2$$

at the point A $(1, 1)$. Show that this tangent meets the curve again at a point B whose abscissa is $\frac{1}{6}$, and find the ordinate of B.

    (ii) The area bounded by the parabola $y^2 = 16x$ and the lines $x = 0$ and $y = 8$ is rotated about the y-axis. Show that the volume so formed is $128\pi/5$.

97. A curve is given by the equation

$$y = x(x - 1)(px + q)$$

where p and q are constants. If the tangents to the curve at $(0, 0)$, $(1, 0)$ are the lines $y = -x$, $y = 3x - 3$, respectively, find the values of p and q and sketch the curve.

Find the area bounded by the curve and that portion of the x-axis between the points $(0, 0)$, $(1, 0)$.

98. Find (i) the gradients of the tangents to the curve whose equation is

$$y^2 = x^2(1 - x^2)$$

at the points $(0, 0)$ and $(1, 0)$, and (ii) the coordinates of the turning-points on the curve.

Sketch the curve.

A surface is formed by rotating the whole of this curve through four right angles about the x-axis. Find the volume enclosed by the surface.

99. Find the approximate value of $x^3 - 3x^2 + 4x + 2$ when $x = 1{\cdot}003$.

100. Find the approximate value of $\dfrac{1}{x^4}$ when $x = 2{\cdot}04$.

101. Find the roots of $4x^3 + 18x^2 + 24x + 9 = 0$, correct to one decimal place.

102. Sketch the graph of $y = \tan x - 2x$ for $0 < x < \dfrac{\pi}{2}$. Use Newton's method to find the root of $\tan x = 2x$ in this interval, correct to two decimal places.

103. Use Simpson's rule with five ordinates to evaluate approximately $\displaystyle\int_0^1 \dfrac{1}{2 + x^2}\, dx$.

Also evaluate this integral directly (as an inverse tangent, cf. p. 93), and compare your answers.

104. Sketch the graphs of (i) $3 \cos x + 4 \cos 3x$, (ii) $\cos 3x + \sin 5x$.

105. (*a*) Draw accurately on squared paper the graphs of (i) $\cos 4x$, (ii) $\cos 5x$, (iii) $\cos 4x + \cos 5x$.

(*b*) (P) Find out what you can about Resonance.

106. Draw accurately, using the same axes, the graphs of $\sin x$, $\sin x - \frac{1}{2} \sin 2x$, $\sin x - \frac{1}{2} \sin 2x + \frac{1}{3} \sin 3x$; take values of x from $-\pi$ to π. (It can be proved, beyond our present scope, that in this range as more terms are taken in the 'Fourier series' $\sin x - \frac{1}{2} \sin 2x + \frac{1}{3} \sin 3x \ldots$, so a closer approximation is reached to the graph of $\frac{1}{2}x$.)

107. Find the average value of $\sin^2 t$ over the interval $0 \leqslant t \leqslant \dfrac{\pi}{2}$.

Papers A to H

A

A1. (i) Differentiate the expression $3x^{\frac{1}{2}} + 4 - 5x^{-\frac{1}{2}}$, and integrate the expression $x^2 - 3 - 2x^{-2}$, both with respect to x.

(ii) Find a maximum and a minimum value of the expression $x^3 - 3x$, and give a rough sketch of the curve $y = x^3 - 3x$.

A2. Find the equation of the tangent to the curve

$$y = x^2 - 4x + 5$$

at the point $(5, 10)$, and also the equation of the line through $(5, 10)$ which is perpendicular to the tangent.

Find the coordinates of the point on the curve at which the tangent will be parallel to the line $y = 2x$.

A3. Evaluate the integrals

$$\int_{-1}^{1} x^2 \, dx, \qquad \int_{-1}^{1} x^4 \, dx, \qquad \int_{1}^{2} \frac{dx}{x^2}.$$

How could you show graphically, without evaluation, that the value of the first integral is greater than the value of the second?

A4. If y is a cubic function of x of the form $y = ax^3 + bx^2 + c$, complete the table below:

x	0	-1	-2
y	-4		
$\dfrac{dy}{dx}$		-3	
$\dfrac{d^2y}{dx^2}$			-6

Describe the nature of the points on the graph of this function at which (i) $x = 0$, (ii) $x = -1$, (iii) $x = -2$.

A5. (i) Sketch the curve $y = x^2(3 - x)$.

Find the equation of the tangent to the curve at the point $(3, 0)$. Find also the equation of the parallel tangent.

(ii) A piece of lead is cast into a solid right circular cylinder. Find the ratio of the cylinder height to the radius of its base if its total surface area is as small as possible.

B

B1. (i) Find the maximum value of $(5 + x)(1 - x)$.

(ii) Find the indefinite integral $\int (x^3 + 1) \, dx$, and evaluate

$$\int_{1}^{2} \frac{x^2 + 1}{x^2} \, dx.$$

B2. (i) Differentiate $\sqrt[3]{x}$, and integrate $x + 2\sqrt{x} + 1$, in each case with respect to x.

(ii) A point is travelling along a straight line with a velocity which is given by $(3t^2 - 4t + 3)$ m s^{-1}, where t seconds is the time measured from the instant when the point is 2 m in the positive direction from a fixed point O on the line. Find its distance from O after 1 s, and also after 3 s. Find also its minimum velocity.

B3. A rectangular box without a lid is made of cardboard of negligible thickness. The sides of the base are $2x$ cm and $3x$ cm, and the height is y cm. If the total area of the cardboard is 200 cm^2, prove that $y = \dfrac{20}{x} - \dfrac{3x}{5}$.

Find the dimensions of the box when its volume is a maximum.

B4. Calculate the area between the curve

$$y = 1 + 2x + 3x^2,$$

the x-axis, and the ordinates $x = -1$, $x = 3$ by Simpson's rule for three ordinates. Verify by integration that the rule gives this area exactly.

B5. Sketch the curve whose equation is

$$y^2 = x^3(1 - x)$$

paying attention to the gradients at $(0, 0)$ and $(1, 0)$.

This curve is rotated through four right angles about the x-axis. A uniform solid is bounded by the surface thus generated. Find the volume and the coordinates of the centre of gravity of this solid.

C

C1. (i) Differentiate $x + x^{\frac{1}{2}} + x^{-\frac{1}{2}} + x^{-1}$ with respect to x.

(ii) From the equations $z = x^3 + y^3 - 15x$ and $x - y = 3$ find an equation connecting z and x. Find then the maximum and minimum value of z.

C2. The velocity of a train starting from rest is proportional to t^2, where t is the time which has elapsed since it started. If the distance it has covered at the end of 6 seconds is 6 m, find the velocity and the rate of acceleration at that instant.

C3. State *without proof* sufficient conditions that a function $f(x)$ has (i) a maximum value at $x = a$ and (ii) a minimum value at $x = b$.

Find the maximum and minimum values of the function

$$\frac{x + 2}{(x^2 + 1)^{\frac{1}{2}}}.$$

C4. Find the equation of the normal to the curve $100y = x^4$ at the point $(5, \frac{125}{4})$

The line $x = c$ is chosen so that an area 10 is cut off between the curve, the line, and the axis of x. Find the value of c to one place of decimals.

C5. The curve whose equation is

$$y = x(x - a)(x - b),$$

where $0 < a < b$, cuts the axis of x at O (the origin), A and B in this order. Calculate the area contained between the curve and the portion OA of the axis. If the two areas enclosed between the curve and the axis are equal in magnitude, find the ratio of a to b.

D

D1. Find, *from first principles*, the differential coefficient of $\dfrac{1}{x^3}$ with respect to x.

Differentiate the following expressions with respect to x, giving your results as simply as possible:

(i) $\dfrac{x}{1-x}$,   (ii) $\dfrac{x^{\frac{1}{2}}}{1-x^{\frac{1}{2}}}$,   (iii) $\sin^2 3x$,   (iv) $\tan^{-1}\left(\dfrac{2x}{1-x^2}\right)$.

D2. A car starts from rest with an acceleration proportional to the time. It travels 12 m in the first 3 seconds. Calculate its velocity and acceleration at the end of this time.

Find the distance travelled up to the instant when the velocity and acceleration are numerically equal.

D3. A vessel has the shape formed by rotating the curve $4y = x^2$ through 2π radians about the axis of y which is vertical. Find the volume of liquid which will fill the vessel to a depth of h cm.

If a hole at the lowest point of the vessel allows liquid to escape at the rate of 4π cm^3 s^{-1}, prove that the rate at which the level falls when the depth is h cm is $1/h$ cm s^{-1}. Find the time taken for the depth to decrease from 8 cm to 6 cm.

D4. Find the equations of the tangents to the curve $y = x^3 + 2$ which are parallel to the line $y = 3x + 2$.

Find the area in the first quadrant bounded by the coordinate axes, the line $x = 1$, and the curve. Show that the line $y = 3x$ divides this area in the ratio $1:2$.

D5. A sports field is to have the shape of a rectangular area $ABCD$ with semi-circular areas at opposite ends on BC and AD as diameters. Its perimeter is to be 400 metres long and the area of the rectangle $ABCD$ is to be a maximum. Find the dimensions of the rectangle.

E

E1. Explain what is meant by the *derivative* of a function $f(x)$ with respect to x.

Differentiate, reducing results to their simplest form:

$$\frac{x^2 - 4}{x^2(x+4)}\;;\qquad \sec^2 x - 2x \tan x.$$

Show that the expression
$$\sqrt{(x^2)} - \sqrt{(x^5)},$$
where x is positive, and the positive values of the square roots are taken, has a maximum value, and find it to three significant figures.

E2. A spherical balloon is being inflated so that, at the time when the radius is 5 cm, the radius is increasing at the rate of 0·15 cm s^{-1}. Find the rate of increase of the volume at this time.

If the rate of increase of the volume thereafter remains constant, find the rate at which the radius is increasing when the balloon has a radius of 6 cm.

E3. (i) Illustrate with a sketch-graph that, if a is an approximate root of an equation $f(x) = 0$, then $a - \dfrac{f(a)}{f'(a)}$ is not always a closer approximation.

(ii) Use Newton's method to find the real root of the equation $x^3 + 4x = 6$ correct to one decimal place.

E4. Find the equations of the tangent and normal to the curve

$$y = x(x - 1)^2$$

at the origin O. Find also the coordinates of the point P in which the tangent meets the curve again. Calculate the area bounded by the arc OP and the chord OP.

E5. The curved surface of an open bowl with a flat circular base may be traced out by the complete revolution of a portion of the curve $ay = x^2$ about the vertical axis Oy. The radius of the top rim of the bowl is twice that of the base, and the capacity of the bowl is $\frac{5}{6}\pi a^3$ cu. units. Find the vertical height of the bowl.

F

F1. Differentiate the following functions with respect to x:

$$\text{(i)} \ \frac{x^2}{x - 1} ; \qquad \text{(ii)} \ \frac{(1 + x^2)^{\frac{1}{2}} - x}{(1 + x^2)^{\frac{1}{2}} + x}.$$

Prove that, if $y = x \sin x$, then

$$x^2 \frac{d^2y}{dx^2} - 2x \frac{dy}{dx} + (2 + x^2)y = 0.$$

F2. (a) Sketch the graphs of

$$(x - 1)(x - 2) \text{ and } \frac{1}{(x - 1)(x - 2)}.$$

(The coordinates of turning points are not required.)

(b) If $y = x^3 + 7x - 10$, show by considering the signs of y and dy/dx that y is zero for one and only one real value of x. Calculate this value of x correct to one decimal place.

F3. The loss of heat from a closed full hot-water tank is proportional to its surface area. A cylindrical tank has flat ends. If its volume is fixed, determine the ratio of the length to the radius for the loss of heat to be a minimum.

Find also whether such a tank would retain heat more efficiently than a cubical one of equal volume.

F4. The part of the curve $y = +x^{\frac{1}{2}}(x - 1)^{\frac{1}{2}}$ between the lines $x = 1$ and $x = 3$ is rotated through four right angles about the axis of x. Prove that the volume enclosed is $44\pi\sqrt{2}/15$.

Calculate the approximate volume directly from the integral using Simpson's rule with ordinates at $x = 1$, $1\frac{1}{2}$, 2, $2\frac{1}{2}$, and 3.

F5. A particle moves along the axis of x so that its distance from the origin t seconds after starting is given by the formula $x = a \cos pt$. Prove that the velocity of the particle changes direction once, and once only, between the times $t = 0$ and $t = 2\pi/p$ and that the change of direction occurs at the point $x = -a$.

The direction from the origin of a second particle is given by the formula

$$x = a \cos pt + \tfrac{1}{2}a \cos 2pt.$$

Write down expressions for its velocity and acceleration at time t. Show that between $t = 0$ and $t = 2\pi/p$ the velocity of the particle changes direction three times, and find the values of x at which these changes occur.

G

G1. Find, *from first principles*, the differential coefficient of sin x with respect to x.

[You may assume that $\dfrac{\sin \theta}{\theta} \to 1$ as $\theta \to 0$.]

Differentiate the following functions of x with respect to x, expressing your results as simply as possible:

$$\text{(i) } \frac{3x-1}{2x^2-1}, \qquad \text{(ii) } \sin^2 x \tan x, \qquad \text{(iii) } \left(\frac{x}{a-x}\right)^{\frac{1}{4}}.$$

G2. The gradient at any point (x, y) of a curve is given by

$$\frac{dy}{dx} = -3x^2 + 3,$$

and the curve passes through the point $(2, 0)$; find its equation and *sketch* the graph, indicating the turning-points.

Find the distance from the y-axis of the centre of gravity of a uniform lamina bounded by the curve and the positive halves of the x and y axes.

G3. Sketch the curve $y^2 = 1 - x$.

Find the area of the region bounded by the y-axis and the part of the curve for which x is positive. Find also the volumes generated when this region is rotated through two right-angles about (i) the x-axis; (ii) the y-axis.

G4. Points P and Q lie on the axis Ox at distances p, q respectively from O, where

$$\frac{1}{p} + \frac{1}{q} = \frac{1}{2}.$$

If P moves a small distance δp along Ox, find, to a first approximation, the corresponding distance δq moved by Q. Hence find the ratio of the velocities of P and Q when $p = 3$.

$$\left[\text{Hint: } \frac{\delta\left(\dfrac{1}{p}\right)}{\delta p} \cdot \delta p \approx -\frac{1}{p^2}\,\delta p.\right]$$

G5. A beam of rectangular cross-section is to be cut from a cylindrical log of radius a. The stiffness of the beam is proportional to xy^3 where x is the breadth and y the depth of the section. Find the cross-sectional area of the beam

    (i) of greatest volume;

    (ii) of greatest stiffness;

that can be cut from the log.

H

H1. (i) Using Simpson's rule with five ordinates find the mean value of the function $y = \sqrt{(x^3 + 1)}$ over the interval $x = 0$ to $x = 4$.

(ii) The area enclosed between the curve $y = x(4 - x)$ and the straight line $y = 2x$ is rotated through four right angles about the axis of x. Prove that the volume generated is $32\pi/5$.

H2. Prove that, if u and v are functions of x,

$$\frac{d}{dx}\left(\frac{u}{v}\right) = \frac{1}{v^2}\left(v\frac{du}{dx} - u\frac{dv}{dx}\right).$$

Differentiate the following expressions with respect to x, giving your results as simply as possible:

(i) $\dfrac{x^2 + 1}{x}$, (ii) $(a^2 - x^2)^{\frac{3}{2}}$, (iii) $\sin^{-1}\left(\dfrac{x}{1 + x}\right)$

H3. State, without proof, rules for finding the maximum and minimum values of a function by means of the differential calculus.

An isosceles triangle ABC, in which $AB = AC$, is inscribed in a circle of given radius a. Denoting $\angle BAC$ by 2θ, express the sum of the perpendiculars from the three vertices to the opposite sides in terms of a and $\sin\theta$. Prove that the sum is a maximum (*not* a minimum) when the triangle is equilateral.

H4. A point moves in a straight line so that its distance at time t from a given point O of the line is x, where

$$x = t^2 \sin t + 6t \cos t - 12 \sin t.$$

Find its velocity at time t, and prove that the acceleration is then

$$-t^2 \sin t - 2t \cos t + 2 \sin t.$$

Determine the times $(t > 0)$ at which the acceleration has a turning value, distinguishing between maxima and minima.

H5. Prove that the slope of the curve whose equation is

$$y = 1 + x + \tfrac{1}{2}x^2 + \tfrac{1}{6}x^3$$

is always positive.

Show that the curve has a point of inflexion where $x = -1$, the slope there being $\tfrac{1}{2}$.

Prove also that the tangent at the point $(0, 1)$ meets the curve again at the point $(-3, -2)$.

Sketch the curve, indicating clearly the point of inflexion and the tangents at the points $(-1, \tfrac{1}{3})$ and $(0, 1)$.

7
Methods of Integration

All things change.
HERACLITUS

The rules given in the previous chapter reduce differentiation largely to a matter of routine. For integration, however, it is sometimes more of a problem to find the most promising line of attack. Some methods are given in this chapter, and we return again to this topic in Chapter 10.

Integration 'by Inspection'

The simplest method of integration is to know the answer, or at least be able to guess its pattern. Any polynomial can, of course, be integrated on sight, e.g. $\int (4x^2 - 6x + 3)\,dx = \frac{4}{3}x^3 - 3x^2 + 3x + c$. Again,

$$\int (3x + 4)^7\,dx = \tfrac{1}{24}(3x + 4)^8 + c$$

(cf. Example 6.3, p. 86).

EXAMPLE 7.1   *To evaluate* $I = \int \dfrac{1}{\sqrt{(3x + 1)}}\,dx.$

$\dfrac{1}{\sqrt{(3x + 1)}} = (3x + 1)^{-\frac{1}{2}}$, and it is a fair guess that the integral is a multiple of $(3x + 1)^{+\frac{1}{2}}$, since $\int x^{-\frac{1}{2}}\,dx = -2x^{\frac{1}{2}}$. Now $\dfrac{d}{dx}(3x + 1)^{\frac{1}{2}} = \frac{1}{2}(3x + 1)^{-\frac{1}{2}}.3$ (using the chain rule), $= \frac{3}{2}(3x + 1)^{-\frac{1}{2}}.$

Therefore, $I = \int (3x + 1)^{-\frac{1}{2}}\,dx = \frac{2}{3}(3x + 1)^{\frac{1}{2}} + c.$

The table on page 119 can be constructed using previous results (cf. p. 93).

We recall also that certain integrals can be evaluated with the help of standard trigonometric formulae, e.g.

(a) $\displaystyle\int 2\cos^2 x\,dx = \int (1 + \cos 2x)\,dx = x + \tfrac{1}{2}\sin 2x + c;$

(b) $\displaystyle\int 2\cos 4x \cos 3x\,dx = \int (\cos 7x + \cos x)\,dx = \tfrac{1}{7}\sin 7x + \sin x + c;$

(c) $\displaystyle\int \tan^2 x\,dx = \int (\sec^2 x - 1)\,dx = \tan x - x + c.$

In the table below, and occasionally elsewhere, we have suppressed the 'arbitrary constant'. It should, however, always be inserted in worked examples, to avoid 'results' such as the following one.

To 'prove' that $\frac{1}{2} = 1$.

$$\int \sin 2x \, dx = -\tfrac{1}{2} \cos 2x = -\tfrac{1}{2}(2 \cos^2 x - 1) = \tfrac{1}{2} - \cos^2 x;$$

also $\int \sin 2x \, dx = \int 2 \sin x \cos x \, dx = \sin^2 x$. Thus $\frac{1}{2} - \cos^2 x = \sin^2 x$, so that $\frac{1}{2} = \sin^2 x + \cos^2 x = 1$. It is left to the reader to unmask the fallacy.

$f(x)$	$\int f(x)\,dx$
$\sin x$	$-\cos x$
$\cos x$	$\sin x$
$\sec^2 x$	$\tan x$
$\operatorname{cosec}^2 x$	$-\cot x$
$\sec x \tan x$	$\sec x$
$\dfrac{1}{\sqrt{(1 - x^2)}}$	$\sin^{-1} x$
$\dfrac{1}{1 + x^2}$	$\tan^{-1} x$

EXAMPLE 7.2 $\int \sec^2 (ax + b) \, dx$

$$\frac{d}{dx} \tan (ax + b) = a \sec^2 (ax + b);$$

thus

$$\int \sec^2 (ax + b) \, dx = \frac{1}{a} \tan (ax + b) + c.$$

EXAMPLE 7.3 $\int \dfrac{1}{4 + 9x^2} \, dx.$

$$\frac{d}{dx}\left(\tan^{-1} \frac{3x}{2}\right) = \frac{1}{1 + \left(\dfrac{3x}{2}\right)^2} \times \frac{3}{2} = \frac{6}{4 + 9x^2}.$$

Thus $\int \dfrac{1}{4 + 9x^2} \, dx = \tfrac{1}{6} \tan^{-1} \dfrac{3x}{2} + c.$

EXAMPLE 7.4   *To evaluate* $I = \int \dfrac{2x}{\sqrt{(x^2 + 1)}}\, dx.$

Here the key lies in the fact that $2x$ is the derivative of $x^2 + 1$. Since

$$\frac{d}{dx}(x^2 + 1)^{\frac{1}{2}} = \tfrac{1}{2}(x^2 + 1)^{-\frac{1}{2}} \cdot 2x = \frac{x}{\sqrt{(x^2 + 1)}},$$

it follows that $I = 2(x^2 + 1)^{\frac{1}{2}} + c.$

The idea of the last example can be generalized thus:

$$\int \frac{f'(x)}{\{f(x)\}^n}\, dx = \frac{1}{-n + 1}\{f(x)\}^{-n+1}, \qquad (n \neq 1),$$

but it will be more profitable to work through Exercise XXX, numbers 27 to 33, rather than attempt to memorize such a formula. It can also be verified (by differentiating both sides) that

$$\int f(x)f'(x)\, dx = \tfrac{1}{2}\{f(x)\}^2,$$

e.g. $\displaystyle\int \frac{\tan^{-1} x}{1 + x^2}\, dx = \tfrac{1}{2}(\tan^{-1} x)^2.$ This should be checked by differentiating $(\tan^{-1} x)^2$ as a 'function of a function'.

The reader who is appalled by Examples 7.1 to 7.4 (and feels he will never be an expert at 'inspecting') may be consoled to know that each of these integrals can be done slightly more painstakingly by the method described in the next section.

EXERCISE XXX

In questions 1 to 35, integrate the given function with respect to x.

1. $x^3 - 6x^2 + 4x + 2.$

2. $(x + 1)^4.$

3. $(4x + 3)^7.$

4. $\dfrac{1}{(1 - 2x)^4}.$

5. $\dfrac{1}{\sqrt{(3x + 1)}}.$

6. $(6 - 7x)^{\frac{3}{2}}.$

7. $\sin 3x.$

8. $\cos(4x + 1).$

9. $\operatorname{cosec}^2 4x.$

10. $\tan^2 4x.$

11. $\sec 3x \tan 3x.$

12. $\cos^2 x.$

13. $\cos^4 x.$ [Use $\cos^2 x = \tfrac{1}{2}(1 + \cos 2x)$]

14. $\dfrac{1}{\sqrt{(1 - x^2)}}.$

15. $\dfrac{1}{\sqrt{(1 - 4x^2)}}.$

16. $\dfrac{1}{\sqrt{(9 - 4x^2)}}.$

17. $\dfrac{1}{\sqrt{(3 - 2x^2)}}$.

18. $\dfrac{1}{1 + x^2}$.

19. $\dfrac{1}{1 + 4x^2}$.

20. $\dfrac{1}{9 + 4x^2}$.

21. $\dfrac{1}{3 + 2x^2}$.

22. $x + 4 - \dfrac{1}{\sqrt{(1 - 3x^2)}}$.

23. $\sin 7x \cos 4x$.

24. $\sin 6x \sin 5x$.

25. $\cos^2 4x$.

26. $\sin^2 7x$.

27. $\dfrac{x}{(1 + x^2)^4}$.

28. $\dfrac{3 + 10x}{(1 + 3x + 5x^2)^3}$.

29. $\dfrac{x}{\sqrt{(1 + x^2)}}$.

30. $\dfrac{x^2}{\sqrt{(1 + x^3)}}$.

31. $\dfrac{4 + 3x^2}{(1 + 4x + x^3)^7}$.

32. $\dfrac{x + 2}{(x^2 + 4x + 1)^{\frac{1}{3}}}$.

33. $\dfrac{7 \cos x - 4 \sin x}{(7 \sin x + 4 \cos x)^2}$.

34. $\dfrac{\sin^{-1} x}{\sqrt{(1 - x^2)}}$.

35. $\dfrac{x + 1}{\sqrt{(1 - x^2)}}$.

36. $\displaystyle\int_0^{\frac{\pi}{2}} \cos 6x \cos 4x \, dx$.

37. Evaluate   (i) $\dfrac{d}{dx} (\cos^4 x)$,   (ii) $\int \cos^3 x \sin x \, dx$.

38. $\int \sin^5 x \cos x \, dx$.

39. $\int \sin^4 x \cos^4 x \, dx$. [Use formula for $\sin 2x$]

40. $\displaystyle\int \dfrac{x^4}{1 + x^2} \, dx$. [Write $x^4 = 1 - (1 - x^4)$]

41. $\displaystyle\int_0^1 \dfrac{x^2(x^2 + 2)}{1 + x^2} \, dx$.

Integration by Substitution

EXAMPLE 7.5   *To evaluate $I = \displaystyle\int \dfrac{1}{\sqrt{(1 - x^2)}} \, dx$.*

Put $x = \sin t$. Then $\dfrac{dx}{dt} = \cos t$; write this (monstrously, for the moment) as $dx = \cos t \, dt$. Then

$$I = \int \frac{1}{\sqrt{(1 - x^2)}} \, dx = \int \frac{1}{\sqrt{(1 - \sin^2 t)}} \cos t \, dt$$

$$= \int \frac{\cos t}{\cos t} \, dt = \int 1 \, dt = t = \sin^{-1} x.$$

EXAMPLE 7.6   *To evaluate* $I = \int (3x + 2)^7 \, dx$.

Put $3x + 2 = t$. Then $3 \dfrac{dx}{dt} = 1$, or $dx = \tfrac{1}{3} dt$. Thus

$$I = \int t^7 . \tfrac{1}{3} dt = \tfrac{1}{3} . \tfrac{1}{8} t^8 = \tfrac{1}{24}(3x + 2)^8.$$

All that can be said so far for the method used in these examples is that it gives the right answers; we now give a justification for it. Let $f(x)$ be continuous, and so integrable (cf. p. 59), and denote $\int f(x) \, dx$ by y; then $\dfrac{dy}{dx} = f(x)$. Now if x is a differentiable function of another variable t, then

$$\frac{dy}{dt} = \frac{dy}{dx} . \frac{dx}{dt} = f(x)\frac{dx}{dt}. \text{ Thus } y = \int f(x)\frac{dx}{dt} \, dt. \text{ As this last integral is with}$$

respect to t, it is implied that $f(x)$ is expressed in terms of t; in fact we should properly write $f(x)$ as, say, $g(t)$. Thus

$$\int f(x) \, dx = \int g(t)\frac{dx}{dt} \, dt \tag{1}$$

We shall require that $\dfrac{dx}{dt}$ is a continuous function of t, so that $g(t)\dfrac{dx}{dt}$ is also continuous; this ensures the existence of the right-hand integral.

Equation (1) is the formula for *integration by substitution*. It justifies the procedure used in the above examples, e.g. in Example 7.5 $f(x) = \dfrac{1}{\sqrt{(1 - x^2)}}$, $g(t) = \dfrac{1}{\sqrt{(1 - \sin^2 t)}}$, and the 'monstrous' step of writing $\dfrac{dx}{dt} = \cos t$ in the form $dx = \cos t \, dt$ is valid in this context, since in (1) dx is replaced by $\dfrac{dx}{dt} \, dt$, i.e. in this example by $\cos t \, dt$.

EXAMPLE 7.7   *To evaluate*   (i) $\int x\sqrt{(x + 2)} \, dx$;   (ii) $\int x\sqrt{(x^2 + 1)} \, dx$;

(iii) $\displaystyle\int \frac{1}{(1 + x^2)^{\frac{3}{2}}} \, dx$; (iv) $\displaystyle\int \frac{1}{x^2 + x + 1} \, dx$; (v) $\int \sin^3 x \, dx$.

(i) Let $I = \int x\sqrt{(x + 2)} \, dx$. Put $x + 2 = y^2$, then $dx = 2y \, dy$ and

$$I = \int (y^2 - 2)y . 2y \, dy = \int (2y^4 - 4y^2) \, dy = \tfrac{2}{5}y^5 - \tfrac{4}{3}y^3 + c$$

$$= \tfrac{2}{5}(x + 2)^{\frac{5}{2}} - \tfrac{4}{3}(x + 2)^{\frac{3}{2}} + c.$$

(ii) Let $I = \int x\sqrt{(x^2 + 1)} \, dx$. Put $x^2 + 1 = y^2$, then $2x \, dx = 2y \, dy$ and

$$I = \int y . y \, dy = \int y^2 \, dy = \tfrac{1}{3}y^3 + c = \tfrac{1}{3}(x^2 + 1)^{\frac{3}{2}} + c.$$

(iii) Let $I = \int \dfrac{1}{(1 + x^2)^{\frac{3}{2}}}\, dx$. Put $x = \tan\theta$, then $dx = \sec^2\theta\, d\theta$ and

$$I = \int \frac{1}{(1 + \tan^2\theta)^{\frac{3}{2}}} \sec^2\theta\, d\theta = \int \frac{\sec^2\theta}{\sec^3\theta}\, d\theta$$

$$= \int \cos\theta\, d\theta = \sin\theta + c = \frac{x}{\sqrt{(1 + x^2)}} + c.$$

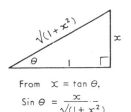

From $x = \tan\theta$,

$\sin\theta = \dfrac{x}{\sqrt{(1 + x^2)}}$

(The reader should try the alternative substitution $1 + x^2 = y^2$ and see why it does not help.

Generally, a quadratic expression for x calls for a trigonometric substitution.)

(iv) Let $I = \int \dfrac{1}{x^2 + x + 1}\, dx$. [The integrand, $\dfrac{1}{x^2 + x + 1}$, is reminiscent

of $\dfrac{1}{1 + x^2}$, which suggests that the answer may well be an inverse

tangent.] Completing the square, for $x^2 + x$, we have

$$x^2 + x + 1 = (x + \tfrac{1}{2})^2 + \tfrac{3}{4} = (x + \tfrac{1}{2})^2 + \left(\frac{\sqrt{3}}{2}\right)^2,$$

thus $I = \displaystyle\int \frac{1}{(x + \tfrac{1}{2})^2 + \left(\dfrac{\sqrt{3}}{2}\right)^2}\, dx$. Now put $x + \tfrac{1}{2} = \dfrac{\sqrt{3}}{2}\tan\theta$, then

$dx = \dfrac{\sqrt{3}}{2} \sec^2\theta\, d\theta$ and

$$I = \int \frac{1}{\left(\dfrac{\sqrt{3}}{2}\right)^2(1 + \tan^2\theta)} \cdot \frac{\sqrt{3}}{2} \sec^2\theta\, d\theta = \frac{2}{\sqrt{3}} \int 1\, d\theta = \frac{2}{\sqrt{3}}\theta + c$$

$$= \frac{2}{\sqrt{3}} \tan^{-1}\left\{\frac{2}{\sqrt{3}}(x + \tfrac{1}{2})\right\} + c.$$

(v) Let $I = \int \sin^3 x\, dx = \int \sin^2 x \cdot \sin x\, dx = \int (1 - \cos^2 x)\sin x\, dx$. Put $\cos x = y$, then $-\sin x\, dx = dy$ and

$$I = -\int (1 - y^2)\, dy = \tfrac{1}{3}y^3 - y + c = \tfrac{1}{3}\cos^3 x - \cos x + c.$$

Exercise XXXI

1. (D) Use suitable substitutions to evaluate the integrals given in Examples 7.1 to 7.4, and reconcile your answers with those obtained 'by inspection'.

2. **(D)** Guess suitable substitutions for the following integrals.

(i) $\int x(3x + 4)^7 \, dx$;　(ii) $\int \dfrac{x^2}{\sqrt{(x + 1)}} \, dx$;　(iii) $\int \dfrac{1}{x^2 + 2x + 3} \, dx$;

(iv) $\int \dfrac{1}{\sqrt{(3 + 2x - x^2)}} \, dx$.

In the remainder of this exercise, integrate the given functions with respect to x.

3. $x(x + 4)^7$.

4. $(x - 1)(3x + 1)^6$.

5. $\dfrac{x}{(3x - 1)^4}$.

6. $\dfrac{x}{(3x^2 + 4)^2}$.

7. $x^2 \sqrt{(x + 1)}$.

8. $\dfrac{x^3}{(x + 1)^{\frac{3}{2}}}$.

9. $\dfrac{1}{4 + 25x^2}$.

10. $\dfrac{1}{\sqrt{(1 - 4x^2)}}$.

11. $\dfrac{1}{\sqrt{(9 - 4x^2)}}$.

12. $x\sqrt{(x^2 + 2)}$.

13. $x^3 \sqrt{(1 + x^2)}$.

14. $x(1 + x^2)^3$.

15. $\sqrt{(1 - x^2)}$.

16. $\sin^3 x \cos x$.

17. $\sin^4 x \cos^3 x$. [Write $\cos^3 x = \cos^2 x \cdot \cos x = (1 - \sin^2 x) \cos x$]

18. $\cos^5 x \sin x$.

19. $\cos^4 x \sin^3 x$.

20. $\dfrac{1}{x\sqrt{(x^2 - 1)}}$.

21. $\dfrac{1}{(1 + x^2)^2}$.

22. $\dfrac{1}{(x + 3)\sqrt{(1 + x)}}$.

23. $\dfrac{1}{x^2 - 4x + 13}$.

24. $\dfrac{1}{\sqrt{(3 - 2x - x^2)}}$.

25. $\dfrac{1}{\sin^2 x + 4 \cos^2 x}$. [Divide numerator and denominator by $\cos^2 x$, then put $\tan x = t$]

26. $\sqrt{\left(\dfrac{x}{1 - x}\right)}$. [Put $x = \sin^2 \theta$]

27. $\sqrt{\left(\dfrac{1 + x}{1 - x}\right)}$. [Put $x = \cos 2\theta$]

28. $\dfrac{x^2}{1 + x^6}$. [Put $x^3 = y$]

29. $\dfrac{\cos x}{4 - \cos^2 x}$. [Write $\cos^2 x = 1 - \sin^2 x$ and $\sin x = y$]

30. $\dfrac{\cos^3 x}{\sin^2 x}$. [Put $\sin x = y$]   32. $\sin^5 x$.

31. $\tan^3 x \sec^2 x$.

We now give an alternative justification for the method of substitution, using definite integrals. Let $I = \displaystyle\int_a^b f(x)\,dx = \lim_{\delta x \to 0} \Sigma_{x=a}^b f(x)\,\delta x$. Also let x, $\dfrac{dx}{d\theta}$ be continuous functions of θ and $f(x) = g(\theta)$; further, suppose that when $x = a$, $\theta = \alpha$; when $x = b$, $\theta = \beta$. Then

$$I = \lim_{\delta\theta \to 0} \Sigma_{\theta=\alpha}^\beta g(\theta)\,\frac{\delta x}{\delta\theta}\,\delta\theta = \int_\alpha^\beta g(\theta)\,\frac{dx}{d\theta}\,d\theta.^*$$

x and θ will, in practice, be differentiable functions of each other; for $\dfrac{dx}{d\theta}$ to exist, we require $\dfrac{d\theta}{dx} \neq 0$. To ensure that this last condition is satisfied, it is necessary to check that *as x increases from a to b, so θ either increases or decreases steadily from α to β.*

EXAMPLE 7.8   *To evaluate* $I = \displaystyle\int_0^{\frac{1}{3}} \dfrac{1}{\sqrt{(4 - 9x^2)}}\,dx.$

The substitution $3x = 2 \sin\theta$ is indicated, as this will reduce the denominator of the integrand to $2 \cos\theta$. Then when $x = 0$, we can take $\theta = 0$; when $x = \frac{1}{3}$, $\sin\theta = \frac{1}{2}$, $\theta = \dfrac{\pi}{6}$. It is easy to see that as x increases from 0 to $\frac{1}{3}$, so θ increases all the way, from 0 to $\dfrac{\pi}{6}$. We therefore proceed as follows.

Put $x = \frac{2}{3} \sin\theta$, so that $dx = \frac{2}{3} \cos\theta\,d\theta$, and

$$I = \int_0^{\frac{\pi}{6}} \frac{1}{2\cos\theta} \cdot \frac{2}{3}\cos\theta\,d\theta = \frac{1}{3}\Big[\theta\Big]_0^{\frac{\pi}{6}} = \frac{\pi}{18}.$$

EXERCISE XXXII

1. $\displaystyle\int_0^1 x(x + 7)^{11}\,dx.$

2. $\displaystyle\int_2^3 \dfrac{x}{\sqrt{(x + 1)}}\,dx.$

3. $\displaystyle\int_0^1 \dfrac{1}{1 + x^2}\,dx.$

4. $\displaystyle\int_0^{\frac{1}{3}} \dfrac{1}{\sqrt{(1 - 9x^2)}}\,dx.$

---

* This step is not 'obvious'. It follows from the definition of the integral as the limit of a sum with the help of the 'first mean-value theorem' (p. 45), since within each interval $\delta\theta$, $\dfrac{dx}{d\theta}$ is somewhere equal to $\dfrac{\delta x}{\delta\theta}$.

5. $\displaystyle\int_0^{\frac{\pi}{2}} \sin^3 x \, dx.$

6. $\displaystyle\int_0^{\frac{\pi}{2}} \cos^3 x \sin^2 x \, dx.$

7. $\displaystyle\int_0^{\frac{\pi}{3}} \frac{1}{\sin^2 x + 3 \cos^2 x} \, dx.$

Integration by Parts

This method is often useful for integrating the product of two different types of function, e.g. for $\int (3x - 1) \sin x \, dx$. The rule is

$$\int f(x) \cdot g(x) \, dx = \left\{ \int f(x) \, dx \right\} \cdot g(x) - \int \left\{ \int f(x) \, dx \right\} \cdot g'(x) \, dx \qquad (1)$$

EXAMPLE 7.9　*To evaluate* $I = \int x \cos x \, dx.$

Taking $\cos x$ as 'the first', or $f(x)$, we have

$$I = \sin x \cdot x - \int \sin x \cdot 1 \, dx = x \sin x + \cos x + c.$$

The rule (1), for *integration by parts*, is an analogue of the formula for the differentiation of a product,

$$\frac{d}{dx} \{h(x) \cdot g(x)\} = h(x) \cdot g'(x) + g(x) \cdot h'(x) \qquad (2)$$

In fact, integrating both sides of eqn. (2) with respect to x, we have

$$h(x) \cdot g(x) = \int h(x) \cdot g'(x) \, dx + \int g(x) \cdot h'(x) \, dx,$$

or

$$\int h'(x) \cdot g(x) \, dx = h(x) \cdot g(x) - \int h(x) \cdot g'(x) \, dx.$$

Now writing $h'(x) = f(x)$, so that $h(x) = \int f(x) \, dx$, this reduces to the rule (1), above.*

There is no extra difficulty in applying integration by parts to definite integrals.

EXAMPLE 7.10　*To evaluate* $I = \displaystyle\int_0^{\frac{\pi}{4}} x \cos 2x \, dx.$

$$I = \left[\tfrac{1}{2} \sin 2x \cdot x \right]_0^{\frac{\pi}{4}} - \tfrac{1}{2} \int_0^{\frac{\pi}{4}} \sin 2x \, dx$$

$$= \frac{\pi}{8} + \frac{1}{4} \left[\cos 2x \right]_0^{\frac{\pi}{4}} = \frac{\pi}{8} - \frac{1}{4}.$$

---

* Feeling runs high on the best way to remember the rule. The reader may care to improve on 'the integral of a product is the integral-of-the-first times the second, minus the integral of the-said-integral-times-the-derivative-of-the-other'.

It is not always obvious which of the functions to choose for integration first, and sometimes it is in any case necessary to repeat the process.

EXAMPLE 7.11 $\int x^2 \sin x \, dx.$

$$I = -\cos x \cdot x^2 + \int \cos x \cdot 2x \, dx$$

$$= -x^2 \cos x + \sin x \cdot 2x - \int 2 \sin x \, dx$$

$$= -x^2 \cos x + 2x \sin x + 2 \cos x + c.$$

Taking $f(x) = 1$ sometimes leads to a successful integration.

EXAMPLE 7.12 $\int \sin^{-1} x \, dx.$

Writing $I = \int 1 \times \sin^{-1} x \, dx$, we have $I = x \sin^{-1} x - \int x \cdot \dfrac{1}{\sqrt{(1 - x^2)}} \, dx.$

Now $\int \dfrac{x}{\sqrt{(1 - x^2)}} \, dx = -(1 - x^2)^{\frac{1}{2}}$, so $I = x \sin^{-1} x + \sqrt{(1 - x^2)} + c.$

EXERCISE XXXIII

In questions 1 to 12, integrate the given functions with respect to x.

1. $x \sin x.$
2. $(3x + 1) \cos 2x.$
3. $x^2 \cos x.$
4. $x^3 \sin x.$
5. $\cos^{-1} x.$
6. $x \sin x \cos x.$
7. $x \sin 2x \cos 3x.$
8. $x \cos^2 x.$
9. $x^2 \sin^2 x.$
10. $x \sin^{-1} x.$ [Start by putting $x = \sin y$]
11. $x \tan^{-1} x.$
12. $x^3 \tan^{-1} x$ [cf. Exercise XXX, No. 40]
13. $\displaystyle\int_0^{\frac{\pi}{2}} x \cos x \, dx.$
14. $\displaystyle\int_0^{\frac{1}{2}} x \sin^{-1} x \, dx.$
15. $\displaystyle\int_0^{\frac{\pi}{4}} (x^2 + 5x + 2) \sin 2x \, dx.$

Definite Integrals*

We recall from Chapter 4 (cf. pp. 60 to 62) that if $f(x), g(x)$ are continuous functions of x, then

(1) $$\int_a^b f(x) \, dx = \int_a^c f(x) \, dx + \int_c^b f(x) \, dx.$$

(2) $$\int_b^a f(x) \, dx = -\int_a^b f(x) \, dx.$$

---
* This section can be omitted on the first reading.

(3) If $f(x) > g(x)$ in the interval $a \leqslant x \leqslant b$, then $\displaystyle\int_a^b f(x)\,dx > \int_a^b g(x)\,dx$. This result still holds if $f(x) = g(x)$ in part of the interval. The reader should illustrate this graphically.

(4) If $f(x)$ is an odd function, $\displaystyle\int_{-a}^a f(x)\,dx = 0$.

(5) If $g(x)$ is an even function, $\displaystyle\int_{-a}^a g(x)\,dx = 2\int_0^a g(x)\,dx$.

We now add

(6) $\displaystyle\int_a^b f(x)\,dx = \int_a^b f(y)\,dy$. This is self-evident, from the meaning of definite integration, but a verification is afforded by the trivial substitution $x = y$.

(7) $$\int_0^a f(x)\,dx = \int_0^a f(a - x)\,dx.$$

The reader should illustrate this result graphically, and also supply a proof, using the substitution $x = a - y$.

(8) $$\frac{d}{dx}\left\{\int_a^x f(t)\,dt\right\} = f(x).$$

By definition,

$$\frac{d}{dx}\left\{\int_a^x f(t)\,dt\right\} = \lim_{h \to 0} \frac{\displaystyle\int_a^{x+h} f(t)\,dt - \int_a^x f(t)\,dt}{h}$$

$$= \lim_{h \to 0} \frac{\displaystyle\int_x^{x+h} f(t)\,dt}{h}$$

$$= \lim_{h \to 0} \frac{hf(\xi)}{h}$$

where $x < \xi < x + h$, by result (3), p. 60

$$= f(x),$$

as required.

EXAMPLE 7.13　*To evaluate* $I = \displaystyle\int_0^\pi \frac{x \sin x}{1 + \cos^2 x}\,dx.$

Setting $x = \pi - y$,

$$I = -\int_\pi^0 \frac{(\pi - y)\sin y}{1 + \cos^2 y}\,dy = \int_0^\pi \frac{(\pi - x)\sin x}{1 + \cos^2 x}\,dx$$

$$= \pi\int_0^\pi \frac{\sin x}{1 + \cos^2 x}\,dx - I.$$

Thus

$$2I = \pi \int_0^\pi \frac{\sin x}{1 + \cos^2 x}\, dx = \pi \int_1^{-1} \left(-\frac{1}{1 + z^2}\right) dz,$$

$$\text{putting } \cos x = z, \ -\sin x\, dx = dz$$

$$= \pi \int_{-1}^1 \frac{1}{1 + z^2}\, dz$$

$$= \pi \left[\tan^{-1} z\right]_{-1}^1 = \pi \left\{\frac{\pi}{4} - \left(-\frac{\pi}{4}\right)\right\} = \frac{\pi^2}{2},$$

and so $I = \dfrac{\pi^2}{4}$.

Notation. For brevity, $\displaystyle\int \frac{1}{f(x)}\, dx$ can be written as $\displaystyle\int \frac{dx}{f(x)}$ and $\displaystyle\int \frac{f(x)}{g(x)}\, dx$

as $\displaystyle\int \frac{f(x)\, dx}{g(x)}$, e.g.

$$\int \frac{1}{\sqrt{(4 - 9x^2)}}\, dx = \int \frac{dx}{\sqrt{(4 - 9x^2)}},$$

$$\int \frac{x}{(1 + x^2)^2}\, dx = \int \frac{x\, dx}{(1 + x^2)^2}.$$

EXAMPLE 7.14 *To evaluate* $I = \displaystyle\int_{-1}^1 \frac{dx}{\sqrt{(1 - 2\alpha x + \alpha^2)}}$ $(\alpha \neq 0)$.

By inspection,

$$\int \frac{dx}{\sqrt{(1 - 2\alpha x + \alpha^2)}} = -\frac{1}{\alpha}(1 - 2\alpha x + \alpha^2)^{\frac{1}{2}}.$$

Now $(1 - 2\alpha x + \alpha^2)^{\frac{1}{2}} \geqslant 0$, and care must be taken to ensure that the right values of the square roots are taken when the limits ± 1 for x are inserted. For example, $(1 - 2\alpha + \alpha^2)^{\frac{1}{2}} = \alpha - 1$ or $1 - \alpha$ according as $\alpha > 1$ or $\alpha < 1$.

We have

$$I = -\frac{1}{\alpha}\left[(1 - 2\alpha x + \alpha^2)^{\frac{1}{2}}\right]_{-1}^1$$

$$= -\frac{1}{\alpha}\left\{(1 - 2\alpha + \alpha^2)^{\frac{1}{2}} - (1 + 2\alpha + \alpha^2)^{\frac{1}{2}}\right\}$$

$$= -\frac{1}{\alpha}\left\{|1 - \alpha| - |1 + \alpha|\right\}.$$

If

$$\alpha > 1, \quad I = -\frac{1}{\alpha}\{\alpha - 1 - (1 + \alpha)\} = \frac{2}{\alpha};$$

if

$$-1 \leqslant \alpha \leqslant 1 \quad (\alpha \neq 0), \quad I = -\frac{1}{\alpha}\{1 - \alpha - (1 + \alpha)\} = 2;$$

if

$$\alpha < -1, \quad I = -\frac{1}{\alpha}\{1 - \alpha - (-1 - \alpha)\} = -\frac{2}{\alpha}.$$

Exercise XXXIV

1. $\displaystyle\int_{-\frac{\pi}{2}}^{\frac{\pi}{2}} \sin^7 x \, dx.$

   2. $\displaystyle\int_{-\frac{\pi}{2}}^{\frac{\pi}{2}} \sin^9 x \cos^4 x \, dx.$

3. Show that $\displaystyle\int_0^{\frac{\pi}{2}} f(\sin x) \, dx = \int_0^{\frac{\pi}{2}} f(\cos x) \, dx.$

4. Show that $\displaystyle\int_0^{\frac{\pi}{2}} \cos^{2n+1} x \, dx = -\int_{\frac{\pi}{2}}^{\pi} \cos^{2n+1} x \, dx$, and hence that

$$\int_0^{\pi} \cos^{2n+1} x \, dx = 0.$$

5. Evaluate $\displaystyle\int_0^{\pi} \cos^7 x \, dx.$

6. Show that $\displaystyle\int_a^b f(x) \, dx = \int_a^b f(a + b - x) \, dx.$

7. Show that $\displaystyle\int_0^{\frac{\pi}{2}} \frac{\sin x}{\sin x + \cos x} \, dx = \int_0^{\frac{\pi}{2}} \frac{\cos x}{\sin x + \cos x} \, dx$ (cf. question 3 above); deduce the common value of these integrals.

8. $\displaystyle\int_0^{\pi} \frac{\sin \theta \, d\theta}{\sqrt{(1 - 2a \cos \theta + a^2)}}.$

We conclude this chapter with an application of result (3), p. 128, which will introduce the subject-matter of Chapter 8.

If $x > 0$, we have

$$\sin x < x \tag{1}$$

so that

$$\int_0^x \sin t \, dt < \int_0^x t \, dt,$$

i.e.

$$\left[-\cos t\right]_0^x = 1 - \cos x < \frac{x^2}{2}$$

or

$$\cos x > 1 - \frac{x^2}{2}. \tag{1'}$$

Using the same result again,

$$\int_0^x \cos t \, dt > \int_0^x \left(1 - \frac{t^2}{2}\right) dt$$

or

$$\sin x > x - \frac{x^3}{3!}. \tag{2}$$

Therefore

$$\int_0^x \sin t \, dt > \int_0^x \left(t - \frac{t^3}{3!}\right) dt$$

or

$$1 - \cos x > \frac{x^2}{2!} - \frac{x^4}{4!}$$

i.e.

$$\cos x < 1 - \frac{x^2}{2!} + \frac{x^4}{4!}. \tag{2}$$

So

$$\int_0^x \cos t \, dt < \int_0^x \left(1 - \frac{t^2}{2!} + \frac{t^4}{4!}\right) dt$$

or

$$\sin x < x - \frac{x^3}{3!} + \frac{x^5}{5!}. \tag{3}$$

By continuing this process, it can be seen that, for any integer n, $\sin x$ lies between

$$x - \frac{x^3}{3!} + \frac{x^5}{5!} - \ldots - \frac{x^{4n-1}}{(4n-1)!}$$

and

$$x - \frac{x^3}{3!} + \frac{x^5}{5!} - \ldots - \frac{x^{4n-1}}{(4n-1)!} + \frac{x^{4n+1}}{(4n+1)!}.$$

Now the difference between these expressions, namely $\frac{x^{4n+1}}{(4n+1)!}$, tends to zero as $n \to \infty$ (cf. Appendix 2), so that

$$\sin x = x - \frac{x^3}{3!} + \frac{x^5}{5!} - \frac{x^7}{7!} + \ldots. \tag{A}$$

Writing $\sin x$ as equal to this 'infinite series' means that the sum of k terms of the series tends to $\sin x$ as $k \to \infty$.

If $x < 0$, (A) still holds, the argument being slightly modified, e.g.

$$\sin x > x$$

therefore

$$\int_x^0 \sin t \, dt > \int_x^0 t \, dt$$

so

$$\int_0^x \sin t \, dt < \int_0^x t \, dt$$

so that

$$1 - \cos x < \frac{x^2}{2},$$ etc.

The reader should prove for himself (by considering the inequalities $(1')$, $(2')$, . . . above, instead of (1), (2), . . .) that, for all values of x,

$$\cos x = 1 - \frac{x^2}{2!} + \frac{x^4}{4!} - \frac{x^6}{6!} + \dots .$$

EXERCISE XXXV

1. Draw accurately, on the same sheet of graph paper, the graphs of (i) $\sin x$, (ii) x, (iii) $x - \frac{x^3}{3!}$, (iv) $x - \frac{x^3}{3!} + \frac{x^5}{5!}$, for $0 \leqslant x \leqslant \pi$.

2. Using the inequality $x - \frac{x^3}{3!} < \sin x < x$, find the maximum possible error obtained by replacing $\sin x$ by x when $x = \frac{\pi}{10}$. Also find the error by looking up $\sin 18°$ in your tables.

Miscellaneous questions (in sets of 3)

3. Integrate the following with respect to x: (i) $(1 + 9x)^4$; (ii) $\frac{1}{1 + 9x^2}$; (iii) $\frac{x}{(1 + 9x^2)^2}$; (iv) $\frac{1}{\sqrt{(1 + 9x)}}$.

4. Evaluate (i) $\int_0^\pi \cos^3 x \, dx$; (ii) $\int_{-\frac{\pi}{3}}^{\frac{\pi}{3}} \sin^7 x \, dx$; (iii) $\int_0^{\frac{\pi}{2}} \sin^7 x \cos x \, dx$.

5. Differentiate with respect to x: $\int_1^{x^2+1} \frac{1}{2 + t^2} \, dt$. [Use the chain rule, first differentiating with respect to $x^2 + 1$.]

6. Write down, by inspection, the values of (i) $\int \frac{14x - 9x^2}{(4 + 7x^2 - 3x^3)^7} \, dx$; (ii) $\int \tan 3x \sec^2 3x \, dx$; (iii) $\int \frac{x}{\sqrt{(1 - x^2)}} \, dx$.

7. Integrate with respect to x: (i) $x\sqrt{(x+1)}$; (ii) $\dfrac{1}{x^2 + 4x + 9}$;
   (iii) $\dfrac{x+4}{\sqrt{(1-x^2)}}$; (iv) $\sin^4 x$.

8. Using the fact that $0 < x^5 < x^2$ in the range $(0, 1)$, find limits between which $\displaystyle\int_0^1 \dfrac{dx}{1+x^5}$ lies. Also find the value of this integral obtained by using Simpson's rule with five ordinates.

9. $\displaystyle\int_0^{\frac{\pi}{2}} x^3 \sin x \, dx$.

10. Find the indefinite integrals:

   (i) $\displaystyle\int \dfrac{dx}{(1-3x)^3}$; (ii) $\displaystyle\int \tan^2 2x \, dx$;

   (iii) $\displaystyle\int 2x \tan^{-1} x \, dx$; (iv) $\displaystyle\int \dfrac{\cos x \, dx}{2 - \cos^2 x}$.

11. Prove that

$$\tfrac{1}{2}\int_0^1 x^4(1-x)^4 \, dx < \int_0^1 \dfrac{x^4(1-x)^4}{1+x^2} \, dx < \int_0^1 x^4(1-x)^4 \, dx.$$

   Verify the identity

$$x^4(1-x)^4 = (1+x^2)(4 - 4x^2 + 5x^4 - 4x^5 + x^6) - 4,$$

   and, by using this identity in the second of the three integrals, prove that

$$\dfrac{22}{7} - \dfrac{1}{630} < \pi < \dfrac{22}{7} - \dfrac{1}{1260}.$$

8
Taylor-Maclaurin Expansions

Change and decay in all around I see.
LYTE

In this chapter two important topics are introduced informally, (i) Taylor's theorem and (ii) the exponential function. To give a general picture quickly, a number of assumptions will be made, which must subsequently be challenged. A more rigorous treatment of exponentials and logarithms is given in Chapter 9, and of Taylor's theorem in Appendix 3.

We first consider *polynomials*, that is, expressions of the form

$$a_0 + a_1 x + a_2 x^2 + \ldots + a_{n-1} x^{n-1} + a_n x^n,$$

$a_0, a_1, \ldots, a_n$ here being real constants.

EXAMPLE 8.1 *To determine the polynomial f(x) of degree 3 whose value and that of its first, second and third derivatives are respectively* 1, 3, 4 *and* -2 *when* $x = 0$.*

Let $f(x) = a_0 + a_1 x + a_2 x^2 + a_3 x^3$; we are given that $f(0) = 1, f'(0) = 3,$ $f''(0) = 4, f'''(0) = -2$.

Putting $x = 0$, we have immediately $1 = f(0) = a_0$. Also

$$f'(x) = a_1 + 2a_2 x + 3a_3 x^2$$

so that $3 = f'(0) = a_1$. Differentiating again, $f''(x) = 2a_2 + 6a_3 x$, so that

$$4 = f''(0) = 2a_2, \qquad a_2 = 2$$

and finally $f'''(x) = 6a_3$ so that $-2 = f'''(0) = 6a_3, a_3 = -\frac{1}{3}$.

Substituting the values of a_0, a_1, a_2, a_3 which we have found, we have as the required polynomial

$$f(x) = 1 + 3x + 2x^2 - \tfrac{1}{3}x^3.$$

It is remarkable that, *if*

$$f(x) = a_0 + a_1 x + \ldots + a_n x^n, \tag{1}$$

---

* The more sophisticated reader may care to investigate more generally the number of 'clues' required to identify a polynomial of given degree. E.g. if $y = a_0 + a_1 x + a_2 x^2 + a_3 x^3$, *four* values of y corresponding to given values of x will provide four simultaneous equations for $a_0, a_1,$ a_2, a_3, just sufficient to determine them uniquely. (Any exceptions?)

the coefficients $a_0, a_1, \ldots, a_n$ are all determined if the value of $f(x)$ and all its derivatives are given for one particular value of x. (For simplicity, we choose this as $x = 0$, but the argument can be modified for any other value.)

This statement is easily justified, by a generalization of Example 8.1. We have immediately $f(0) = a_0$. Also $f'(x) = a_1 + 2a_2x + 3a_3x + \ldots + na_nx^{n-1}$, so that $f'(0) = a_1$.

Differentiating again,

$$f''(x) = 2a_2 + 3.2a_3x + 4.3a_4x^2 + \ldots + n(n-1)a_nx^{n-2}$$

so that $f''(0) = 2a_2$.

Also $f'''(x) = 3.2.1a_3 + 4.3.2a_4x + \ldots + n(n-1)(n-2)a_nx^{n-3}$, whence $f'''(0) = 3!\,a_3$, and similarly it can be seen that $f^{(r)}(0) = r!\,a_r$, i.e. $a_r = \dfrac{f^{(r)}(0)}{r!}$. Substituting the values of $a_0, a_1, \ldots, a_n$ in (1), we thus have

$$f(x) = f(0) + xf'(0) + \frac{x^2}{2!}f''(0) + \ldots + \frac{x^r}{r!}f^{(r)}(0) + \ldots + \frac{x^n}{n!}f^{(n)}(0). \quad (2)$$

We next seek to extend this idea to cover functions other than polynomials. For a start, we examine how far the pattern of (2) agrees with the 'infinite series' given at the end of the previous chapter,

$$\sin x = x - \frac{x^3}{3!} + \frac{x^5}{5!} - \frac{x^7}{7!} + \ldots \quad (3)$$

If $f(x) = \sin x$, we have

$f(x)$	$\sin x$	$f(0)$	0	$f(0)$	0
$f'(x)$	$\cos x$	$f'(0)$	1	$xf'(0)$	x
$f''(x)$	$-\sin x$	$f''(0)$	0	$\frac{x^2}{2!}f''(0)$	0
$f'''(x)$	$-\cos x$	$f'''(0)$	-1	$\frac{x^3}{3!}f'''(0)$	$-\frac{x^3}{3!}$
$f^{(iv)}(x)$	$\sin x$	$f^{(iv)}(0)$	0	$\frac{x^4}{4!}f^{(iv)}(0)$	0
$f^{(v)}(x)$	$\cos x$	$f^{(v)}(0)$	1	$\frac{x^5}{5!}f^{(v)}(0)$	$\frac{x^5}{5!}$

and so on.

By comparing this table with (3), it can be seen that, if $f(x) = \sin x$, we have

$$f(x) = f(0) + xf'(0) + \frac{x^2}{2!}f''(0) + \frac{x^3}{3!}f'''(0) + \ldots \quad (4)$$

EXERCISE XXXVI

1. If $f(x) = 3 + 4x - 5x^2 + 2x^3$, find the values of $f(0)$, $f'(0)$, $f''(0)$ and $f'''(0)$.
Verify that $f(x) = f(0) + xf'(0) + \dfrac{x^2}{2!}f''(0) + \dfrac{x^3}{3!}f'''(0)$.

2. Use the relationship (2) above to verify that (i) $(1 + x)^3 = 1 + 3x + 3x^2 + x^3$; (ii) $(1 + x)^n = 1 + {}_nC_1x + {}_nC_2x^2 + \ldots + {}_nC_rx^r + \ldots + x^n$, n being a positive integer (cf. Appendix 1, p. 322).

3. Verify that the relationship (4) holds when $f(x) = \cos x$ (cf. p. 132).
(4) is known as *Maclaurin's expansion** for a function $f(x)$, the right-hand side being a *Maclaurin series*. We defer for the moment a statement of Maclaurin's theorem, which gives the conditions under which (4) holds.

EXAMPLE 8.2   *To find the Maclaurin series for* $(1 + x)^n$, *where n is not a positive integer or zero.*

$f(x)$	$(1 + x)^n$	$f(0)$	1
$f'(x)$	$n(1 + x)^{n-1}$	$f'(0)$	n
$f''(x)$	$n(n - 1)(1 + x)^{n-2}$	$f''(0)$	$n(n - 1)$
$f'''(x)$	$n(n - 1)(n - 2)(1 + x)^{n-3}$	$f'''(0)$	$n(n - 1)(n - 2)$

etc.
Substituting in (4), we have the *binomial series*

$$(1 + x)^n = 1 + nx + \frac{n(n - 1)}{1 \cdot 2}x^2 + \frac{n(n - 1)(n - 2)}{1 \cdot 2 \cdot 3}x^3 + \ldots \quad (5)$$

This result must be regarded with suspicion, as we have not yet *proved* (4) (except in the special case when $f(x) = \sin x$) or investigated conditions under which it is valid. Plainly one necessary condition is that the right-hand side should *converge*, that is, the sum of the first p terms of the series should tend to a limit as $p \to \infty$. A necessary condition for this in turn is that the p-th term should tend to zero as $p \to \infty$.†

Now the p-th term of the right-hand side of (5) does not always tend to zero. For example, putting $n = -1$, $x = -2$ in (5) would yield the absurd conclusion that

$$-1 = 1 + 2 + 4 + 8 + \ldots.$$

---

* The result was, in fact, discovered by Brook Taylor (1712), but partial justice has been done, as Taylor's name is given to an apparently more general result; see below.
† For consider the series $s_p = u_1 + u_2 + u_3 + \ldots + u_p$, and suppose that as $p \to \infty$, $s_p \to s$, so that we can write $s = u_1 + u_2 + u_3 + \ldots$, as an infinite series. Then we have $u_p = s_p - s_{p-1} = (s_p - s) + (s - s_{p-1})$, which tends to zero as $p \to \infty$.

On the other hand, putting $n = -1$ and $x = \frac{1}{2}$ gives

$$(1 + \tfrac{1}{2})^{-1} = 1 + (-1)\tfrac{1}{2} + \frac{(-1)(-2)}{1 \cdot 2} \cdot (\tfrac{1}{2})^2 + \frac{(-1)(-2)(-3)}{1 \cdot 2 \cdot 3} \cdot (\tfrac{1}{2})^3 + \ldots$$

i.e. $1 - \frac{1}{2} + (\frac{1}{2})^2 - (\frac{1}{2})^3 + \ldots = \frac{2}{3}$, which is correct.* We must at present be content with the statement 'it can be shown that'

$$(1 + x)^n = 1 + nx + \frac{n(n-1)}{1 \cdot 2}x^2 + \ldots \text{ for } |x| < 1.$$

Maclaurin's theorem may be stated as follows.

If $f^{(n-1)}(x)$ is a continuous function of x over the range $|x| \leqslant |h|$ and $f^{(n)}(x)$ exists for $|x| < |h|$, then for all x such that $|x| \leqslant |h|$ we have

$$f(x) = f(0) + xf'(0) + \frac{x^2}{2!}f''(0) + \ldots + \frac{x^{n-1}}{(n-1)!}f^{(n-1)}(0) + R_n,$$

where $R_n = \dfrac{x^n}{n!}f^{(n)}(\theta x)$, θ being some number such that $0 < \theta < 1$.

If, further, $R_n \to 0$ as $n \to \infty$, then $f(x)$ may be expanded as an infinite series, namely $f(x) = f(0) + xf'(0) + \dfrac{x^2}{2!}f''(0) + \ldots$.

The reader should at some stage refer to Appendix 3 for a proof of the above, in conjunction with Taylor's theorem (introduced below), but this should be postponed at least till the end of this chapter. It is, in any case, important to study the *statement* of the theorem, to appreciate the conditions under which a function can be expanded as a Maclaurin series.

The importance of the theorem lies in the fact that it not only gives a succession of polynomial approximations to a given function but also provides information on the possible error involved.

EXAMPLE 8.3   *To show that $(1 + x)^{\frac{1}{2}} \approx 1 + \frac{1}{2}x$ if x is small, and to investigate the error when $x = 0.02$.*

Taking $n = 2$ in Maclaurin's theorem, we have

$$f(x) = f(0) + xf'(0) + \frac{x^2}{2!}f''(\theta x),$$

where $0 < \theta < 1$. If $f(x) = (1 + x)^{\frac{1}{2}}$, $f'(x) = \frac{1}{2}(1 + x)^{-\frac{1}{2}}$ and $f''(x) = -\frac{1}{4}(1 + x)^{-\frac{3}{2}}$ so

$$(1 + x)^{\frac{1}{2}} = 1 + \frac{1}{2}x + \frac{x^2}{2!} \cdot -\frac{1}{4}(1 + \theta x)^{-\frac{3}{2}} \approx 1 + \frac{1}{2}x \text{ if } x \text{ is small.}$$

---

* $a + ar + ar^2 + \ldots = \dfrac{a}{1-r}$ if $|r| < 1$. Putting $a = 1, r = -\frac{1}{2}$ gives $1 - \dfrac{1}{2} + \dfrac{1}{2^2} - \ldots = \dfrac{2}{3}$.

Putting $x = 0.02$ yields

$$\sqrt{(1.02)} = 1 + 0.01 - 0.00005(1 + 0.02\theta)^{-\frac{3}{2}}$$

Now $(1 + 0.02\theta)^{-\frac{3}{2}} < 1$, so that $\sqrt{(1.02)}$ lies between 1.01 and 1.00995.

EXAMPLE 8.4   *To evaluate $\sqrt[3]{(64.03)}$ correct to 6 decimal places.*

To use the binomial series for $(1 + x)^{\frac{1}{3}}$, *in which we must have* $|x| < 1$, we first write $64.03 = 64(1 + \frac{3}{6400})$, so that $\sqrt[3]{(64.03)} = 4(1 + \frac{3}{6400})^{\frac{1}{3}}$. Now if $|x| < 1$,

$$(1 + x)^{\frac{1}{3}} = 1 + \tfrac{1}{3}x + \frac{\frac{1}{3}\cdot-\frac{2}{3}}{1\cdot2}x^2 + \ldots$$

So that

$$\left(1 + \frac{3}{6400}\right)^{\frac{1}{3}} = 1 + \frac{1}{3}\cdot\frac{3}{6400} + \frac{\frac{1}{3}\cdot-\frac{2}{3}}{1\cdot2}\left(\frac{3}{6400}\right)^2 + \frac{\frac{1}{3}\cdot-\frac{2}{3}\cdot-\frac{5}{3}}{1\cdot2\cdot3}\left(\frac{3}{6400}\right)^3 + \ldots$$

The third term in this expansion is $-(\frac{1}{6400})^2$ which is numerically less than 4×10^{-8}. The subsequent terms are smaller still, and we shall be safe in taking only the first two. (The strict reason for our safety is that the remainder after these two terms is $\dfrac{\frac{1}{3}\cdot-\frac{2}{3}}{1\cdot2}(\frac{3}{6400})^2(1 + \frac{3}{6400})^{-\frac{5}{3}}$ where $0 < \theta < 1$, and this is numerically less than 4×10^{-8}.) We thus have

$$\sqrt[3]{(64.03)} \approx 4(1 + \tfrac{1}{3}\cdot\tfrac{3}{6400}) = 4 \times 1.00015625 = 4.000625(0),$$

the error from neglecting the third term being numerically less than $4 \times 4 \times 10^{-8}$, so that $\sqrt[3]{(64.03)} = 4.000625$ to 6 decimal places.

EXAMPLE 8.5   *To find the first three terms in the Maclaurin expansion of*

$$\frac{\cos x}{(1 + x)^{\frac{2}{3}}}, \qquad (0 < x < 1).$$

It would be very laborious to differentiate the given expression repeatedly, so instead we write it as $\cos x . (1 + x)^{-\frac{2}{3}}$. Now $\cos x = 1 - \dfrac{x^2}{2!} + \dfrac{x^4}{4!} - \ldots$

and $(1 + x)^{-\frac{2}{3}} = 1 - \tfrac{2}{3}x + \dfrac{-\frac{2}{3}\cdot-\frac{5}{3}}{1\cdot2}x^2 + \ldots = 1 - \tfrac{2}{3}x + \tfrac{5}{9}x^2 + \ldots$

Thus, *assuming the series can be multiplied together,* we have

$$\cos x(1 + x)^{-\frac{2}{3}} = \left(1 - \frac{x^2}{2} + \frac{x^4}{24} - \ldots\right)(1 - \tfrac{2}{3}x + \tfrac{5}{9}x^2 + \ldots)$$

$$= 1 - \tfrac{2}{3}x + \tfrac{1}{18}x^2 + \tfrac{1}{3}x^3 + \ldots,$$

so that the first three terms are $1 - \tfrac{2}{3}x + \tfrac{1}{18}x^2$.

It is not, however, always permissible to 'multiply' two infinite series together in this manner (cf. Exercise XXXVIII, question 3). We may proceed more cautiously as follows.

$$\cos x = 1 - \frac{x^2}{2} + \frac{x^4}{24}\cos(\theta x), \qquad (0 < \theta < 1)$$

and

$$(1 + x)^{-\frac{2}{3}} = 1 - \tfrac{2}{3}x + \frac{-\tfrac{2}{3} \cdot -\tfrac{5}{3}}{1 \cdot 2} x^2$$

$$+ \frac{-\tfrac{2}{3} \cdot -\tfrac{5}{3} \cdot -\tfrac{8}{3}}{1 \cdot 2 \cdot 3} x^3 (1 + \varphi x)^{-\frac{11}{3}}, \qquad (0 < \varphi < 1).$$

Now, since $1 + \varphi x > 1$, $(1 + \varphi x)^{-\frac{11}{3}} < 1$, and so neglecting x^3 and higher powers of x, we have

$$\cos x (1 + x)^{-\frac{2}{3}} \approx \left(1 - \frac{x^2}{2}\right)(1 - \tfrac{2}{3}x + \tfrac{5}{9}x^2)$$

$$\approx 1 - \tfrac{2}{3}x + \tfrac{1}{18}x^2.$$

Taylor's Theorem

The *first mean-value theorem* was introduced on p. 45. It may be stated in the form

$$f(a + h) = f(a) + hf'(a + \theta h) \tag{1}$$

or

$$f'(a + \theta h) = \frac{f(a + h) - f(a)}{h}, \tag{2}$$

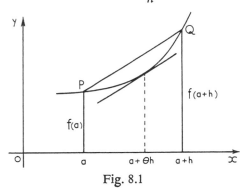

Fig. 8.1

which is intuitively clear from the graph, Fig. 8.1. The right-hand side of eqn. (2) is the gradient of the chord PQ and the theorem states in effect that this is equal to the gradient of the curve somewhere in between, say where $x = a + \theta h$ ($0 < \theta < 1$). Putting $a = 0$, $h = x$ in eqn. (1) gives

$$f(x) = f(0) + hf'(\theta x),$$

i.e. Maclaurin's theorem with $n = 1$ (cf. p. 137).

By analogy with

$$f(x) = f(0) + xf'(0) + \frac{x^2}{2!}f''(0) + \ldots,$$

this suggests the possibility of an expansion in the form

$$f(a + h) = f(a) + hf'(a) + \frac{h^2}{2!}f''(a) + \ldots.$$

In fact, Taylor's theorem may be stated as follows.

If $f^{(n-1)}(x)$ is continuous for $|x - a| \leqslant |h|$ and $f^{(n)}(x)$ exists for $|x - a| < |h|$, then

$$f(a + h) = f(a) + hf'(a) + \ldots + \frac{h^{n-1}}{(n-1)!}f^{(n-1)}(a) + R_n,$$

where $R_n = \dfrac{h^n}{n!}f^{(n)}(a + \theta h)$, θ being some number such that $0 < \theta < 1$.

If, further, $R_n \to 0$ as $n \to \infty$, then $f(a + h) = f(a) + hf'(a) + \dfrac{h^2}{2!}f''(a) + \ldots.$

This should be compared with Maclaurin's theorem (p. 137); these theorems are essentially equivalent (cf. Appendix 3).

EXAMPLE 8.6 *Prove that the function $f(x)$ has a maximum value at $x = a$ when $f'(a) = 0$ and $f''(a)$ is negative.*

This can be demonstrated graphically (cf. p. 96), but we can now give a more formal treatment. By Taylor's theorem, $f(a + h) = f(a) + hf'(a) + \dfrac{h^2}{2!}f''(a + \theta h)$, $0 < \theta < 1$, i.e. $f(a + h) - f(a) = \dfrac{h^2}{2!}f''(a + \theta h)$, if $f'(a) = 0$. Now, *assuming that $f''(x)$ is continuous at $x = a$*, if $f''(a) < 0$ then $f''(a + \theta h) < 0$ for all h within a certain range containing 0. Thus, within this range for h, $f(a + h) - f(a) < 0$ and so $f(x)$ has a maximum value at $x = a$.

L'Hôpital's Rule

The 'first mean-value theorem', namely

$$f(a + h) = f(a) + hf'(a + \theta h), \qquad (0 < \theta < 1),$$

leads to a rule for evaluating $\lim\limits_{x \to a} \dfrac{f(x)}{g(x)}$ when $f(a) = g(a) = 0$ $\left(\text{e.g. } \lim\limits_{x \to 0} \dfrac{\sin x}{x}\right)$.

For in this case

$$\lim_{x \to a} \frac{f(x)}{g(x)} = \lim_{x \to a} \frac{f(x) - f(a)}{g(x) - g(a)}$$

$$= \lim_{h \to 0} \frac{f(a + h) - f(a)}{g(a + h) - g(a)}$$

(writing $x = a + h$),

$$= \lim_{h \to 0} \frac{f'(a + \theta h)}{g'(a + \varphi h)}$$

where $0 < \theta < 1, 0 < \varphi < 1$

$$= \frac{f'(a)}{g'(a)}.$$

This is *L'Hôpital's rule*.

This breaks down when $f'(a) = g'(a) = 0$; it can then be proved that $\lim_{x \to a} \frac{f(x)}{g(x)} = \frac{f^{(n)}(a)}{g^{(n)}(a)}$, where $g^{(n)}(a)$ is the first non-vanishing derivative of $g(x)$ at $x = a$, but this is beyond our present scope.

EXERCISE XXXVII

In questions 1 to 4, find the first four terms in the Maclaurin expansion of the given function, stating in each case the range of values of x for which the expansion is valid. In questions 5 to 9, give also the general term.

1. $(1 + x)^{-3}$.

2. $(1 - x)^{\frac{3}{2}}$.

3. $(1 + 3x)^{-\frac{1}{3}}$.

4. $(2 + x)^{-4}$. $\left[\text{Write as } 2^{-4}\left(1 + \frac{x}{2}\right)^{-4}\right]$

5. $(1 - x)^{-2}$.

6. $\sin 6x$.

7. $\cos 5x$.

8. $\sin^2 x$.

9. $\sin 4x \cos 5x$.

10. Investigate the values of x, h for which it is true that

$$\frac{1}{x + h} = \frac{1}{x} - \frac{h}{x^2} + \frac{h^2}{x^3} - \frac{h^3}{x^4} + \cdots.$$

11. Show that $(1 + x)^{\frac{1}{3}} \approx 1 + \frac{1}{3}x - \frac{1}{9}x^2$ if x is small. Evaluate the right-hand side when $x = 0 \cdot 01$, and investigate the possible error in taking this as an approximation for $\sqrt[3]{(1 \cdot 01)}$.

12. Find the square root of $1 \cdot 05$, correct to 5 decimal places.

13. If x is small, show that $\sin x(1 - 2x)^{\frac{7}{2}} \approx x - 7x^2 + \frac{52}{3}x^3$.

14. Show that the tangent at $(0, 1)$ to the curve $y = \dfrac{1}{(1 + x)^2}$ is $y = 1 - 2x$. Sketch roughly, using the same axes, the graphs of the functions $\dfrac{1}{(1 + x)^2}$, $1 - 2x$, $1 - 2x + 3x^2$ and $1 - 2x + 3x^2 - 4x^3$.

15. Verify that Maclaurin's expansion yields $(1 - x)^{-1} = 1 + x + x^2 + x^3(1 - \theta x)^{-4}$. Find an expression for θ explicitly in terms of x, and give a direct proof that the value of this expression is between 0 and 1 if $|x| < 1$. Is this also the case when $x = 2$? (If not, can you explain this?)

16. Use L'Hôpital's rule to evaluate (i) $\lim\limits_{\theta \to 0} \dfrac{\sin \theta}{\theta}$; (ii) $\lim\limits_{x \to a} \dfrac{x^n - a^n}{x^m - a^m}$.

17. Evaluate $\lim\limits_{x \to 0} \dfrac{1 - \cos x}{x^2}$.

18. (D) Let $s_p = 1 + \frac{1}{2} + \frac{1}{3} + \ldots + \dfrac{1}{p}$. Show that

$$s_{2^k} = 1 + \tfrac{1}{2} + (\tfrac{1}{3} + \tfrac{1}{4}) + (\tfrac{1}{5} + \tfrac{1}{6} + \tfrac{1}{7} + \tfrac{1}{8}) + \ldots + \left(\frac{1}{2^{k-1} + 1} + \ldots + \frac{1}{2^k}\right)$$

$$> 1 + \tfrac{1}{2} + (\tfrac{1}{4} + \tfrac{1}{4}) + (\tfrac{1}{8} + \tfrac{1}{8} + \tfrac{1}{8} + \tfrac{1}{8}) + \ldots + \left(\frac{1}{2^k} + \ldots + \frac{1}{2^k}\right)$$

$$= 1 + \tfrac{1}{2} + \tfrac{1}{2} + \tfrac{1}{2} + \ldots + \tfrac{1}{2}$$

$$= 1 + k.$$

Deduce that as $p \to \infty$, $s_p \to \infty$, that is, the series $1 + \frac{1}{2} + \frac{1}{3} + \ldots$ does not tend to a finite limit; it *diverges*.

The condition that the p-th term of a series should tend to zero as $p \to \infty$ is necessary for convergence (cf. p. 136), but this last example shows that it is not a sufficient condition to ensure convergence. A general discussion of tests and conditions for convergence of series is beyond our present scope.

The Exponential Function

The reader is probably familiar with the compound interest law, $A = P\left(1 + \dfrac{R}{100}\right)^T$. As a reminder from the arithmetic course, a principal, £P, is invested at R per cent per annum, so that at the end of the first year $P\dfrac{R}{100}$ is added and the new principal is thus $P\left(1 + \dfrac{R}{100}\right)$. After each year, the principal is multiplied by the same 'growth factor' $1 + \dfrac{R}{100}$, so that after T years the amount £A is given by $A = P\left(1 + \dfrac{R}{100}\right)^T$.

Now, if instead of being paid annually, the interest is paid at successive time-intervals of δt years, the amount at the beginning of such an interval being £x will become £$x\left(1 + \dfrac{R}{100} \delta t\right)$ in time δt. Denoting the change in x by δx, we have $\delta x = x\dfrac{R}{100} \delta t$. Letting $\delta t \to 0$, so that the amount of money is growing continuously, we have

$$\frac{dx}{dt} = \left(\frac{R}{100}\right)x \tag{1}$$

This equation expresses the fact that the rate of growth of the amount x is proportional to the amount itself at any given instant. This 'compound interest law', whereby the gradient of a function is proportional to the function itself, crops up persistently in applications; we give a few examples before investigating what function has this property.

EXAMPLE 8.7 *To formulate mathematically: 'If bacteria are free to breed, the rate of increase is proportional to the number of organisms already living.'*

If the bacteria occupy volume N at time t, the given statement can be translated into mathematics as

$$\frac{dN}{dt} = kN, \tag{2}$$

where k is a constant.

EXAMPLE 8.8 *To translate into mathematics Newton's law of cooling: 'If a hot body is placed in a cool room, the rate at which its temperature changes is proportional to the difference between the temperature of the body and that of the surrounding medium.'*

If T is the temperature difference at time t between the body and its surroundings (the latter being constant), we have

$$\frac{dT}{dt} = -kT \tag{3}$$

EXAMPLE 8.9 *To formulate the law of radioactive decay.*

'A gramme of radium contains $2 \cdot 6 \times 10^{21}$ *atoms. As the process of radioactive decay proceeds some of the radium nuclei give off α-particles and are replaced by those of a new element, radon* ($Z = 86$). *There is thus a continual disappearance of the radium, and experiments show that in about 1500 years half of the nuclei of the original radium atoms are transformed. In the next 1500 years the original number is reduced to a quarter, and so on.*

This 'exponential' decay of radium in which, in any time-interval, a definite fraction of the atoms present at the beginning of the interval disappears through radioactive decay, is characteristic of all radioactive substances.'*

If the quantity of radium present at time t is m, then the law given above in English can be translated into mathematics as

$$\frac{dm}{dt} = -km \tag{4}$$

The eqns. (1) to (4) are all of the type

$$\frac{d}{dx}\{f(x)\} = kf(x) \tag{5}$$

---

* Powell and Occhialini, *Nuclear Physics in Photographs* (O.U.P. 1947), p. 18.

We now investigate informally the nature of $f(x)$ if it is to satisfy such an equation. *The discussion which follows is highly suspect, and the reader should make a careful note of all unwarranted assumptions.* We are, in fact, simply exploring at this stage, and nothing is proved.

We start with the special case $k = 1$, so that

$$\frac{d}{dx}\{f(x)\} = f(x) \tag{6}$$

It is clear that if $f(x)$ is a solution of eqn. (6), then so is $cf(x)$, so we might as well find the solution for which $f(0) = 1$. Using Maclaurin's series recklessly for the moment,

$$f(x) = f(0) + xf'(0) + \frac{x^2}{2!}f''(0) + \dots$$

$$= 1 + x + \frac{x^2}{2!} + \frac{x^3}{3!} + \dots$$

since, by virtue of (6), $f(0) = f'(0) = f''(0) = \dots = 1$.

Assuming that we can differentiate an infinite series term by term, we have

$$\frac{d}{dx}\{f(x)\} = 1 + \frac{2x}{2!} + \frac{3x^2}{3!} + \dots = 1 + x + \frac{x^2}{2!} + \dots = f(x).$$

Assuming also the possibility of multiplying together infinite series,

$$f(x) \cdot f(y) = \left(1 + x + \frac{x^2}{2!} + \frac{x^3}{3!} + \dots\right)\left(1 + y + \frac{y^2}{2!} + \frac{y^3}{3!} + \dots\right)$$

$$= 1 + (x + y) + \frac{1}{2!}(x^2 + 2xy + y^2) + \dots$$

$$= 1 + (x + y) + \frac{(x + y)^2}{2!} + \frac{(x + y)^3}{3!} + \dots$$

$$= f(x + y).$$

This last result is reminiscent of the index law. We shall accordingly write

$$e^x = 1 + x + \frac{x^2}{2!} + \dots \tag{7}$$

e being the number obtained by writing $x = 1$, viz.

$$e = 1 + 1 + \frac{1}{2!} + \frac{1}{3!} + \dots *$$

---

* The number e thus defined is irrational, being approximately 2·718 (cf. Chapter 9).

We shall assume, during the remainder of this chapter, that *there is a function e^x which is the sum of the infinite series* $1 + x + \dfrac{x^2}{2!} + \ldots$ *such that* (i) $\dfrac{d}{dx}(e^x) = e^x$ *and* (ii) $e^x \cdot e^y = e^{x+y}$ *for all real x, y*; in fact, e^x is the x-th power of the number e*. This assertion will be justified later, but the reader should work carefully through the following exercise, and convince himself that it is by no means proved as yet.

<h3 style="text-align:center">EXERCISE XXXVIII</h3>

1. (D) Has the convergence of the series $1 + x + \dfrac{x^2}{2!} + \dfrac{x^3}{3!} + \ldots$ been established yet?

2. (D) It can be proved† that the series $\sin 2x + \frac{1}{2}\sin(2^2 x) + \dfrac{1}{2^2}\sin(2^3 x) + \ldots$ converges for all values of x. Examine what happens when you differentiate term by term and then put $x = 0$.

3. (D) It can also be proved† that the series $\dfrac{1}{\sqrt{1}} - \dfrac{1}{\sqrt{2}} + \dfrac{1}{\sqrt{3}} - \dfrac{1}{\sqrt{4}} + \ldots$ converges. 'Multiplying' this series by itself, i.e.

$$\left(\frac{1}{\sqrt{1}} - \frac{1}{\sqrt{2}} + \frac{1}{\sqrt{3}} - \ldots\right)\left(\frac{1}{\sqrt{1}} - \frac{1}{\sqrt{2}} + \frac{1}{\sqrt{3}} - \ldots\right)$$

gives $\dfrac{1}{\sqrt{1}\sqrt{1}} - \left(\dfrac{1}{\sqrt{1}\sqrt{2}} + \dfrac{1}{\sqrt{2}\sqrt{1}}\right) + \left(\dfrac{1}{\sqrt{1}\sqrt{3}} + \dfrac{1}{\sqrt{2}\sqrt{2}} + \dfrac{1}{\sqrt{3}\sqrt{1}}\right)$

$- \left(\dfrac{1}{\sqrt{1}\sqrt{4}} + \dfrac{1}{\sqrt{2}\sqrt{3}} + \dfrac{1}{\sqrt{3}\sqrt{2}} + \dfrac{1}{\sqrt{4}\sqrt{1}}\right) + \ldots$.

Using the fact that $\sqrt{r}\sqrt{(n-r)} \leqslant \dfrac{r+n-r}{2} = \frac{1}{2}n$, show that $\dfrac{1}{\sqrt{1}\sqrt{(n-1)}} + \dfrac{1}{\sqrt{2}\sqrt{(n-2)}} + \ldots + \dfrac{1}{\sqrt{(n-1)}\sqrt{1}} > \dfrac{2(n-1)}{n}$, which does not tend to zero as $n \to \infty$. Deduce that the 'product' of two convergent series is not necessarily itself convergent.

4. (D) e^x being defined by eqn. (7), to what extent has it been proved that $e^x \cdot e^y = e^{x+y}$?

5. (P) Find out what is meant in chemistry by the Law of Mass Action. [For a reaction $X \to Y$ 'of the first order', if a represents the number of molecules of X before the reaction begins, and if x is the number which have changed into Y after a time t, then $(a - x)$ molecules remain, and the rate of change at this time

---

* For the meaning of e^x when x is irrational, cf. Appendix 2.

† Future mathematical specialists will no doubt check up; see for example Binmore, *Mathematical Analysis* (C.U.P.). The series in question 2 is 'absolutely convergent' by comparison with $1 + \dfrac{1}{2} + \dfrac{1}{2^2} + \dfrac{1}{2^3} + \ldots$, whose sum is 2.

is $\dfrac{dx}{dt} = k(a - x)$, cf. Goddard and James, *The Elements of Physical Chemistry*, (Longmans) p. 139. Writing $a - x = y$, the last equation becomes $\dfrac{dy}{dt} = -ky$, which is of the type (5) above.]

EXAMPLE 8.10   (i) $\dfrac{d}{dx}(e^{4x})$; (ii) $\dfrac{d}{dx}(e^{\sin x})$; (iii) $\int e^{5x+2}\,dx$.

(i) To differentiate e^{4x}, we use the chain rule. Differentiating e^y with respect to y gives e^y; thus $\dfrac{d}{d(4x)}(e^{4x}) = e^{4x}$. We have, accordingly,

$$\frac{d}{dx}(e^{4x}) = \frac{d}{d(4x)}\,e^{4x} \times \frac{d(4x)}{dx} = e^{4x} \times 4 = 4e^{4x}.$$

(ii) Similarly, $\dfrac{d}{dx}(e^{\sin x}) = e^{\sin x}\cos x$.

(iii) By guesswork, $\int e^{5x+2}\,dx$ will be a multiple of e^{5x+2}.

Now $\dfrac{d}{dx}(e^{5x+2}) = e^{5x+2} \times 5$, and so the required integral is $\tfrac{1}{5}e^{5x+2}$.

We have found e^x as a solution of the *differential equation* $\dfrac{dy}{dx} = y$. Since $\dfrac{d}{dx}(Ae^{kx}) = Ake^{kx}$, where A, k are any constants, it follows that Ae^{kx} is a solution of $\dfrac{dy}{dx} = ky$, and it will be proved later that this is the 'general' solution, i.e. all solutions are of this form.

EXERCISE XXXIX

In questions 1 to 8, differentiate the given function with respect to x.

1. e^{3x}.

2. e^{4-2x}.

3. e^{x^2+x}.

4. $e^{\cos x}$.

5. $x^2 e^{5x}$.

6. $e^x \sin x$.

7. $(3x + 2)e^{-x}$.

8. $\dfrac{x}{e^x + 1}$.

In questions 9 to 14, integrate the given function with respect to x.

9. e^{2x}.

10. $e^{\frac{x}{2}}$.

11. e^{4-3x}.

12. xe^x. [Use parts]

13. xe^{x^2}.

14. $e^x \sin x$. [Use parts twice]

15. Evaluate (i) $\dfrac{d^4}{dx^4}(e^{3x})$; (ii) $\dfrac{d^n}{dx^n}(e^{ax})$.

16. Verify that $y = 2e^{3x} - 5e^{2x}$ is a solution of the differential equation $\dfrac{d^2y}{dx^2} - 5\dfrac{dy}{dx} + 6y = 0$. Can you find a more general solution?

17. Find the solution of $\dfrac{dy}{dx} = y$ such that $y = 2$ when $x = 0$.

18. Find the general solution of the equation $\dfrac{dy}{dx} = 4y$.

19. Give the first four terms and the general term of the Maclaurin expansion of e^{kx}.

20. Find the first four terms of the Maclaurin expansion of (i) $e^{\sin x}$, (ii) $e^x \sin x$.

21. The tangent at a point P to the curve $y = be^{x/a}$ meets the x-axis at T and N is the foot of the ordinate through P. Prove that $TN = a$, for all positions of P.
    If the curve meets the y-axis at Q and O is the origin, express in terms of a, b and the ordinate PN (i) the area $ONPQ$ and (ii) the volume formed by rotating this area through four right angles about the x-axis.

If $x = e^y$, then $y = \ln x$ or $\log_e x$ ('logarithm of x to base e'), by the definition of a logarithm. ln is used as an abbreviation for 'natural' logarithm.

We have $\dfrac{dx}{dy} = e^y$, so $\dfrac{dy}{dx} = \dfrac{1}{e^y} = \dfrac{1}{x}$. That is, $\dfrac{d}{dx}(\ln x) = \dfrac{1}{x}$.

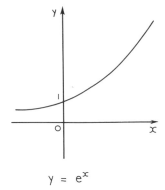

$y = e^x$

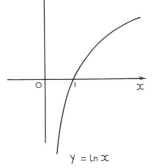

$y = \ln x$

Fig. 8.2

The reader should plot some points and sketch the graphs of $y = 2^x$ and $y = 3^x$ to satisfy himself that the graph of $y = e^x$ is as shown in Fig. 8.2. As the logarithm is the inverse function (p. 89), the graph of $y = \ln x$ is obtained by interchanging x and y.

Since $\ln x$ is only defined for positive x it is impossible to expand $\ln x$ as a

Maclaurin series, $[f(x) = f(0) + \ldots$ breaks down since $f(0)$ is not defined]. However, if $f(x) = \ln(1 + x)$, we have

$f(x)$	$\ln(1 + x)$	$f(0)$	0
$f'(x)$	$(1 + x)^{-1}$	$f'(0)$	1
$f''(x)$	$-(1 + x)^{-2}$	$f''(0)$	$-1!$
$f'''(x)$	$2(1 + x)^{-3}$	$f'''(0)$	$2!$
$f^{(iv)}(x)$	$-3!(1 + x)^{-4}$	$f^{(iv)}(0)$	$-3!$
$\ldots$	$\ldots$	$\ldots$	$\ldots$
$f^{(n)}(x)$	$(-1)^{n-1}(n - 1)!(1 + x)^{-n}$	$f^{(n)}(0)$	$(-1)^{n-1}(n - 1)!$

Thus, subject to the remainder after n terms tending to zero as $n \to \infty$, we have

$$\ln(1 + x) = x - \frac{x^2}{2} + \frac{x^3}{3} - \frac{x^4}{4} + \ldots . \tag{1}$$

It will be shown in the next chapter that this expansion is valid only for $-1 < x \leqslant 1$. In particular, putting $x = 1$, we have

$$\ln 2 = 1 - \tfrac{1}{2} + \tfrac{1}{3} - \tfrac{1}{4} + \ldots$$

On replacing x by $-x$ in eqn. (1), we have

$$\ln(1 - x) = -x - \frac{x^2}{2} - \frac{x^3}{3} - \ldots \qquad (-1 \leqslant x < 1)$$

EXAMPLE 8.11 *To expand* (i) $\ln(2 + x)$, (ii) $\ln(1 + x + x^2)$ *in ascending powers of* x, *stating the necessary restrictions on the values of* x.

(i)

$$\ln(2 + x) = \ln 2\left(1 + \frac{x}{2}\right) = \ln 2 + \ln\left(1 + \frac{x}{2}\right)^{*}$$

$$= \ln 2 + \frac{x}{2} - \frac{1}{2}\left(\frac{x}{2}\right)^2 + \frac{1}{3}\left(\frac{x}{2}\right)^3 - \ldots,$$

provided that $-1 < \dfrac{x}{2} \leqslant 1$, i.e. $-2 < x \leqslant 2$.

(ii)

$$\ln(1 + x + x^2) = \ln\left(\frac{1 - x^3}{1 - x}\right) = \ln(1 - x^3) - \ln(1 - x)$$

$$= -x^3 - \tfrac{1}{2}(x^3)^2 - \tfrac{1}{3}(x^3)^3 - \ldots + x + \tfrac{1}{2}x^2 + \tfrac{1}{3}x^3 + \ldots,$$

---

* We are using here the rule $\ln ab = \ln a + \ln b$. This follows from our other, *so far unproved*, assumptions about e^x. For let $\ln a = p$, $\ln b = q$. Then $a = e^p$, $b = e^q$, whence $ab = e^{p+q}$ and so $\ln ab = p + q = \ln a + \ln b$. The whole theory will be established more firmly in the next chapter.

the coefficient of x^k being $\dfrac{1}{k}$ if $k \neq 3n$ and $-\dfrac{2}{k}$ if $k = 3n$. The range of values for x is $-1 \leqslant x < 1$.

EXERCISE XL

In questions 1 to 8, differentiate the given function with respect to x.

1. $\ln (2x + 1)$.

5. $\dfrac{\ln x}{x}$.

2. $\ln (3 - x)$.

6. $\dfrac{x}{\ln x}$.

3. $\ln \sin x$.

7. $\ln 5x$.

4. $x^3 \ln x$.

8. $\ln kx$.

    (Explain why the answers to numbers 7 and 8 are the same.)

9. $\dfrac{d^n}{dx^n} \{\ln (1 + 2x)\}$.

In questions 10 to 12, give the first four terms in the Maclaurin expansion of the function, and state the range of values of x for which it is valid.

10. $\ln \left(1 + \dfrac{x}{2}\right)$.

12. $\ln (1 + e^x)$.

11. $\ln (3 - x)$.

Leibniz's Theorem

This theorem is concerned with the n-th derivative of a product, that is, $\dfrac{d^n}{dx^n} (uv)$ where u, v are functions of x. With its help, a number of further functions can be expressed as Maclaurin expansions.

    For compactness, we shall use suffixes to denote differentiation, so that $\dfrac{d^n y}{dx^n}$ will be written as y_n; as usual, $_nC_r$ denotes the number of ways of choosing r things from n.

    Leibniz's theorem states:

$$(uv)_n = u_n v + {}_nC_1 u_{n-1} v_1 + {}_nC_2 u_{n-2} v_2 + \ldots + {}_nC_r u_{n-r} v_r + \ldots + uv_n.$$

    Before proving the theorem, we shall look at some special cases, with a view to making things look clearer. When $n = 1$, the result reads $(uv)_1 = u_1 v + uv_1$, that is, $\dfrac{d}{dx} (uv) = \dfrac{du}{dx} v + u \dfrac{dv}{dx}$, which has been established previously. Differentiating this again, we have $\dfrac{d^2}{dx^2} (uv) = \dfrac{d^2u}{dx^2} v + \dfrac{du}{dx} \dfrac{dv}{dx} + \dfrac{du}{dx} \dfrac{dv}{dx} + u \dfrac{d^2v}{dx^2}$, or $(uv)_2 = u_2 v + 2u_1 v_1 + uv_2$, that is, Leibniz's theorem with

$n = 2$. We shall now prove the theorem by induction (the reader should refer to Appendix 1 if necessary).

We assume the result true for a particular value of n, say $n = k$. Thus

$$(uv)_k = u_k v + {}_kC_1 u_{k-1} v_1 + \ldots + {}_kC_{r-1} u_{k+1-r} v_{r-1} + {}_kC_r u_{k-r} v_r + \ldots + uv_k \tag{1}$$

The general term here is ${}_kC_r u_{k-r} v_r$.

Differentiating once more, there will be contributions to the term in $u_{k+1-r} v_r$ from both ${}_kC_{r-1} u_{k+1-r} v_{r-1}$ and ${}_kC_r u_{k-r} v_r$, in fact

$$(uv)_{k+1} = u_{k+1} v + \ldots + ({}_kC_r + {}_kC_{r-1}) u_{k+1-r} v_r + \ldots + uv_{k+1} \tag{2}$$

But ${}_kC_r + {}_kC_{r-1} = {}_{k+1}C_r$ (cf. Appendix 1), so the general term in eqn. (2) is ${}_{k+1}C_r u_{k+1-r} v_r$. It follows that if the result is true for $n = k$, it is also true for $n = k + 1$. But we have seen that it is true for $n = 1$, and so, by the Principle of Mathematical Induction, the theorem is proved for all positive integral values of n.

EXAMPLE 8.12   *To find the n-th derivative of $x^3 e^x$.*

Using Leibniz's theorem, $\dfrac{d^n}{dx^n}(x^3 e^x) = e^x x^3 + ne^x \cdot 3x^2 + {}_nC_2 e^x \cdot 6x + {}_nC_3 e^x \cdot 6$ (subsequent derivatives of x^3 vanishing), $= e^x\{x^3 + 3nx^2 + 3n(n-1)x + n(n-1)(n-2)\}$.

EXAMPLE 8.13   *Given that $y = \sin^{-1} x$, prove that*

$$(1 - x^2)\frac{d^2 y}{dx^2} - x\frac{dy}{dx} = 0.$$

Hence prove that

$$(1 - x^2)y_{n+2} - (2n + 1)xy_{n+1} - n^2 y_n = 0,$$

where y_n denotes $\dfrac{d^n y}{dx^n}$.

Expand y as a Maclaurin series in x as far as the term in x^5. (Assume y is defined as in Fig. 6.5 (a).)

If $y = \sin^{-1} x$, $\dfrac{dy}{dx} = \dfrac{1}{\sqrt{(1 - x^2)}} = (1 - x^2)^{-\frac{1}{2}}$

and

$$\frac{d^2 y}{dx^2} = -\tfrac{1}{2}(1 - x^2)^{-\frac{3}{2}} \cdot -2x = x(1 - x^2)^{-\frac{3}{2}}$$

so that

$$(1 - x^2)\frac{d^2 y}{dx^2} - x\frac{dy}{dx} = x(1 - x^2)^{-\frac{1}{2}} - x(1 - x^2)^{-\frac{1}{2}} = 0,$$

as required. Writing this result as

$$(1 - x^2)y_2 - xy_1 = 0$$

and differentiating n times, by Leibniz's theorem, we have

$$(1 - x^2)y_{n+2} + n(-2x)y_{n+1} + \frac{n(n-1)}{1 \cdot 2}(-2)y_n - xy_{n+1} - ny_n = 0$$

or $\qquad\qquad (1 - x^2)y_{n+2} - (2n + 1)xy_{n+1} - n^2y_n = 0 \qquad\qquad$ (1)

If $y = f(x) = \sin^{-1} x$, putting $x = 0$ in (1) yields

$$f^{(n+2)}(0) - n^2f^{(n)}(0) = 0 \qquad\qquad (2)$$

Now $f(0) = 0$, and $f''(0) = 0$ so, from eqn. (2), $f^{(2n)}(0) = 0$.
   Also, since $f'(0) = 1$, we have $f'''(0) = 1^2, f^{(v)}(0) = 3^2, f^{(vii)}(0) = 5^2 . 3^2$ etc.
   Using the Maclaurin expansion $f(x) = f(0) + xf'(0) + \ldots$ (the question
implies that existence can be assumed), we have $\sin^{-1} x = x + \dfrac{x^3}{3!} +$
$\dfrac{3^2x^5}{5!} + \ldots$, so that as far as x^5 the series is $x + \dfrac{x^3}{6} + \dfrac{3x^5}{40}$.

EXAMPLE 8.14   *Show that*

$$\frac{d}{dx}(e^{ax} \sin bx) = (a^2 + b^2)^{\frac{1}{2}} e^{ax} \sin (bx + \alpha),$$

where $\tan \alpha = b/a$.
   Deduce that

$$\frac{d^n}{dx^n}(e^{ax} \sin bx) = (a^2 + b^2)^{n/2}e^{ax} \sin (bx + n\alpha).$$

Apply Leibniz's theorem to obtain another expression for $\dfrac{d^n}{dx^n}(e^{ax} \sin bx)$,
and by comparing the two results prove that

$$2^{n/2} \sin \frac{n\pi}{4} = \binom{n}{1} - \binom{n}{3} + \binom{n}{5} - \ldots .$$

$\left[\dbinom{n}{r} \text{ is the coefficient of } x^r \text{ in the binomial expansion of } (1 + x)^n.\right]$

$\dfrac{d}{dx}(e^{ax} \sin bx) = ae^{ax} \sin bx + be^{ax} \cos bx$

$$= (a^2 + b^2)^{\frac{1}{2}} e^{ax} \left\{ \frac{a}{\sqrt{a^2 + b^2}} \sin bx + \frac{b}{\sqrt{a^2 + b^2}} \cos bx \right\}$$

$$= (a^2 + b^2)^{\frac{1}{2}} e^{ax} \sin (bx + \alpha), \quad \text{where } \tan \alpha = \frac{b}{a}.$$

Thus, on successive differentiations, a factor of $(a^2 + b^2)^{\frac{1}{2}}$ is introduced and α is added to the 'sine' expression. It follows that

$$\frac{d^n}{dx^n}(e^{ax}\sin bx) = (a^2 + b^2)^{\frac{n}{2}}e^{ax}\sin(bx + n\alpha). \tag{1}$$

To apply Leibniz's theorem to $e^{ax}\sin bx$, we require the result that

$$\frac{d^n}{dx^n}(\sin bx) = b^n\sin\left(bx + \frac{n\pi}{2}\right),$$

which can be proved by induction. It is also readily seen that

$$\frac{d^n}{dx^n}(e^{ax}) = a^n e^{ax},$$

so we have

$$\frac{d^n}{dx^n}(e^{ax}\sin bx)$$

$$= a^n e^{ax}\sin bx + \binom{n}{1}a^{n-1}e^{ax}b\sin\left(bx + \frac{\pi}{2}\right)$$

$$+ \binom{n}{2}a^{n-2}e^{ax}b^2\sin(bx + \pi) + \ldots + e^{ax}b^n\sin\left(bx + \frac{n\pi}{2}\right). \tag{2}$$

Comparing eqns. (1) and (2), [with an eye on the answer, e.g. we presumably require $(a^2 + b^2)^{\frac{n}{2}}$ to become $2^{\frac{n}{2}}$], if we put $a = b = 1$, $x = 0$ so that $\tan\alpha = \frac{b}{a} = 1$ and $\sin n\alpha = \sin\frac{n\pi}{4}$, we have

$$2^{\frac{n}{2}}\sin\frac{n\pi}{4} = \binom{n}{1}\sin\frac{\pi}{2} + \binom{n}{2}\sin\pi + \binom{n}{3}\sin\frac{3\pi}{2} + \ldots$$

$$= \binom{n}{1} - \binom{n}{3} + \binom{n}{5} - \ldots,$$

as required.

EXERCISE XLI

(Miscellaneous examination questions involving the contents of this chapter.)

1. Use Maclaurin's series to obtain the expansion (in ascending powers of x) of $\ln(1 + x)$ as far as the term is x^5. Deduce that

$$\frac{1}{2}\ln\left(\frac{1 + x}{1 - x}\right) = x + \frac{x^3}{3} + \frac{x^5}{5} + \ldots.$$

and by substituting $x = 0 \cdot 2$, estimate the value of $\ln 1 \cdot 5$ to 4 decimal places.

2. Differentiate with respect to x:

(i) $\dfrac{1 + x^2}{1 - x}$;   (ii) $\sqrt{x} \ln x$;   (iii) $e^{x \tan x} \cos x$.

Prove by means of Maclaurin's theorem that the first four terms in the expansion of $\cos x + \sin^{-1} x$ in ascending powers of x are

$$1 + x - \tfrac{1}{2}x^2 + \tfrac{1}{6}x^3.$$

3. State Maclaurin's theorem for the expansion of a function $f(x)$ in a series of ascending powers of x, giving the coefficient of x^n.

If $y = e^{\cos x}$, prove that

$$\frac{d^2y}{dx^2} + \frac{dy}{dx} \sin x + y \cos x = 0$$

and that, if x is so small that x^5 and higher powers of x can be neglected, then

$$y = e\left(1 - \frac{x^2}{2} + \frac{1}{6}x^4\right).$$

4. Find the maximum and minimum values of $x^2 e^{-x}$ and sketch the graph of this function.

Find the equation of the tangent to the graph at the point at which $x = 1$.

5. Prove that, if $f'(a) = 0$ and $f''(a)$ is positive, then the graph of the function $f(x)$ has a minimum at the point whose abscissa is a.

A variable isosceles triangle is circumscribed about a circle of given radius. Prove that the area of the triangle is a minimum (and not a maximum) when the triangle is equilateral.

6. Given that $y = (\sec x + \tan x)^p$ where p is constant, prove that

$$\cos x \frac{dy}{dx} = py.$$

Find, by using Leibniz's theorem or otherwise, the values at $x = 0$ of the first four derivatives of y.

Using Maclaurin's theorem, show that $(\sec x + \tan x)^p$ is equal to $e^{px} + \tfrac{1}{6}px^3(1 + px)$ if powers of x above the fourth are neglected.

7. Prove from first principles that $\dfrac{d}{dx} \sec x = \sec x \tan x$. If $y = e^x \tan x$, show that

$$\frac{d^2y}{dx^2} - 2(1 + \tan x) \frac{dy}{dx} + (1 + 2 \tan x)y = 0.$$

Find the first three non-zero terms of the Maclaurin expansion of $e^x \tan x$.

8. Find the values of x for which the function $e^{-x/a} \sin x$, where a is a positive constant, has maximum and minimum values and show that these values of the function form a geometric progression. Distinguish between the maximum and minimum values.

9. If $y = e^{e^x}$, prove that

$$e^{-x} \frac{d^3y}{dx^3} = \frac{d^2y}{dx^2} + 2 \frac{dy}{dx} + y.$$

Use Maclaurin's theorem to find a series for e^{e^x} as far as the term in x^3, and hence find the value of the function to 3 significant figures when $x = 0{\cdot}1$, given that $e \approx 2{\cdot}718$.

10. Differentiate with respect to x:

(i) $\dfrac{(1-x)^2}{(1+x)}$;   (ii) $x^2 \tan^{-1}(x^2)$;   (iii) $\ln(1 + \tan x)$.

By means of Taylor's theorem, or otherwise, prove that when $x - a$ is small the expression $\sqrt{(x + 3a)} - \sqrt{(5a - x)}$ is represented approximately by

$$\frac{1}{2a^{\frac{1}{2}}}(x - a) + \frac{1}{256a^{\frac{5}{2}}}(x - a)^3.$$

9
Special Functions

I begin to smell a rat.
CERVANTES

The exponential and logarithmic functions play an important part in the applications of calculus and we must now put their theory on a more rigorous basis. The introductory approach of the previous chapter can be made respectable with the help of more advanced convergence theory but, to avoid the latter, we shall instead make a fresh start. All the assumptions and results of Chapter 8 must accordingly, for the time being, be forgotten.

We consider first

$$f(t) = \int_1^t \frac{1}{x}\, dx \qquad (t > 0)$$

This integral is of interest as giving the area under the rectangular hyperbola (Fig. 9.1) $y = \dfrac{1}{x}$ from $x = 1$ to $x = t$ and also as the exceptional case when $\int x^n\, dx \neq \dfrac{x^{n+1}}{n+1}$, $n + 1$ being zero.

We have
$$f(ab) = \int_1^{ab} \frac{1}{x}\, dx = \int_1^a \frac{1}{x}\, dx + \int_a^{ab} \frac{1}{x}\, dx$$

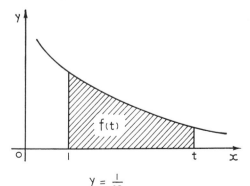

$$y = \frac{1}{x}$$

Fig. 9.1

155

so
$$f(ab) = f(a) + \int_a^{ab} \frac{1}{x}\, dx.$$

Now, putting $x = az$, $dx = a\, dz$,

$$\int_a^{ab} \frac{1}{x}\, dx = \int_1^b \frac{1}{az} \cdot a\, dz = \int_1^b \frac{1}{z}\, dz = f(b).$$
Thus

$$f(ab) = f(a) + f(b) \tag{1}$$

(The reader will now guess that $f(t)$ turns out to be $\ln t$, but we will not change notation until this conjecture is justified.)

Also

$$f(a^p) = \int_1^{a^p} \frac{1}{x}\, dx,$$

p being any real number,*

$$= \int_1^a \frac{1}{y^p}\, py^{p-1}\, dy$$

(putting $x = y^p$, $dx = py^{p-1}\, dy$)

$$= p \int_1^a \frac{1}{y}\, dy$$

$$= pf(a) . \tag{2}$$

Now we *define* e as the number such that $f(e) = \int_1^e \frac{1}{x}\, dx = 1$. From eqn. (2), we have

$$f(e^x) = xf(e) = x. \tag{3}$$

Now, by the definition of a logarithm,

$$\ln (e^x) = x, \tag{4}$$

so comparing eqns. (3) and (4) we can now identify f with $\ln$.

Thus

$$\ln t = \int_1^t \frac{1}{x}\, dx.$$

In particular, $\ln 1 = \int_1^1 \frac{1}{x}\, dx = 0$.

We note that $\ln x$ is defined only for $x > 0$.

* Cf. Appendix 2.

Equation (1) can now be re-written as $\ln (ab) = \ln a + \ln b$, and eqn. (2) as $\ln (a^p) = p \ln a$. Putting $p = -1$, $\ln \left(\dfrac{1}{a}\right) = -\ln a$, so that $\ln 1 = 0$, as is otherwise evident since $e^0 = 1$. Writing $b = \dfrac{1}{c}$,

$$\ln \left(\frac{a}{c}\right) = \ln a + \ln \left(\frac{1}{c}\right) = \ln a - \ln c.$$

If $t > 1$,
$$\int_1^t \frac{1}{t}\,dx < \int_1^t \frac{1}{x}\,dx < \int_1^t dx$$

or
$$\frac{1}{t}\Big[x\Big]_1^t < \ln t < \Big[x\Big]_1^t$$

that is,
$$\frac{t-1}{t} < \ln t < t - 1 \tag{5}$$

Writing $t = 1 + \dfrac{x}{n}$, we have $\dfrac{\dfrac{x}{n}}{1 + \dfrac{x}{n}} < \ln \left(1 + \dfrac{x}{n}\right) < \dfrac{x}{n}$ so that

$$\frac{x}{1 + \dfrac{x}{n}} < n \ln \left(1 + \frac{x}{n}\right) = \ln \left(1 + \frac{x}{n}\right)^n < x.$$

As $n \to \infty$, $\dfrac{x}{1 + \dfrac{x}{n}} \to x$, so $\lim\limits_{n \to \infty} \ln \left(1 + \dfrac{x}{n}\right)^n = \ln \lim\limits_{n \to \infty} \left(1 + \dfrac{x}{n}\right)^n = x,$ *

whence
$$\lim_{n \to \infty} \left(1 + \frac{x}{n}\right)^n = e^x. \tag{6}$$

Since $\ln x = \displaystyle\int_1^x \frac{du}{u}$, it follows from the definition

$$f'(x) = \lim_{h \to 0} \frac{f(x + h) - f(x)}{h}$$

that
$$\frac{d}{dx}(\ln x) = \lim_{h \to 0} \frac{\displaystyle\int_1^{x+h} \frac{du}{u} - \int_1^x \frac{du}{u}}{h} = \lim_{h \to 0} \frac{\displaystyle\int_x^{x+h} \frac{du}{u}}{h} = \lim_{h \to 0} \frac{h \cdot \dfrac{1}{\xi}}{h},$$

where ξ is between x and $x + h$ (cf. p. 60), $= \dfrac{1}{x}$.

---

* The inversion of lim and ln illustrates an important general property of *continuous* functions.

Now if $x = \ln y$, so that $y = e^x$, $\dfrac{dx}{dy} = \dfrac{1}{y}$ and so $\dfrac{dy}{dx} = y$, i.e.

$$\frac{d}{dx}(e^x) = e^x.$$

Having now proved this result, we may use Maclaurin's theorem in the form

$$f(x) = f(0) + xf'(0) + \ldots + \frac{x^{n-1}}{(n-1)!}\, f^{(n-1)}(0) + \frac{x^n}{n!} f^{(n)}(\theta x),$$

giving $\qquad e^x = 1 + x + \dfrac{x^2}{2!} + \ldots + \dfrac{x^{n-1}}{(n-1)!} + R_n,$

where $\qquad R_n = \dfrac{x^n}{n!} e^{\theta x},\ (0 < \theta < 1).$

Now $e^{\theta x} < 1$ if $x \leqslant 1$ and $< e^x$ if $x > 1$, while $\dfrac{x^n}{n!} \to 0$ as $n \to \infty$ (cf. Appendix 2). It follows that $R_n \to 0$ as $n \to \infty$ for all x, and so

$$e^x = 1 + x + \frac{x^2}{2!} + \frac{x^3}{3!} + \ldots \text{ for all } x.$$

EXAMPLE 9.1 $\quad \dfrac{d}{dx}(a^x).$

Let $y = a^x$, then $\ln y = x \ln a$. Differentiating with respect to x, we have $\dfrac{d}{dx}(\ln y) = \dfrac{d}{dy}(\ln y) \cdot \dfrac{dy}{dx}$, so that $\dfrac{1}{y}\dfrac{dy}{dx} = \ln a$.

Thus $\dfrac{d}{dx}(a^x) = a^x \cdot \ln a$. (Cf. Exercise XLII, question 7.)

EXERCISE XLII (D)

1. (i) Illustrate graphically the result (5), namely $\dfrac{t-1}{t} < \ln t < t - 1$, if $t > 1$.

   (ii) By writing $s = \dfrac{1}{t}$, show that (5) also holds for $t < 1$. [If $t < 1$, $s > 1$. $\dfrac{s-1}{s} < \ln s < s - 1$ becomes $1 - t < -\ln t < \dfrac{1-t}{t}$. If $a > b$, then $-a < -b$.]

2. Use the result (5), with $t^{\frac{1}{2}}$ instead of t, to show that $\lim_{t \to \infty} \dfrac{\ln t}{t} = 0$. [Divide through by t.]

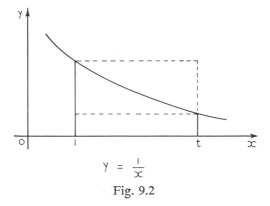

$$y = \frac{1}{x}$$

Fig. 9.2

3. (*Irrationality of e*) Cf. Example 2.5, p. 20. Suppose $e = 1 + \frac{1}{1!} + \frac{1}{2!} + \frac{1}{3!} + \ldots = \frac{a}{b}$ where a, b are integers. Taking a value of $n > b$, so that $n! \, e$ is an integer, we have

$$n! \, e = n!\left(1 + \frac{1}{1!} + \frac{1}{2!} + \ldots + \frac{1}{n!}\right) + \frac{1}{n+1} + \frac{1}{(n+1)(n+2)}$$

$$+ \frac{1}{(n+1)(n+2)(n+3)} + \ldots.$$

(i) Show that

$$\frac{1}{n+1} + \frac{1}{(n+1)(n+2)} + \frac{1}{(n+1)(n+2)(n+3)} + \ldots$$

to k terms is less than

$$\frac{1}{n+1} + \frac{1}{(n+1)^2} + \frac{1}{(n+1)^3} + \ldots = \frac{1}{n+1} \cdot \frac{1}{1 - \frac{1}{n+1}} = \frac{1}{n},$$

for all k.

(ii) Deduce that $\frac{1}{n+1} + \frac{1}{(n+1)(n+2)} + \frac{1}{(n+1)(n+2)(n+3)} + \ldots$ is a fraction less than 1.

(iii) Complete the *reductio ad absurdum* proof that e is irrational.

4. (i) Show that, for all n,

$$2 < 1 + \frac{1}{1!} + \frac{1}{2!} + \frac{1}{3!} + \ldots + \frac{1}{n!}$$

$$< 1 + 1 + \frac{1}{2} + \frac{1}{2^2} + \frac{1}{2^3} + \ldots = 3.$$

(ii) Deduce that $2 < e < 3$.
($e = 2 \cdot 7182818285 \ldots$)

5. (i) The compound interest law with yearly payment of interest is $A = P\left(1 + \dfrac{R}{100}\right)^T$ (p. 142). Let $\dfrac{R}{100} = x$, $P = 1$, and suppose that payment at the rate of $R = 100x$ per cent per annum is now made n times in the year instead of once; thus $A = \left(1 + \dfrac{x}{n}\right)^{nT}$. Putting $T = 1$, and letting $n \to \infty$, deduce that the amount after a year if interest at $100x$ per cent per annum is paid continuously is $\lim\limits_{n \to \infty} \left(1 + \dfrac{x}{n}\right)^n$.

(ii) By comparing with result (6) above, find to the nearest penny the amount after one year if interest on £1 is paid continuously at the rate of 100 per cent.

6. Show graphically that $1 + \frac{1}{2} + \frac{1}{3} + \ldots + \dfrac{1}{n-1} > \ln n > \frac{1}{2} + \frac{1}{3} + \frac{1}{4} + \ldots + \dfrac{1}{n}$. [See Exercise XVI, question 8 only if necessary.]

7. (a) Show that $a = e^{\ln a}$.
   (b) Differentiate a^x by first writing it in the form $e^{x \ln a}$.
   (c) Differentiate with respect to x (i) 2^x, (ii) x^x.

Continuing to put the results of the last chapter on a proper footing, we next reconsider the expansion for $\ln (1 + x)$. Instead of investigating the remainder R_n as $n \to \infty$ in the Maclaurin expansion, it is easier to proceed directly as follows.

Since $\ln x = \displaystyle\int_1^x \frac{ds}{s}$,

$$\ln (1 + x) = \int_1^{1+x} \frac{ds}{s} = \int_0^x \frac{dt}{1 + t}$$

(putting $s = 1 + t$)

$$= \int_0^x \left\{1 - t + t^2 - t^3 + \ldots + (-1)^{n-1} t^{n-1} + \frac{(-1)^n t^n}{1 + t}\right\} dt$$

$$= x - \frac{x^2}{2} + \frac{x^3}{3} - \ldots + (-1)^{n-1} \frac{x^n}{n} + R_n,$$

where

$$R_n = \int_0^x \frac{(-1)^n t^n}{1 + t}\, dt.$$

(i) $0 < x \leqslant 1$.

$$\int_0^x \frac{t^n}{1 + t}\, dt < \int_0^x t^n\, dt = \frac{x^{n+1}}{n + 1} \to 0 \qquad \text{as } n \to \infty.$$

(ii) $x > 1$. The series $x - \dfrac{x^2}{2} + \dfrac{x^3}{3} - \dots$ is not convergent, since $\left|\dfrac{x^n}{n}\right|$ does

not tend to zero as $n \to \infty$ (cf. p. 136 and Appendix 2). So far we have

established that $\ln(1 + x) = x - \dfrac{x^2}{2} + \dfrac{x^3}{3} - \dfrac{x^4}{4} + \dots$ if $0 < x \leqslant 1$,

but the series is not valid if $x > 1$.

(iii) $-1 < x < 0$. We have

$$|R_n| = \left|\int_0^x \frac{t^n}{1+t}\,dt\right| = \int_0^y \frac{s^n}{1-s}\,ds$$

(writing $t = -s$ and $x = -y$, so that $0 < y < 1$)

$$< \int_0^y \frac{s^n}{1-y}\,ds = \frac{1}{1-y} \cdot \frac{y^{n+1}}{n+1} \to 0 \qquad \text{as } n \to \infty.$$

The series therefore also holds good for $-1 < x < 0$.

(iv) $x = -1$. The expansion is not valid, as $1 + \frac{1}{2} + \frac{1}{3} + \frac{1}{4} + \dots$ is divergent (cf. Exercise XXXVII, question 18).

(v) $x < -1$. Again the series concerned is not convergent, since $\left|\dfrac{x^n}{n}\right|$ does not tend to zero as $n \to \infty$.

To sum up, *the expansion* $\ln(1 + x) = x - \dfrac{x^2}{2} + \dfrac{x^3}{3} - \dfrac{x^4}{4} + \dots$ *is valid if and only if* $-1 < x \leqslant 1$.

Using the following hints only as necessary, the reader should prove

similarly that $\tan^{-1} x = x - \dfrac{x^3}{3} + \dfrac{x^5}{5} - \dfrac{x^7}{7} + \dots$ *if and only if* $|x| \leqslant 1$.

$$\left[\int_0^x \frac{dt}{1+t^2} = \int_0^x \left\{1 - t^2 + t^4 - \dots + (-1)^{n-1}t^{2n-2} + \frac{(-1)^n t^{2n}}{1+t^2}\right\} dt;\right.$$

$$\text{if } |x| \leqslant 1, \qquad \int_0^x \frac{t^{2n}}{1+t^2}\,dt < \int_0^x t^{2n}\,dt;$$

$$\left.\text{if } |x| > 1, \qquad \int_1^x \frac{t^{2n}}{1+t^2}\,dt > \int_1^x \frac{t^{2n}}{1+x^2}\,dt.\right]$$

Applications to Integration

If $ax + b > 0$, $\dfrac{d}{dx} \ln(ax + b) = \dfrac{a}{ax+b}$, so that

$$\int \frac{dx}{ax+b} = \frac{1}{a} \ln(ax+b) \tag{1}$$

Also, if $ax + b < 0$,

$$\int \frac{dx}{ax+b} = -\int \frac{1}{-(ax+b)}\,dx = +\frac{1}{a} \ln\{-(ax+b)\}. \tag{2}$$

Equations (1) and (2) can be combined in the statement

$$\int \frac{dx}{ax + b} = \frac{1}{a} \ln |ax + b|.$$

EXAMPLE 9.2 $I = \int \dfrac{2x}{x^2 + 1} \, dx$

Putting $x^2 + 1 = y$, we have $2x \, dx = dy$ so that

$$I = \int \frac{dy}{y} = \ln y = \ln (x^2 + 1).$$

Seeking for a generalization, $\dfrac{d}{dx} \{\ln f(x)\} = \dfrac{1}{f(x)} \times f'(x)$, by the chain rule. Thus

$$\int \frac{f'(x)}{f(x)} \, dx = \ln \{f(x)\}.$$

EXAMPLE 9.3 $I = \int \cot x \, dx.$

$I = \displaystyle\int \dfrac{\cos x}{\sin x} \, dx.$ Now $\dfrac{d}{dx} (\sin x) = \cos x$, thus $I = \ln \sin x.$

We assume that the reader is by this time familiar with partial fractions.

EXAMPLE 9.4

$$\int \frac{dx}{(x - 1)(x + 2)}$$

We have

$$\int \frac{dx}{(x - 1)(x + 2)} = \int \left(\frac{\frac{1}{3}}{x - 1} - \frac{\frac{1}{3}}{x + 2} \right) dx$$

$$= \tfrac{1}{3} \ln |x - 1| - \tfrac{1}{3} \ln |x + 2| + c$$

$$= \tfrac{1}{3} \ln \left| \frac{x - 1}{x + 2} \right| + c.$$

EXAMPLE 9.5

$$I = \int \frac{dx}{(x - 1)(x + 2)^2}.$$

$$I = \int \left(\frac{\frac{1}{9}}{x - 1} - \frac{\frac{1}{9}}{x + 2} - \frac{\frac{1}{3}}{(x + 2)^2} \right) dx$$

$$= \tfrac{1}{9} \ln \left| \frac{x - 1}{x + 2} \right| + \frac{1}{3(x + 2)} + c.$$

EXAMPLE 9.6 $I = \displaystyle\int \frac{dx}{(x-1)(x^2+2)}$

$$I = \frac{1}{3} \int \left(\frac{1}{x-1} - \frac{x+1}{x^2+2} \right) dx$$

$$= \frac{1}{3} \int \left(\frac{1}{x-1} - \frac{x}{x^2+2} - \frac{1}{x^2+2} \right) dx$$

Now $\displaystyle\int \frac{x}{x^2+2}\, dx = \frac{1}{2} \ln(x^2+2)$ and $\displaystyle\int \frac{1}{x^2+2}\, dx = \frac{1}{\sqrt{2}} \tan^{-1}\left(\frac{x}{\sqrt{2}}\right)$.

Thus $I = \frac{1}{3} \ln|x-1| - \frac{1}{6} \ln(x^2+2) - \dfrac{1}{\sqrt{2}} \tan^{-1}\left(\dfrac{x}{\sqrt{2}}\right) + c.$

EXERCISE XLIII

1. Show that $\frac{1}{2} \ln \left(\dfrac{1+x}{1-x}\right) = x + \dfrac{x^3}{3} + \dfrac{x^5}{5} + \ldots$, stating the range of values for x for which this expansion is valid.

2. Show that $\dfrac{\pi}{4} = 1 - \frac{1}{3} + \frac{1}{5} - \frac{1}{7} + \ldots$.

In questions 3 to 22, integrate the given function with respect to x.

3. $\dfrac{1}{2x}$.

4. $\dfrac{1}{1-3x}$.

5. $\dfrac{1}{x(x+1)}$.

6. $\dfrac{1}{x^2-1}$.

7. $\dfrac{1}{x(x-1)(x-2)}$.

8. $\dfrac{1}{x^2(x-1)}$.

9. $\dfrac{1}{(x^2+1)(x-1)}$.

10. $\dfrac{1}{x(3x-1)(2x+5)}$.

11. $\dfrac{1}{(1-x)(1+2x)}$.

12. $\ln x$. [Use parts.]

13. $x \ln x$.

14. xe^x.

15. $(\ln x)^2$. [Put $x = e^y$.]

16. $\dfrac{2x}{x^2-1}$.

17. $\dfrac{x^2}{x^3+1}$.

18. $\dfrac{21x^2+10x-2}{7x^3+5x^2-2x+1}$.

19. $\cot x$.

20. $\tan 2x$.

21. $\dfrac{e^x}{1+e^x}$.

22. $\dfrac{\cos x}{1+\sin x}$.

23. (i) $\dfrac{d}{dx}\{\ln(\sec x + \tan x)\}$; (ii) $\int \sec x\, dx$.

24. $\displaystyle\int_0^{\frac{\pi}{4}} \tan x\, dx$ to 3 decimal places, given that $\ln 2 = 0.6931$.

25. (D) Differentiate $x^3 e^x \sin x$ with respect to x. [Compare whether it is quicker to treat as a triple product or to proceed as follows. Let $y = x^3 e^x \sin x$, then $\ln y = 3 \ln x + x + \ln \sin x$. Differentiating with respect to x, $\dfrac{1}{y}\dfrac{dy}{dx} = \dfrac{3}{x} + 1 + \cot x$, so $\dfrac{dy}{dx} = x^2 e^x\{(3 + x)\sin x + x \cos x\}$.]

26. $\dfrac{d}{dx}\{x^5 e^x \ln x\}$.

Hyperbolic Functions

We define the hyperbolic functions

$$\cosh x = \frac{e^x + e^{-x}}{2}$$

$$\sinh x = \frac{e^x - e^{-x}}{2}$$

Then $\tanh x = \dfrac{\sinh x}{\cosh x}$, $\coth x = \dfrac{\cosh x}{\sinh x}$, $\operatorname{sech} x = \dfrac{1}{\cosh x}$, $\operatorname{cosech} x = \dfrac{1}{\sinh x}$.

Three reasons for introducing these new functions are:

(i) A number of integrals can be expressed in terms of them.

(ii) A uniform chain hanging under gravity takes up the shape $y = c \cosh \dfrac{x}{c}$ (cf. Chapter 12).

(iii) If $x = a \cosh\theta$, $y = b \sinh\theta$, then

$$\frac{x^2}{a^2} - \frac{y^2}{b^2} = \cosh^2\theta - \sinh^2\theta = \left(\frac{e^x + e^{-x}}{2}\right)^2 - \left(\frac{e^x - e^{-x}}{2}\right)^2$$

$$= \tfrac14\{e^{2x} + 2 + e^{-2x}\} - \tfrac14\{e^{2x} - 2 + e^{-2x}\}$$

$$= 1 \quad (\text{see Fig. 9.3})$$

(iii) shows how the hyperbolic functions got their name. $x = a \cosh\theta$, $y = b \sinh\theta$ are parametric equations for the hyperbola $\dfrac{x^2}{a^2} - \dfrac{y^2}{b^2} = 1$, or rather only that part for which $x > 0$ (if $a > 0$), since $\cosh\theta$ is always positive.

There is a close analogy with the circular functions, $\sin\theta$, $\cos\theta$, etc. For example,

$$\cos^2\theta + \sin^2\theta = 1 \qquad \cosh^2\theta - \sinh^2\theta = 1 \text{ (as shown above).}$$

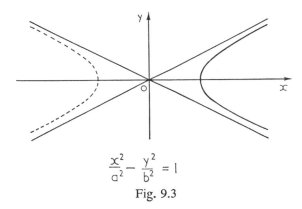

$$\frac{x^2}{a^2} - \frac{y^2}{b^2} = 1$$

Fig. 9.3

Corresponding to each of the 'usual' trigonometric formulae, for cos $(A + B)$, tan $2A$, etc. there is one for hyperbolic functions. These must be proved by going back to the definitions in terms of e^x, but can be remembered by *Osborn's rule*: to obtain a 'hyperbolic' formula from a 'circular' one, *change* cos *into* cosh *and* sin *into* sinh, *and change the sign for a product of two sines.* Thus, sin A sin B is changed to $-\sinh A \sinh B$, $\sin^2 \theta$ $(= \sin \theta . \sin \theta)$ becomes $-\sinh^2 \theta$ and $\tan^2 \theta$ $\left(= \dfrac{\sin^2 \theta}{\cos^2 \theta} \right)$ becomes $-\tanh^2 \theta$.

EXAMPLE 9.7   *To prove that* cosh $2x = \cosh^2 x + \sinh^2 x$.

We have
$$\cosh^2 x + \sinh^2 x = \left(\frac{e^x + e^{-x}}{2}\right)^2 + \left(\frac{e^x - e^{-x}}{2}\right)^2$$
$$= \tfrac{1}{4}(e^{2x} + 2 + e^{-2x} + e^{2x} - 2 + e^{-2x})$$
$$= \frac{e^{2x} + e^{-2x}}{2}$$
$$= \cosh 2x.$$

EXAMPLE 9.8   *If* $\theta = \ln \tan \left(\dfrac{\pi}{4} + \dfrac{\varphi}{2} \right)$, *show that* tanh $\theta = \sin \varphi$.

If $\theta = \ln \tan \left(\dfrac{\pi}{4} + \dfrac{\varphi}{2} \right)$,

$$e^\theta = \tan \left(\frac{\pi}{4} + \frac{\varphi}{2} \right) = \frac{1 + \tan \frac{\varphi}{2}}{1 - \tan \frac{\varphi}{2}} = \frac{\cos \frac{\varphi}{2} + \sin \frac{\varphi}{2}}{\cos \frac{\varphi}{2} - \sin \frac{\varphi}{2}}$$

so

$$\tanh\theta = \frac{e^\theta - e^{-\theta}}{e^\theta + e^{-\theta}}$$

$$= \left(\frac{\cos\frac{\varphi}{2} + \sin\frac{\varphi}{2}}{\cos\frac{\varphi}{2} - \sin\frac{\varphi}{2}} - \frac{\cos\frac{\varphi}{2} - \sin\frac{\varphi}{2}}{\cos\frac{\varphi}{2} + \sin\frac{\varphi}{2}}\right) \bigg/ \left(\frac{\cos\frac{\varphi}{2} + \sin\frac{\varphi}{2}}{\cos\frac{\varphi}{2} - \sin\frac{\varphi}{2}} + \frac{\cos\frac{\varphi}{2} - \sin\frac{\varphi}{2}}{\cos\frac{\varphi}{2} + \sin\frac{\varphi}{2}}\right)$$

$$= \frac{\left(\cos\frac{\varphi}{2} + \sin\frac{\varphi}{2}\right)^2 - \left(\cos\frac{\varphi}{2} - \sin\frac{\varphi}{2}\right)^2}{\left(\cos\frac{\varphi}{2} + \sin\frac{\varphi}{2}\right)^2 + \left(\cos\frac{\varphi}{2} - \sin\frac{\varphi}{2}\right)^2}$$

$$= \frac{4\sin\frac{\varphi}{2}\cos\frac{\varphi}{2}}{2\left(\cos^2\frac{\varphi}{2} + \sin^2\frac{\varphi}{2}\right)} = 2\sin\frac{\varphi}{2}\cos\frac{\varphi}{2} = \sin\varphi.$$

EXAMPLE 9.9   *To solve the equation* $3\sinh x - \cosh x = 1$.

We have

$$\frac{3(e^x - e^{-x})}{2} - \frac{e^x + e^{-x}}{2} = 1$$

or $2e^x - 4e^{-x} = 2, \quad e^x - 2e^{-x} = 1$

Writing $e^x = y, \quad y - \dfrac{2}{y} = 1$

giving $y^2 - y - 2 = 0; (y - 2)(y + 1) = 0$.

Thus $e^x = 2$ or -1. But e^x cannot be negative, so the only solution is given by $e^x = 2$, i.e. $x = \ln 2$.

[N.B. This method is quicker and more reliable than attempting an analogy with the 'usual methods' for $3\sin x - \cos x = 1$.]

EXERCISE XLIV

In questions 1 to 8, write down the corresponding formula for hyperbolic functions, and prove a selection of them.

1. $\sin(A - B) = \sin A \cos B - \cos A \sin B$.
2. $\sin 2\theta = 2\sin\theta\cos\theta$.
3. $\cos(A + B) = \cos A \cos B - \sin A \sin B$.
4. $\sec^2\varphi = 1 + \tan^2\varphi$.

5. $\cos X + \cos Y = 2 \cos \dfrac{X + Y}{2} \cos \dfrac{X - Y}{2}$.

6. $\cos X - \cos Y = -2 \sin \dfrac{X + Y}{2} \sin \dfrac{X - Y}{2}$.

7. $\sin 2\theta = \dfrac{2 \tan \theta}{1 + \tan^2 \theta}$. 8. $\cos A = \dfrac{1 - \tan^2 \dfrac{A}{2}}{1 + \tan^2 \dfrac{A}{2}}$.

9. Show that sinh x is an odd function and cosh x an even one.

10. Write down the Maclaurin expansions (giving the general term) for (i) cosh x; (ii) sinh x; (iii) sinh $3x$.

11. Show that, if $\theta = \ln \tan \varphi$, then $\tanh \theta = -\cos 2\varphi$.

12. Solve the equation 7 cosh $x - 5$ sinh $x = 5$.

$$\frac{d}{dx} (\cosh x) = \frac{d}{dx} \left(\frac{e^x + e^{-x}}{2} \right) = \frac{e^x - e^{-x}}{2} = \sinh x.$$

Similarly, $\dfrac{d}{dx} (\sinh x) = \cosh x.$

$$\frac{d}{dx} (\tanh x) = \frac{d}{dx} \left(\frac{\sinh x}{\cosh x} \right) = \frac{\cosh^2 x - \sinh^2 x}{\cosh^2 x} = \operatorname{sech}^2 x.$$

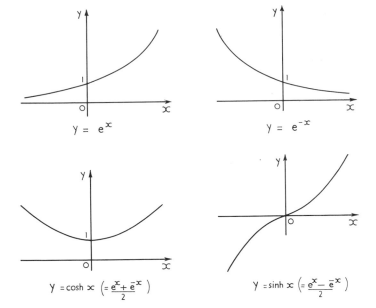

Fig. 9.4

For $y = \tanh x = \dfrac{e^x - e^{-x}}{e^x + e^{-x}}$ (Fig 9.5), we observe that (i) $|y| < 1$ for all x, (ii) $y \to 1$ as $x \to \infty$, (iii) $y \to -1$ as $x \to -\infty$, (iv) when $x = 0$, $y = 0$, (v) tanh x is an odd function, (vi) when $x > 0$, $y > 0$, (viii) $\dfrac{dy}{dx} = \operatorname{sech}^2 x, > 0.$

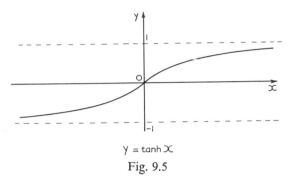

$y = \tanh x$

Fig. 9.5

Inverse Hyperbolic Functions

cosh x is a one-to-one function if $x \geqslant 0$ and we define $y = \cosh^{-1} x$ as the *positive* value of y such that $x = \cosh y$. Thus $\cosh^{-1} x$ is a function of x.

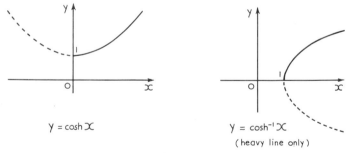

$y = \cosh x$ $y = \cosh^{-1} x$

(heavy line only)

Fig. 9.6

If $y = \cosh^{-1} x$,

$$x = \cosh y = \frac{e^y + e^{-y}}{2}.$$

To find an alternative expression for $\cosh^{-1} x$, we find y explicitly in terms of x from this relationship. We have

$$e^y + e^{-y} = 2x$$

or, writing $e^y = z$,

$$z + \frac{1}{z} = 2x,$$

whence

$$z^2 - 2xz + 1 = 0,$$

so that

$$e^v = z = \frac{2x \pm \sqrt{(4x^2 - 4)}}{2}$$

$$= x \pm \sqrt{(x^2 - 1)}$$

and so

$$\cosh^{-1} x = y = \ln (x \pm \sqrt{(x^2 - 1)}).$$

The right-hand side may be written as $\pm\ln (x + \sqrt{(x^2 - 1)})$, since $\{x + \sqrt{(x^2 - 1)}\}\{x - \sqrt{(x^2 - 1)}\} = 1$, remembering that $\ln \left(\frac{1}{a}\right) = -\ln a$. The positive value must be taken to conform with our definition. Thus

$$\cosh^{-1} x = \ln \{x + \sqrt{(x^2 - 1)}\}.$$

Similar investigation yields that $\sinh^{-1} x = \ln \{x \pm \sqrt{(x^2 + 1)}\}$; in this case, the positive sign must be taken for the square root since the logarithm of a negative number has no meaning for us. Thus

$$\sinh^{-1} x = \ln \{x + \sqrt{(x^2 + 1)}\};$$

as $\sinh x$ is one-to-one, $\sinh^{-1} x$ is defined for all real x.

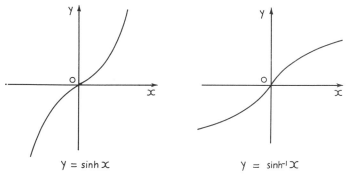

$y = \sinh x$ $y = \sinh^{-1} x$

Fig. 9.7

The reader should now investigate $\tanh^{-1} x$ for himself (cf. Exercise XLV, question 3).

EXAMPLE 9.10   (i) $\dfrac{d}{dx} (\sinh^{-1} x)$; (ii) $\dfrac{d}{dx} (\tanh^{-1} x)$.

(i) If $y = \sinh^{-1} x$, then $\sinh y = x$, so $\cosh y \dfrac{dy}{dx} = 1$. Thus

$$\frac{dy}{dx} = \frac{1}{\cosh y} = \frac{1}{\sqrt{(1 + \sinh^2 y)}} = \frac{1}{\sqrt{(1 + x^2)}} \quad \left(\frac{dy}{dx} > 0\right).$$

(ii) If $z = \tanh^{-1} x$, then $\tanh z = x$, so $\operatorname{sech}^2 z \, \dfrac{dz}{dx} = 1$. Thus

$$\frac{dz}{dx} = \frac{1}{\operatorname{sech}^2 z} = \frac{1}{1 - \tanh^2 z} = \frac{1}{1 - x^2}.$$

EXAMPLE 9.11   (i) $I = \displaystyle\int \frac{dx}{\sqrt{(x^2 - 1)}}$, (ii) $J = \displaystyle\int \frac{dx}{\sqrt{(1 + 3x^2)}}$,

(iii) $K = \int \sqrt{(1 + x^2)} \, dx$.

(i) Putting $x = \cosh \theta$, so that $dx = \sinh \theta \, d\theta$, we have

$$I = \int \frac{dx}{\sqrt{(x^2 - 1)}} = \int \frac{\sinh \theta \, d\theta}{\sqrt{(\cosh^2 \theta - 1)}} = \int \frac{\sinh \theta}{\sinh \theta} \, d\theta = \theta + c =$$

$\cosh^{-1} x + c$.*

(ii) Putting $x = \dfrac{1}{\sqrt{3}} \sinh \theta$ [so that $3x^2 = \sinh^2 \theta$ and $1 + 3x^2 = \cosh^2 \theta$],

$dx = \dfrac{1}{\sqrt{3}} \cosh \theta \, d\theta$, we have

$$J = \int \frac{\dfrac{1}{\sqrt{3}} \cosh \theta}{\cosh \theta} \, d\theta = \frac{1}{\sqrt{3}} \theta + c = \frac{1}{\sqrt{3}} \sinh^{-1} (\sqrt{3} \, x) + c.$$

(iii) Putting $x = \sinh \varphi$ [so that $\sqrt{(1 + x^2)} = \cosh \varphi$], $dx = \cosh \varphi \, d\varphi$, we have $K = \int \cosh^2 \varphi \, d\varphi = \frac{1}{2}\int (1 + \cosh 2\varphi) \, d\varphi = \frac{1}{2}(\varphi + \frac{1}{2} \sinh 2\varphi)$

$$= \tfrac{1}{2}(\varphi + \sinh \varphi \cosh \varphi) = \tfrac{1}{2}\{\sinh^{-1} x + x\sqrt{(1 + x^2)}\}.$$

EXERCISE XLV

1. Sketch the graphs of $\operatorname{sech} x$, $\operatorname{cosech} x$, $\coth x$.
2. Differentiate with respect to x (i) $\sinh 3x$, (ii) $\tanh 4x$, (iii) $\operatorname{sech} x$, (iv) $\sinh^{-1} 2x$, (v) $\cosh^{-1} (4x + 3)$.
3. Show that $\tanh^{-1} x = \frac{1}{2} \ln \left(\dfrac{1 + x}{1 - x} \right)$, $(|x| < 1)$.

In questions 4 to 12, integrate the given function with respect to x.

4. $\dfrac{1}{\sqrt{(x^2 - 1)}}$.

5. $\dfrac{1}{\sqrt{(3x^2 - 2)}}$.

6. $\dfrac{1}{\sqrt{(1 + 3x^2)}}$.

7. $\dfrac{1}{\sqrt{(4 + 25x^2)}}$.

8. $\sqrt{(1 + 4x^2)}$.

---

* Strictly speaking, $I = \cosh^{-1} |x| + c$, but we shall often suppress such modulus signs from now on, assuming that x is within the appropriate range.

9. $\dfrac{1}{\sqrt{(x^2 + 2x + 2)}}$. [Write $x^2 + 2x + 2 = (x - 1)^2 + 1$.]

10. $\dfrac{1}{\sqrt{(x^2 + 4x + 5)}}$.   11. $\dfrac{1}{\sqrt{(x^2 - 6x + 7)}}$.

12. $\sinh^{-1} x$. [Write as $1 \cdot \sinh^{-1} x$ and use parts.]

EXAMPLE 9.12   *The equation for radioactive decay of radium is* $\dfrac{dm}{dt} = -km$

(p. 143), m being measured in grammes and t in years. To calculate the value of k, the 'half-life' of radium (that is, the time taken for a given amount m to be reduced by one-half) being 1500 years.

Method 1. Since $\dfrac{dm}{dt} = -km$, $m = Ae^{-kt}$ (cf. p. 146). Let $m = m_0$ at t_0 and $m = \dfrac{m_0}{2}$ at $t_0 + 1500$.

Then

$$m_0 = Ae^{-kt_0}$$
$$\frac{m_0}{2} = Ae^{-k(t_0+1500)}$$

By division,

$$2 = e^{1500k},$$

whence $1500k = \ln 2$, $k = \dfrac{\ln 2}{1500}$.

Method 2. $\dfrac{dm}{dt} = -km$ can be written $\dfrac{dt}{dm} = -\dfrac{1}{km}$, so

$$[t] = -\frac{1}{k} \int_{m_0}^{\frac{1}{2}m_0} \frac{1}{m}\, dm = -\frac{1}{k}\left[\ln m \right]_{m_0}^{\frac{1}{2}m_0}$$

Thus $1500 = -\dfrac{1}{k} (\ln \tfrac{1}{2}m_0 - \ln m_0) = -\dfrac{1}{k} \ln \tfrac{1}{2} = \dfrac{1}{k} \ln 2$,

and so $k = \dfrac{\ln 2}{1500}$.

EXERCISE XLVI

Miscellaneous questions (in sets of 3)

1. Differentiate with respect to x (i) 3^x; (ii) $\ln \tan \dfrac{x}{2}$.

2. Integrate with respect to x (i) $\dfrac{1}{(x - 1)(x + 3)}$, (ii) $\dfrac{x}{1 + x^2}$, (iii) $\dfrac{1}{\sqrt{(1 + 4x^2)}}$.

3. In a certain chemical reaction the rate of decomposition of a substance at any time t is proportional to the amount m that remains. In any small change, the

percentage increase in the pressure p inside the vessel in which the reaction takes place is proportional to the percentage decrease in the amount m of the substance.

Express the above statements in calculus notation, using a and b as the respective constants of proportionality.

If m and p have the values m_0, p_0 when $t = 0$, find

(i) m in terms of m_0, a and t,

(ii) p in terms of p_0, m_0, b and m,

and hence obtain p in terms of p_0, a, b and t.

$$\left[\text{Hint for (ii): } \lim_{\delta p \to 0} \Sigma \frac{1}{p} \, \delta p = -b \lim_{\delta m \to 0} \Sigma \frac{1}{m} \, \delta m. \right]$$

4. (i) Prove the identity

$$1 + 2x + 2x^2 + x^3 = \frac{(1 + x)(1 - x^3)}{1 - x}$$

and hence expand

$$\ln (1 + 2x + 2x^2 + x^3)$$

in ascending powers of x as far as the term in x^6, stating the necessary restrictions on the values of x.

(ii) Write down the series for e^x and e^{-x} in ascending powers of x.

Prove that, if x^4 and higher powers of x are neglected, then

$$\frac{x}{1 - e^{-2x}} - \frac{1}{2} = \tfrac{1}{2}x + \tfrac{1}{6}x^2.$$

5. Prove that $\cosh 2x = 2 \cosh^2 x - 1$.

If

$$\theta = \ln \tan \left(\frac{3\pi}{8} \right),$$

prove that $3 \tanh 2\theta = 2\sqrt{2}$.

6. A particle moving in a straight line has velocity v when it is at distance s from a fixed point in the line. Prove that its acceleration is then

$$\frac{d}{ds} \left(\frac{1}{2} v^2 \right)$$

A particle is projected vertically upwards with initial velocity u in a resisting medium, the equation of motion while the particle is moving upwards being

$$\frac{d}{ds} \left(\frac{1}{2} v^2 \right) = -g - kv^2$$

where k is a constant. Prove that the particle reaches a height

$$\frac{1}{2k} \ln \left(1 + \frac{ku^2}{g} \right)$$

above the point of projection. [Write $v^2 = y$.]

7. Express

$$\frac{x + 3}{(x + 2)(x + 1)^2}$$

as the sum of three partial fractions.

Obtain the value, correct to three significant figures, of

$$\int_0^1 \frac{(x + 3)\,dx}{(x + 2)(x + 1)^2}$$

8. Show that the function $y = e^{px}$, where p is a positive constant, increases as x increases and sketch roughly the graph of the function.

   Find the equation of the tangent to this curve which passes through the origin. Deduce that the equation

   $$e^{px} = x$$

   has 0, 1 or 2 real roots according as ep is greater than, equal to, or less than unity.

9. A flask having an inlet at the top and a tap at the base holds initially one litre of a uniform salt solution containing m grammes of salt. Another such solution containing n grammes per litre ($n > m$) begins at time $t = 0$ to enter through the inlet and flows in subsequently at the constant rate of U litres per minute; at the same instant the tap is opened and the mixture flows out at the same rate, so that the flask always contains one litre of solution which may be assumed to be thoroughly mixed. If x grammes per litre is the concentration of salt in the flask after t minutes, show that in the short interval δt minutes a mass $xU\,\delta t$ grammes (approximately) of salt leaves the flask, and state the corresponding amount which enters.

   Hence derive an equation expressing dx/dt in terms of x and the constants n and U.

   Find by integration an expression for x in terms of t, U, m and n.

10. Define $\cosh x$ and $\sinh x$ in terms of exponential functions and prove that $\cosh 2x = 1 + 2\sinh^2 x$. Find the values of $\cosh x$ which satisfy the equation

    $$10\cosh x + 2\sinh x = 11.$$

11. Draw the graph of $y = e^{-x}\sin x$ for $x > 0$.

    Obtain the coordinates of all maxima and minima in this range, and prove that the values of y at successive maxima form a geometric progression.

    Obtain also the coordinates of all points of inflexion.

12. A population of insects is placed in an experimental environment and allowed to grow for several days. Its net time-rate of increase is proportional to its size, x insects, at any time t days, and its increases during the fourth day and the fifth day are estimated (from counts of newly hatched and dead insects) as 3566 and 5143 insects respectively.

    By constructing and solving a suitable differential equation, determine the initial size and initial time-rate of increase of the population.

10
Systematic Integration

Our little systems have their day.

TENNYSON

We return now to methods of integration, and shall revise and extend the subject-matter of Chapter 7. Our basic armoury consists of inspection, substitution, 'parts' (Chapter 7) and partial fractions (Chapter 9); but there are a number of subsidiary devices, which will be illustrated below by examples. The reader should make his own table of standard integrals. To those on p. 119 may be added

$f(x)$	$\int f(x)\, dx$
e^{ax}	$\dfrac{1}{a}\, e^{ax}$
$\dfrac{1}{ax + b}$	$\dfrac{1}{a} \ln |ax + b|$
$\dfrac{g'(x)}{g(x)}$	$\ln \{g(x)\}$
$\tan x$	$\ln |\sec x|$
$\sinh x$	$\cosh x$
$\cosh x$	$\sinh x$
$\dfrac{1}{\sqrt{(1 + x^2)}}$	$\sinh^{-1} x$
$\dfrac{1}{\sqrt{(x^2 - 1)}}$	$\cosh^{-1} x$

We shall suppress the arbitrary constant throughout this chapter, but it should be inserted automatically in written work.

EXAMPLE 10.1

$$I = \int x^3 e^{-x^2}\, dx$$

174

Put $x^2 = y$, so that $2x\,dx = dy$ and

$$I = \int x^2 e^{-x^2} \cdot x\,dx = \frac{1}{2}\int ye^{-y}\,dy$$

$$= \frac{1}{2}\left\{-e^{-y}y + \int e^{-y}\,dy\right\} \text{ (by parts)}$$

$$= -\tfrac{1}{2}ye^{-y} - \tfrac{1}{2}e^{-y}$$

$$= -\tfrac{1}{2}e^{-x^2}(x^2 + 1).$$

EXAMPLE 10.2
$$I = \int \frac{3x + 1}{(4x - 3)^{\frac{3}{2}}}\,dx$$

Put $4x - 3 = y^2$, $4\,dx = 2y\,dy$, so that $3x + 1 = \tfrac{3}{4}(y^2 + 3) + 1 = \tfrac{1}{4}(3y^2 + 13)$ and

$$I = \frac{1}{4}\int \frac{3y^2 + 13}{y^3} \cdot \frac{1}{2}y\,dy$$

$$= \frac{1}{8}\int \left(3 + \frac{13}{y^2}\right) dy$$

$$= \frac{3y}{8} - \frac{13}{8y}$$

$$= \tfrac{3}{8}(4x - 3)^{\frac{1}{2}} - \tfrac{13}{8}(4x - 3)^{-\frac{1}{2}}.$$

EXAMPLE 10.3 $I = \int \ln x\,dx.$

By parts (writing the integrand as $1 \cdot \ln x$),

$$I = x \ln x - \int x \cdot \frac{1}{x}\,dx = x \ln x - x.$$

EXAMPLE 10.4
$$\text{(i) } I = \int \frac{x^2}{x^3 + 2}\,dx, \quad \text{(ii) } J = \int \frac{x^2}{(x^3 + 2)^{\frac{3}{2}}}\,dx.$$

Both can be done 'by inspection'.

(i) $I = \tfrac{1}{3}\ln(x^3 + 2)$; (ii) $J = -\tfrac{2}{9}(x^3 + 2)^{-\frac{3}{2}}$

EXAMPLE 10.5
$$I = \int \frac{x^3 + 4}{(x - 1)(x + 2)}\,dx$$

$$\frac{x^3 + 4}{(x - 1)(x + 2)} = x - 1 + \frac{3x + 2}{(x - 1)(x + 2)} \quad \text{(by division)}$$

$$= x - 1 + \frac{\frac{5}{3}}{x - 1} + \frac{\frac{4}{3}}{x + 2}$$

(using partial fractions).

Thus $I = \tfrac{1}{2}x^2 - x + \tfrac{5}{3}\ln|x - 1| + \tfrac{4}{3}\ln|x + 2|.$

Example 10.6

$$I = \int \frac{2x + 7}{x^2 + x + 1} \, dx$$

If only the '7' had been a '1', we could have written the answer, by inspection, as a logarithm. We accordingly write

$$I = \int \left\{ \frac{2x + 1}{x^2 + x + 1} + \frac{6}{x^2 + x + 1} \right\} dx$$

$$= \ln (x^2 + x + 1) + 6 \int \frac{dx}{x^2 + x + 1} \, .$$

Now

$$J = \int \frac{dx}{x^2 + x + 1} = \int \frac{dx}{(x + \frac{1}{2})^2 + \left(\frac{\sqrt{3}}{2}\right)^2}$$

and this is of the 'tan^{-1}' type. By trial and error $\left(\text{or the substitution } x + \frac{1}{2} = \frac{\sqrt{3}}{2} \tan \theta \right)$,

$$J = \frac{2}{\sqrt{3}} \tan^{-1} \left(\frac{x + \frac{1}{2}}{\frac{\sqrt{3}}{2}} \right) = \frac{2\sqrt{3}}{3} \tan^{-1} \left(\frac{2x + 1}{\sqrt{3}} \right),$$

and so

$$I = \ln (x^2 + x + 1) + 4\sqrt{3} \tan^{-1} \left(\frac{2x + 1}{\sqrt{3}} \right)$$

Exercise XLVII

Integrate the given function with respect to x.

1. $(3x + 4)^{-7}$.

2. $(2x - 1)(5x - 2)^4$.

3. $\sin^{-1} x$.

4. $x \sin x$.

5. $\dfrac{1}{x^2 - 1}$.

6. $\dfrac{2x}{x^2 - 1}$

7. xe^{x^2}.

8. $(x + 5)^{-\frac{1}{2}}$.

9. $\dfrac{1}{4x + 1}$.

10. $\dfrac{7x + 2}{(x + 3)^3}$.

11. $e^x (x^2 + 4x + 1)$.

12. $\dfrac{x}{x + 1}$.

13. $\dfrac{1}{(1 - x)(2 - x)}$.

14. $\dfrac{x}{\sqrt{(x+1)}}$.

15. $\dfrac{1}{x^2 + x + 1}$.

16. $\dfrac{1}{x^2 + 5x + 6}$.

17. $\dfrac{x+1}{x^2 + 2x + 7}$.

18. $\dfrac{3x+1}{x^2 + 3x + 4}$.

19. $\dfrac{x^3}{x^2 - 1}$.

20. $\dfrac{x^3}{x^2 + 1}$.

21. $\dfrac{1}{(x-1)(x^2+1)}$.

22. $(7x)^{\frac{1}{2}}$.

23. $\dfrac{1}{1 - x^4}$.

24. $\dfrac{x}{1 - x^3}$.

25. $\dfrac{\sqrt{(x-1)}}{x}$.

For integrals involving $\sqrt{(ax + b)}$, the usual substitution is $ax + b = y^2$ (or y). When $\sqrt{(px^2 + qx + r)}$ occurs, the procedure is to complete the square and see whether the result is of the form (i) $\sqrt{(a^2 - y^2)}$, (ii) $\sqrt{(y^2 - a^2)}$, or (iii) $\sqrt{(a^2 + y^2)}$. The next move is then generally a substitution, for (i) $y = a \sin \theta$; for (ii) $y = a \cosh \theta$; for (iii) $y = a \sinh \theta$, or possibly $y = a \tan \theta$.

EXAMPLE 10.7

$$I = \int \frac{x}{\sqrt{(4 - x^2)}} \, dx$$

By inspection, $I = -(4 - x^2)^{\frac{1}{2}}$. (This should be checked by differentiating the answer, as the numerical factor can easily go wrong.)

EXAMPLE 10.8

$$I = \int_3^5 \frac{3x + 1}{\sqrt{(7 + 6x - x^2)}} \, dx$$

$\dfrac{d}{dx}(7 + 6x - x^2) = 6 - 2x$, and we first write $3x + 1 = -\frac{3}{2}(6 - 2x) + 10$. Thus

$$I = \int_3^5 \frac{-\frac{3}{2}(6 - 2x)}{\sqrt{(7 + 6x - x^2)}} \, dx + \int_3^5 \frac{10}{\sqrt{(7 + 6x - x^2)}} \, dx$$

The first of these integrals can be written down by inspection as

$$-3[(7 + 6x - x^2)^{\frac{1}{2}}]_3^5 = -3(\sqrt{12} - 4) = 12 - 6\sqrt{3}.$$

For the second, writing

$$J = \int_3^5 \frac{dx}{\sqrt{(7 + 6x - x^2)}} = \int_3^5 \frac{dx}{\sqrt{(4^2 - (x - 3)^2)}},$$

and putting $x - 3 = 4 \sin \theta$, $dx = 4 \cos \theta \, d\theta$, we have

$$J = \int_0^{\frac{\pi}{6}} \frac{4 \cos \theta \, d\theta}{4 \cos \theta} = \left[\theta \right]_0^{\frac{\pi}{6}} = \frac{\pi}{6}.$$

Thus $I = 12 - 6\sqrt{3} + 10J = 12 - 6\sqrt{3} + \frac{5}{3}\pi, \approx 6\cdot8(4)$.

EXAMPLE 10.9

$$I = \int \frac{1}{\sqrt{(x^2 + 4x + 13)}} \, dx$$

$$I = \int \frac{dx}{\sqrt{\{(x + 2)^2 + 3^2\}}}$$

Put $x + 2 = 3 \sinh \theta$, $dx = 3 \cosh \theta \, d\theta$

$$I = \int \frac{3 \cosh \theta \, d\theta}{3 \cosh \theta} = \theta = \sinh^{-1}\left(\frac{x + 2}{3} \right).$$

EXAMPLE 10.10

$$I = \int \frac{dx}{x\sqrt{(x^2 + 2x - 4)}}$$

Put $x = \dfrac{1}{y}$, $dx = -\dfrac{1}{y^2} \, dy$.

Then

$$I = -\int \frac{dy}{y^2 \cdot \frac{1}{y} \sqrt{\left(\frac{1}{y^2} + \frac{2}{y} - 4 \right)}} = -\int \frac{dy}{\sqrt{(1 + 2y - 4y^2)}}$$

$$= -\frac{1}{2} \int \frac{dy}{\sqrt{\{\frac{1}{4} + \frac{1}{2}y - y^2\}}} = -\frac{1}{2} \int \frac{dy}{\sqrt{\left\{ \left(\frac{\sqrt{5}}{4} \right)^2 - (y - \frac{1}{4})^2 \right\}}}$$

$$= -\frac{1}{2} \sin^{-1}\left(\frac{y - \frac{1}{4}}{\frac{\sqrt{5}}{4}} \right) = -\frac{1}{2} \sin^{-1}\left(\frac{4 - x}{\sqrt{5}x} \right) = \frac{1}{2} \sin^{-1}\left(\frac{x - 4}{\sqrt{5}x} \right).$$

The substitution $x - a = \dfrac{1}{y}$ is effective for integrals of the type

$$\int \frac{dx}{(x - a)^n \sqrt{(px^2 + qx + r)}},$$

(n a positive integer).

Exercise XLVIII

Integrate the following functions with respect to x.

1. $\dfrac{x}{\sqrt{(x^2 - 1)}}$.

2. $\dfrac{1}{\sqrt{(1 - 2x^2)}}$.

3. $\dfrac{x + 1}{\sqrt{(1 - x^2)}}$.

4. $\dfrac{1}{\sqrt{(x^2 + 4x + 1)}}$.

5. $\dfrac{1}{\sqrt{(1 + 4x - x^2)}}$.

6. $\dfrac{1}{\sqrt{(2 - 3x - 7x^2)}}$.

7. $\dfrac{1}{x\sqrt{(x^2 - 1)}}$.

8. $\dfrac{x^3}{\sqrt{(x^4 + 1)}}$.

9. $\ln x$.

10. $\sqrt{(3x + 4)}$.

11. $x\sqrt{(x^2 + 1)}$.

12. $x^2\sqrt{(x + 1)}$.

13. $\dfrac{1}{(x - 2)^2\sqrt{(5 + 4x - x^2)}}$.

14. $\dfrac{x^3}{1 - x^4}$.

15. $\dfrac{3x + 2}{(3x^2 + 4x - 1)^{\frac{1}{2}}}$.

EXAMPLE 10.11 (i) $I = \int \sin 5\theta \cos 2\theta \, d\theta$, (ii) $J = \int \sin^5 \theta \cos^2 \theta \, d\theta$.

(i) $\int \sin 5\theta \cos 2\theta \, d\theta = \frac{1}{2}\int (\sin 7\theta + \sin 3\theta) \, d\theta$

$$= -\tfrac{1}{14} \cos 7\theta - \tfrac{1}{6} \cos 3\theta.$$

(ii) Put $\cos \theta = c$, then $-\sin \theta \, d\theta = dc$ and

$$J = \int \sin^4 \theta \cos^2 \theta \sin \theta \, d\theta$$

$$= \int (1 - \cos^2 \theta)^2 \cos^2 \theta \sin \theta \, d\theta = -\int (1 - c^2)^2 c^2 \, dc$$

$$= -\int (c^2 - 2c^4 + c^6) \, dc = -\frac{\cos^3 \theta}{3} + \frac{2 \cos^5 \theta}{5} - \frac{\cos^7 \theta}{7}.$$

Any integral of the form $\int \sin^p \theta \cos^q \theta \, d\theta$ can be done by a similar method if at least one of p, q is odd (if q is odd, put $\sin \theta = s$).

EXAMPLE 10.12 $I = \int \sin^4 \theta \cos^2 \theta \, d\theta.$

We use the identities $\begin{cases} \sin 2\theta = 2 \sin \theta \cos \theta \\ 1 - \cos 2\theta = 2 \sin^2 \theta \end{cases}$

$$I = \frac{1}{8} \int \sin^2 2\theta (1 - \cos 2\theta) \, d\theta$$

$$= \frac{1}{16} \int (1 - \cos 4\theta) \, d\theta - \frac{1}{8} \int \sin^2 2\theta \cos 2\theta \, d\theta$$

$$= \tfrac{1}{16}\theta - \tfrac{1}{64} \sin 4\theta - \tfrac{1}{48} \sin^3 2\theta$$

$$= \tfrac{1}{16}\{\theta - \tfrac{1}{4} \sin 4\theta - \tfrac{1}{3} \sin^3 2\theta\}.$$

EXAMPLE 10.13

$$I = \int \frac{\cos x}{1 + \sin^2 x} \, dx$$

Put $\sin x = s$, then $\cos x \, dx = ds$.

$$I = \int \frac{ds}{1 + s^2} = \tan^{-1} s = \tan^{-1} (\sin x).$$

EXAMPLE 10.14

$$I = \int \frac{d\theta}{\cos^2 \theta + 3 \sin^2 \theta}$$

$$I = \int \frac{\sec^2 \theta \, d\theta}{1 + 3 \tan^2 \theta}$$

Put $\tan \theta = t$, then $\sec^2 \theta \, d\theta = dt$,

$$I = \int \frac{dt}{1 + 3t^2} = \frac{1}{\sqrt{3}} \tan^{-1} (\sqrt{3}t) = \frac{1}{\sqrt{3}} \tan^{-1} (\sqrt{3} \tan \theta).$$

EXAMPLE 10.15 $I = \int \sec \theta \, d\theta.$

This can be done outrageously easily by knowing the answer.

$$I = \int \sec \theta \, d\theta = \int \frac{\sec \theta (\sec \theta + \tan \theta)}{\sec \theta + \tan \theta} \, d\theta$$

$$= \int \frac{\sec^2 \theta + \sec \theta \tan \theta}{\sec \theta + \tan \theta} \, d\theta$$

Now it happens that $\dfrac{d}{d\theta} (\sec \theta + \tan \theta) = \sec \theta \tan \theta + \sec^2 \theta$, and so

$$I = \ln |\sec \theta + \tan \theta|. \tag{1}$$

This integral can also be 'played straight' with the substitution $\tan \dfrac{\theta}{2} = t$, which is a stand-by for trigonometric integrals when all else fails. Put

$$\tan \frac{\theta}{2} = t, \qquad \tfrac{1}{2} \sec^2 \frac{\theta}{2}\, d\theta = dt$$

so

$$d\theta = \frac{2\, dt}{\sec^2 \dfrac{\theta}{2}} = \frac{2\, dt}{1 + t^2}$$

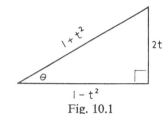

Fig. 10.1

Thus

$$I = \int \sec \theta\, d\theta = \int \frac{1 + t^2}{1 - t^2} \cdot \frac{2}{1 + t^2}\, dt$$

$$= \int \frac{2\, dt}{1 - t^2} = \int \left(\frac{1}{1 - t} + \frac{1}{1 + t} \right) dt$$

$$= \ln \left| \frac{1 + \tan \dfrac{\theta}{2}}{1 - \tan \dfrac{\theta}{2}} \right| \tag{2}$$

Now

$$\sec \theta + \tan \theta = \frac{1 + t^2}{1 - t^2} + \frac{2t}{1 - t^2} = \frac{(1 + t)^2}{1 - t^2}$$

$$= \frac{1 + \tan \dfrac{\theta}{2}}{1 - \tan \dfrac{\theta}{2}} \qquad \left(= \tan \left\{ \frac{\theta}{2} + \frac{\pi}{4} \right\} \right)$$

and so the answers (1) and (2) are reconciled.

EXAMPLE 10.16

$$I = \int \frac{d\theta}{3 \cos \theta + 4 \sin \theta}$$

$$I = \int \frac{d\theta}{5(\tfrac{3}{5} \cos \theta + \tfrac{4}{5} \sin \theta)}$$

$$= \frac{1}{5} \int \frac{d\theta}{\cos (\theta - \alpha)} \qquad \text{where } \tan \alpha = \frac{4}{3},$$

$$= \frac{1}{5} \int \sec (\theta - \alpha)\, d\theta$$

$$= \tfrac{1}{5} \ln |\sec (\theta - \alpha) + \tan (\theta - \alpha)|.$$

EXAMPLE 10.17

$$I = \int \frac{d\theta}{2 + 3 \cos \theta}$$

Put $\tan \dfrac{\theta}{2} = t$, so that $d\theta = \dfrac{2\,dt}{1 + t^2}$ as in Example 10.15. Then

$$I = \int \frac{2\,dt}{(1 + t^2)\left\{2 + \dfrac{3(1 - t^2)}{1 + t^2}\right\}}$$

$$= \int \frac{2\,dt}{2(1 + t^2) + 3(1 - t^2)} = \int \frac{2\,dt}{5 - t^2}$$

$$= \frac{1}{\sqrt{5}} \int \left(\frac{1}{\sqrt{5} - t} + \frac{1}{\sqrt{5} + t}\right) dt$$

$$= \frac{1}{\sqrt{5}} \ln \left| \frac{\sqrt{5} + \tan \dfrac{\theta}{2}}{\sqrt{5} - \tan \dfrac{\theta}{2}} \right|.$$

EXAMPLE 10.18

$$I = \int \frac{2 \sin \theta + 3 \cos \theta}{7 \sin \theta - 5 \cos \theta}\, d\theta$$

We write

$$2 \sin \theta + 3 \cos \theta = a(7 \sin \theta - 5 \cos \theta) + b(7 \cos \theta + 5 \sin \theta),$$

i.e. as the sum of multiples of $7 \sin \theta - 5 \cos \theta$ and its derivative. Comparing coefficients of $\sin \theta$, $\cos \theta$, we have

$$\begin{cases} 2 = 7a + 5b \\ 3 = -5a + 7b \end{cases} \quad \text{whence } a = -\tfrac{1}{74},\ b = \tfrac{31}{74}.$$

Thus

$$I = \int \left\{ -\frac{1}{74}\left(\frac{7 \sin \theta - 5 \cos \theta}{7 \sin \theta - 5 \cos \theta}\right) + \frac{31}{74}\left(\frac{7 \cos \theta + 5 \sin \theta}{7 \sin \theta - 5 \cos \theta}\right) \right\} d\theta$$

$$= -\tfrac{1}{74}\theta + \tfrac{31}{74} \ln |7 \sin \theta - 5 \cos \theta|.$$

Exercise XLIX

Integrate the given functions with respect to x.

1. $\cot x$.

2. $\sec^2 x$.

3. $\csc^2 x$.

4. $\cot^2 x$. $[1 + \cot^2 \theta = \csc^2 \theta]$

5. $\sin 3x \cos 5x$.

6. $\sin^2 7x$.

7. $\sin 3x \sin 4x$.

8. $\cos^5 x \sin x$.

9. $\cos^8 x \sin^3 x$.

10. $\sin^2 x \cos^2 x$.

11. $\dfrac{\sin x}{1 + \cos^2 x}$.

12. $\dfrac{\sec^2 x}{1 + \tan^2 x}$.

13. $\dfrac{1}{2\cos^2 x + \sin^2 x}$.

14. $\dfrac{1}{1 + \cos^2 x}$.

15. $\csc x$.

16. $\dfrac{1}{5\sin x + 12\cos x}$.

17. $\dfrac{1}{\sin x + \cos x}$.

18. $\dfrac{1}{3 + 4\sin x}$.

19. $\dfrac{\sin x}{2\sin x + 3\cos x}$.

20. $\dfrac{x^2}{\sqrt{(1 - x^2)}}$. $[x = \sin \theta]$

21. $\cos x \csc^4 x$.

22. $\sqrt{\left(\dfrac{1 + x}{1 - x}\right)}$. $[x = \cos 2\theta]$

23. $\dfrac{1}{(1 + x^2)^{\frac{3}{2}}}$. $[x = \tan \theta]$

24. $\sec^3 x \tan x$. $[\sec x = y]$

25. $\tan^5 x \sec^4 x$.

26. $\sec^3 x$. [Write as $\sec^2 x$. $\sec x$; parts]

Exercise L

'The Daily Integral'

Occasionally, for a week or so at a time, there should be an 'integral drive', one or two of the following being done daily, to keep in practice.

In questions 1 to 50, integrate the given function with respect to x.

1. $\sin 3x \cos x$.

2. $\dfrac{x}{\sqrt{(1 - x^2)}}$.

3. $x \sin x$.

4. $x^{-\frac{1}{2}}$.

5. $\ln (3 + x)$.

6. $\sec 4x$.

7. $\dfrac{x^2}{\sqrt{(x+1)}}$.

8. $\dfrac{x^3}{x^2-4}$.

9. $(1-\cos x)^2 \sin x$.

10. $(1-\cos x)^2 \cos x$.

11. $\dfrac{1}{x^2+3x+7}$.

12. $\dfrac{1}{\sqrt{(x^2+3x+7)}}$.

13. $\dfrac{x}{(x-1)(x+3)}$.

14. $\dfrac{x+5}{x^2+1}$.

15. $\dfrac{1}{3-\sin x}$.

16. $\dfrac{1}{x\sqrt{(x^2-9)}}$.

17. $\dfrac{x^2-1}{x^2+1}$.

18. $\dfrac{1+e^x}{1-e^x}$.

19. $e^{2x}\cos 3x$.

20. $\dfrac{x^2+1}{x(x+2)}$.

21. $\tan^{-1} 2x$.

22. $x(1+x^2)$.

23. $x(1+x^2)^{\frac{1}{2}}$.

24. $\dfrac{1}{3\cos^2 x + 4\sin^2 x}$.

25. $x\cos^2 x$.

26. $\cos^5 x$.

27. $\dfrac{1}{\sqrt{(4-9x^2)}}$.

28. $\dfrac{1}{x^2+x+1}$.

29. $\cos^{-1} x$.

30. $\sin^5 x \sec^2 x$.

31. $\dfrac{x^5}{x-1}$.

32. $\dfrac{x^5}{\sqrt{(x-1)}}$.

33. $\dfrac{1}{\sqrt{(4-5x-x^2)}}$.

34. $\dfrac{x^5}{(x-1)(x+3)}$.

35. $\dfrac{1}{\sqrt{\{(x-2)(x+1)\}}}$.

36. $\dfrac{1}{(x-1)^2\sqrt{(x^2-4x+1)}}$.

37. $\sec^4 x$.

38. $x^2 \sin x$.

39. $\sin^4 x$.

40. $\dfrac{1}{2+\cos x}$.

41. $\sqrt{\left(\dfrac{x}{1-x}\right)}$.

42. $\dfrac{\sin x}{1+\sin x}$.

43. $x\sin^{-1} x$.

44. $(x+5)(x^2+10x+7)^{-4}$.

45. $\sin^{17} x \cos x$.

46. $x^3 \ln x$.

47. $\dfrac{1}{x^3\sqrt{(x^2-1)}}$.

48. $\sqrt{(1-x^2)}$.

49. $\dfrac{1}{\sin x + \cos x + 1}$.

50. $\dfrac{1}{1 + x^4}$. [Multiply $(1 - \sqrt{2x} + x^2)$ by $(1 + \sqrt{2x} + x^2)$]

51. $\displaystyle\int_1^e \ln x \, dx.$

56. $\displaystyle\int_0^{\frac{\pi}{2}} \sin x \, dx.$

52. $\displaystyle\int_0^{\frac{\pi}{2}} \sin^4 x \cos x \, dx.$

57. $\displaystyle\int_0^{\frac{\pi}{2}} x \sin x \, dx.$

53. $\displaystyle\int_7^8 \dfrac{dx}{(x - 3)(x - 5)}.$

58. $\displaystyle\int_0^{\frac{1}{2}} \sin^{-1} x \, dx.$

54. $\displaystyle\int_3^5 \dfrac{1}{(x - 7)(x - 8)} \, dx.$

59. $\displaystyle\int_0^2 \dfrac{dx}{\sqrt{(1 + x^2)}}$ (to 3 decimal places).

55. $\displaystyle\int_0^{\frac{\pi}{6}} \dfrac{\cos x}{1 + \sin x} \, dx.$

60. $\displaystyle\int_{\frac{\pi}{4}}^{\frac{\pi}{3}} \operatorname{cosec}^2 x \, dx.$

EXERCISE LI

Examination questions

1. Evaluate the definite integrals

(i) $\displaystyle\int_0^\pi \sin^2 2x \, dx;$ (ii) $\displaystyle\int_0^{\frac{\pi}{4}} \tan^2 x \, dx;$ (iii) $\displaystyle\int_{-1}^1 (e^x + 2e^{-x}) \, dx$

2. Integrate with respect to x:

(i) $\dfrac{4x - 1}{2x^2 - x - 3}$,   (ii) $\cos^3 x \sin^2 x.$

Evaluate $\displaystyle\int_0^{\frac{\pi}{2}} x^2 \sin x \, dx.$

3. Find

(i) $\displaystyle\int \sin^3 x \cos^3 x \, dx,$

(ii) $\displaystyle\int \dfrac{dx}{(2x + 3)(x + 2)},$

(iii) $\displaystyle\int_{\frac{\pi}{4}}^{\frac{\pi}{2}} \cot^2 x \, dx.$

4. Integrate the following functions with respect to x:

(i) $\dfrac{x}{(1 - x^2)^2}$;   (ii) $\tan x \sec^2 x;$ (iii) $\dfrac{3(1 - x)}{(1 - 2x)(1 + x)}.$

Evaluate $\displaystyle\int_0^1 xe^{-x} \, dx$

to three significant figures, given that

$$e \approx 2\cdot718.$$

5. (i) Find the integrals
$$\int \frac{dx}{x^2 + 2x - 3}, \quad \int \frac{2x + 1}{x^2 + 2x + 2}\, dx, \quad \int \frac{\cos x}{1 + \sin^2 x}\, dx.$$

(ii) Evaluate
$$\int_0^{\frac{\pi}{4}} \frac{d\theta}{\cos^2 \theta + 4 \sin^2 \theta}, \quad \int_0^{\frac{\pi}{2}} \frac{d\theta}{5 + 3 \cos \theta}.$$

6. Integrate the following functions with respect to x:
$$\text{(i) } \frac{1 - x}{1 + x}; \quad \text{(ii) } x\sqrt{(x - 1)}.$$

Prove that
$$\int_0^{\frac{\pi}{2}} x \sin^2 x\, dx = \tfrac{1}{16}(\pi^2 + 4).$$

7. Integrate with respect to x:
$$\frac{1}{(x - 1)(x^2 + 1)}; \quad \cos^3 x; \quad xe^x.$$

Prove that
$$\int_0^1 \frac{x}{x^2 + x + 1}\, dx = \tfrac{1}{2} \ln 3 - \frac{\pi}{6\sqrt{3}}.$$

8. Integrate the following functions with respect to x:
$$\text{(i) } \left(x + \frac{1}{x}\right)^2; \quad \text{(ii) } \cos 2x \cos x; \quad \text{(iii) } \frac{2x}{(x + 1)(x + 3)}.$$

Evaluate
$$\int_0^{\frac{\pi}{4}} x \sec^2 x\, dx.$$

9. Find the values of
$$\int \frac{dx}{x(x + 1)^3}, \quad \int 4x \sin^{-1} x\, dx.$$

By means of the substitution $x + 1 = 1/t$, or otherwise, prove that
$$\int_0^1 \frac{dx}{(1 + x)\sqrt{(2 + x - x^2)}} = \frac{\sqrt{2}}{3}.$$

10. Evaluate the following integrals
$$\text{(i) } \int_0^1 x\sqrt{(1 - x^2)}\, dx,$$

$$\text{(ii) } \int_0^{\frac{\pi}{4}} \sec^2 x \tan x\, dx,$$

$$\text{(iii) } \int_1^4 \left(\sqrt{x} + \frac{1}{\sqrt{x}}\right) dx.$$

11. (i) Obtain
$$\int \sin 2x \cos \tfrac{1}{2}x\, dx.$$

(ii) Evaluate, correct to three significant figures,

$$(a) \int_0^{\frac{\pi}{4}} \tan x \, dx, \qquad (b) \int_0^{\frac{\pi}{4}} \tan^2 x \, dx.$$

12. Integrate with respect to x:

$$\frac{x^2 + 1}{(x - 1)^2}; \qquad \ln (2x); \qquad \frac{\cos x}{2 - \cos^2 x}.$$

Evaluate

$$\int_0^1 \frac{x + 9}{(x + 2)(3 - 2x)} \, dx.$$

13. Integrate with respect to x:

(i) $\dfrac{5x + 8}{(x - 1)(2x + 3)}$;   (ii) $\sin^2 x$.

By means of the substitution $x^2 = 1/u$, evaluate the integral

$$\int_1^2 \frac{dx}{x^2 \sqrt{(5x^2 - 1)}}.$$

14. Integrate with respect to x:

(i) $\dfrac{1}{x^2 + 4x + 8}$;   (ii) $\sin x \sec^3 x$;   (iii) $x^{\frac{1}{2}} \ln x$.

By making the substitution $x = \pi - y$, or otherwise, prove that

$$\int_0^{\pi} x \sin^3 x \, dx = \tfrac{2}{3}\pi.$$

15. Integrate the following functions with respect to x:

(i) $\dfrac{1}{\sqrt{(2x + 1)}}$;   (ii) $x \cos x$.

Evaluate

$$\int_1^2 \frac{dx}{x + x^4}$$

by means of the substitution $u = x^3$.

16. Integrate with respect to x:

(i) $\cos x \cos 2x$;   (ii) $\dfrac{x^2 - 2x + 4}{x(x^2 + 4)}$.

Evaluate

$$\int_0^2 \frac{2x}{\sqrt{(4 - x^2)}} \, dx.$$

17. Integrate with respect to x:

(i) $\dfrac{1}{\sqrt{(1 - 9x^2)}}$;   (ii) xe^{x^2}.

Evaluate

$$\int_0^{\frac{\pi}{2}} \frac{d\theta}{3 \cos \theta + 4 \sin \theta}.$$

18. (i) Obtain

$$\int \frac{x\, dx}{\sqrt{(2x + 1)}}, \qquad \int (\sin^4 x + \cos^4 x)\, dx.$$

(ii) Evaluate $\displaystyle\int_1^e x \ln x\, dx.$

19. (i) Evaluate $\displaystyle\int_2^4 (1 + x)^3\, dx.$

(ii) Evaluate $\displaystyle\int_0^{\frac{\pi}{2}} \theta \cos 2\theta\, d\theta$ and hence evaluate

$$\int_0^{\frac{\pi}{2}} \theta \cos^2 \theta\, d\theta.$$

(iii) Determine the indefinite integral

$$\int \frac{x\, dx}{x^2 + x + 1}.$$

20. Integrate $\left(1 + \dfrac{3}{x^2}\right)^2$ with respect to x, and evaluate

$$\int_0^{\frac{\pi}{2}} \sin^3 x \cos^2 x\, dx \quad \text{and} \quad \int_0^1 \frac{(1 - x)\, dx}{(x + 1)(x^2 + 1)}.$$

21. Integrate with respect to x:

$$\frac{x + 1}{(x - 1)^2}; \qquad x \ln x; \qquad \sec^3 x \tan x.$$

Prove that

$$\int_0^\pi \frac{\sin x}{4 - \sin^2 x}\, dx = \int_{-1}^1 \frac{1}{u^2 + 3}\, du = \frac{\pi}{3\sqrt{3}}.$$

22. (i) Find the indefinite integrals of

$$\frac{1}{x(x + 1)^2}, \qquad \frac{2x + 3}{x^2 + 2x + 2}.$$

(ii) By substitution, or otherwise, prove that

$$\int_0^{1/\sqrt{2}} x \sin^{-1} x\, dx = \tfrac{1}{8}, \qquad \int_0^1 x^3 \sqrt{(x^2 + 1)}\, dx = \frac{2(1 + \sqrt{2})}{15}.$$

23. Integrate with respect to x:

(i) $\dfrac{\cos^3 x}{\sin^2 x}$,   (ii) $e^x \sin x.$

Evaluate $\displaystyle\int_2^{12} \frac{2x + 5}{(2x - 3)(2x + 1)}\, dx$

to three significant figures, given that $\log_{10} e = 0 \cdot 4343.$

24. Find the indefinite integrals:

$$\int \frac{x^2\, dx}{1 - x^6}; \qquad \int \frac{dx}{\sqrt{\{x(2 - x)\}}}; \qquad \int x \sec^2 x\, dx; \qquad \int \tan^4 x\, dx.$$

25. Integrate with respect to x:

(i) $\dfrac{1}{\sqrt{(5 - 4x - x^2)}}$;   (ii) $x^3 e^{-x^2}$.

If $S = \displaystyle\int \dfrac{\sin x}{a \sin x + b \cos x}\, dx$ and $C = \displaystyle\int \dfrac{\cos x}{a \sin x + b \cos x}\, dx$,

evaluate $aS + bC$ and $aC - bS$. Hence, or otherwise, prove that

$$\int_0^{\frac{\pi}{2}} \dfrac{\sin x}{3 \sin x + 4 \cos x}\, dx = \dfrac{3\pi}{50} + \dfrac{4}{25} \ln\left(\dfrac{4}{3}\right).$$

11
Curves, Surfaces and Solids—(I)

Curve-tracing: Areas, Volumes, Centres of Mass

Applications of calculus to areas, volumes and centres of mass were given in Chapter 5. These can be considerably extended with the extra techniques of integration now at our disposal.

EXAMPLE 11.1   *Show that the curve $a^2y^2 = x^2(a^2 - x^2)$ consists of two loops. Show also that one half of one loop lies within the positive quadrant. Find the area of this half and the coordinates of its centroid.*

(i) The curve is symmetrical about both axes, since if (x, y) satisfies the equation, so do $(-x, y)$, $(x, -y)$ and $(-x, -y)$.

(ii) It passes through the origin, and since in that neighbourhood x is small compared with a, we have $a^2y^2 \approx x^2a^2$; thus near the origin the shape approximates to $y = \pm x$. These are, in fact, the equations of the tangents at the origin, which accordingly is a 'double point'; the curve intersects itself there.

(iii) Since a^2y^2 and x^2 are positive, it follows that $a^2 - x^2 \geqslant 0$, i.e. $|x| \leqslant a$. Also $a^2y^2 = \frac{1}{4}a^4 - (\frac{1}{2}a^2 - x^2)^2 \leqslant \frac{1}{4}a^4$, so $|y| \leqslant \frac{1}{2}a$.

(iv) When $y = 0$, $x = \pm a$ or 0; when $x = 0$, $y = 0$.

We can now sketch the curve (Fig. 11.1), showing that it consists of two loops, one half of one being in the positive quadrant. The area, A, of this half-loop is given by $\lim_{\delta x \to 0} \Sigma_{x=0}^{a} y\, \delta x$ where $ay = +x\sqrt{(a^2 - x^2)}$, so that

$$A = \int_0^a \frac{x}{a}\sqrt{(a^2 - x^2)}\, dx = \left[-\frac{(a^2 - x^2)^{\frac{3}{2}}}{3a} \right]_0^a = \frac{1}{3}a^2.$$

If the coordinates of the centroid are $(\bar{x}, \bar{y})$, we have

$$A\bar{x} = \lim_{\delta x \to 0} \Sigma_{x=0}^{a} xy\, \delta x = \int_0^a yx\, dx = \int_0^a \frac{x^2\sqrt{(a^2 - x^2)}}{a}\, dx$$

$$= \int_0^{\frac{\pi}{2}} a^3 \sin^2 \theta \cos^2 \theta\, d\theta,$$

190

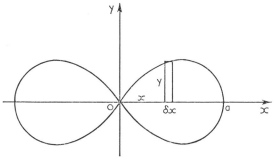

Fig. 11.1

putting $x = a \sin \theta$, $dx = a \cos \theta \, d\theta$,

so
$$A\bar{x} = \tfrac{1}{4}a^3 \int_0^{\frac{\pi}{2}} \sin^2 2\theta \, d\theta$$

$$= \tfrac{1}{8}a^3 \int_0^{\frac{\pi}{2}} (1 - \cos 4\theta) \, d\theta$$

$$= \tfrac{1}{8}a^3 \left[\theta - \tfrac{1}{4} \sin 4\theta \right]_0^{\frac{\pi}{2}} = \frac{\pi a^3}{16}.$$

Thus
$$\bar{x} = \frac{3}{a^2} \times \frac{\pi a^3}{16} = \frac{3\pi a}{16}.$$

Also
$$A\bar{y} = \int_0^a y \cdot \tfrac{1}{2}y \, dx$$

$$= \frac{1}{2a^2} \int_0^a x^2(a^2 - x^2) \, dx$$

$$= \frac{1}{2a^2} \left[\frac{a^2 x^3}{3} - \frac{x^5}{5} \right]_0^a = \frac{a^3}{15},$$

and so
$$\bar{y} = \frac{3}{a^2} \times \frac{a^3}{15} = \frac{a}{5}.$$

Thus the area of the half-loop is $\tfrac{1}{3}a^2$ and the coordinates of its centroid are $\left(\dfrac{3\pi a}{16}, \dfrac{a}{5} \right)$.

As a rough check, the sketch suggests that $a > \bar{x} > \tfrac{1}{2}a$, and this is satisfied by $\bar{x} = \dfrac{3\pi a}{16}$.

EXAMPLE 11.2 *To find the area of the portion of the astroid $x = a \cos^3 t$, $y = a \sin^3 t$, lying in the first quadrant, and the volume of the solid formed by rotating this area through four right angles about the x-axis.*

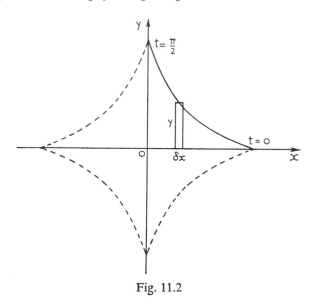

Fig. 11.2

The shape of this curve (Fig. 11.2) was discussed on p. 98. The area concerned is given by

$$A = \lim_{\delta x \to 0} \sum_{x=0}^{a} y \, \delta x$$

$$= \lim_{\delta t \to 0} \sum_{t=\pi/2}^{0} y \frac{\delta x}{\delta t} \, \delta t$$

$$= \int_{\frac{\pi}{2}}^{0} y \frac{dx}{dt} \, dt$$

$$= -\int_{0}^{\frac{\pi}{2}} a \sin^3 t \, (-3a \cos^2 t \sin t) \, dt$$

$$= 3a^2 \int_{0}^{\frac{\pi}{2}} \sin^4 t \cos^2 t \, dt$$

$$= \frac{3a^2}{16} \left[t - \tfrac{1}{3} \sin^3 2t - \tfrac{1}{4} \sin 4t \right]_{0}^{\frac{\pi}{2}} \qquad \text{(cf. p. 180)}$$

$$= \frac{3a^2 \pi}{32}.$$

The volume of the solid of revolution is given by

$$
\begin{aligned}
V &= \int_0^a \pi y^2 \, dx \\
&= \int_{\frac{\pi}{2}}^0 \pi y^2 \frac{dx}{dt} \, dt \\
&= \int_{\frac{\pi}{2}}^0 \pi a^2 \sin^6 t \, (-3a \cos^2 t \sin t) \, dt \\
&= 3\pi a^3 \int_0^{\frac{\pi}{2}} \cos^2 t \sin^7 t \, dt \\
&= 3\pi a^3 \int_0^{\frac{\pi}{2}} \cos^2 t (1 - \cos^2 t)^3 \sin t \, dt \\
&= 3\pi a^3 \int_0^1 z^2 (1 - z^2)^3 \, dz,
\end{aligned}
$$

$$\text{putting } \cos t = z, \ -\sin t \, dt = dz,$$

$$
\begin{aligned}
&= 3\pi a^3 \int_0^1 z^2 (1 - 3z^2 + 3z^4 - z^6) \, dz \\
&= 3\pi a^3 \left[\frac{z^3}{3} - \frac{3z^5}{5} + \frac{3z^7}{7} - \frac{z^9}{9} \right]_0^1 \\
&= \frac{16\pi a^3}{105}.
\end{aligned}
$$

As a rough check, this is much less than $\frac{1}{3}\pi a^3$, the volume of a cone of height and base-radius a. (Such comparisons, however crude, can sometimes avert arithmetical disasters.)

EXAMPLE 11.3   *To find* (i) *the C.M. of an arc of a circle of radius a subtending an angle 2α at the centre,* (ii) *the C.M. of a sector of a circle containing an angle 2α.*

(i) By symmetry, the C.M. is on the central radius of the arc, which we take as the y-axis, as shown in Fig. 11.3. Let the mass per unit length be ρ, so that the total mass is $2\alpha a\rho$, and consider a small element of arc PQ of length $a \, \delta\theta$, where angle $YOP = \theta$. If the ordinate of the C.M. is $\bar{y}$, we have

$$
\begin{aligned}
2\alpha a\rho\bar{y} &= \lim_{\delta\theta \to 0} \sum_{\theta=-\alpha}^{\alpha} a \cos\theta \, . \, \rho a \, \delta\theta \\
&= \int_{-\alpha}^{\alpha} \rho a^2 \cos\theta \, d\theta \\
&= 2\rho a^2 \sin\alpha.
\end{aligned}
$$

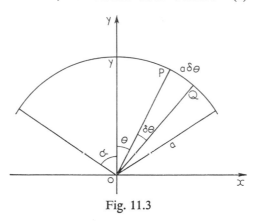

Fig. 11.3

Thus

$$\bar{y} = \frac{a \sin \alpha}{\alpha}.$$

(The reader should check that this lies between a and $a \cos \alpha$, cf. p. 88.)
(ii) The area of the sector may be considered as the sum of a number of thin strips (Fig. 11.4) between radii x and $x + \delta x$, where x ranges

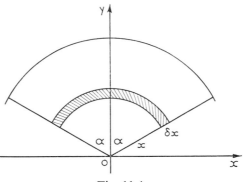

Fig. 11.4

from 0 to a. For such a strip, the ordinate of the C.M. is $\dfrac{x \sin \alpha}{\alpha}$ and the mass may be taken as $\sigma . 2\alpha x \, \delta x^*$ where σ is the mass per unit area. Also the mass of the whole sector is $\dfrac{2\alpha}{2\pi} \times \pi a^2 \sigma = \alpha a^2 \sigma$.

---

* It is, in fact, $2\alpha\sigma(x + \frac{1}{2}\delta x) \, \delta x$, but on integration the limit of the sum of terms with a factor of $(\delta x)^2$ is zero, since, if equal intervals δx are taken, $\lim\limits_{\delta x \to 0} \Sigma f(x)(\delta x)^2 = \lim\limits_{\delta x \to 0} I \, \delta x$ (where $I = \int f(x) \, dx) = 0$. The reader must be prepared to give such an explanation if challenged, but need not normally do so.

Thus if the ordinate of the C.M. of the sector is denoted by $\bar{z}$, we have

$$\alpha a^2 \sigma \bar{z} = \lim_{\delta x \to 0} \sum_{x=0}^{a} \frac{x \sin \alpha}{\alpha} \times 2\alpha\sigma x \, \delta x$$

$$= \int_0^a 2x^2 \sigma \sin \alpha \, dx$$

$$= \tfrac{2}{3}a^3 \sigma \sin \alpha.$$

Thus

$$\bar{z} = \frac{2a \sin \alpha}{3\alpha}.$$

EXERCISE LII

1. Sketch the curve whose equation is
$$y = (x + 1)(5 - x)$$
and find the coordinates of the points A and B at which the line $y = 2x + 5$ cuts the curve.

Find the area bounded by the chord AB, the x-axis and the intermediate arcs of the curve.

Find also the volume produced when this area is rotated about the x-axis through four right angles.

2. Find the area enclosed between the curves
$$ay^2 = 2x^3, \qquad 16ax^2 = y^3,$$
and the coordinates of its centre of gravity.

3. Find the tangents at the origin to the curve
$$y^2 = x^2 \frac{1 - x^2}{1 + x^2}.$$

Sketch the curve and find the area of a loop.

Find the volume of the uniform solid of revolution formed by rotating a loop through four right angles about the x-axis.

4. Show that the area enclosed by the curve
$$x = f(t), \qquad y = g(t),$$
the axis of x, and the ordinates determined by $t = t_1$, $t = t_2$ is
$$\int_{t_1}^{t_2} g(t) f'(t) \, dt.$$

Sketch the arc of the cycloid
$$x = 2(t + \sin t), \qquad y = 2(1 - \cos t)$$
from $x = 0$ to $x = 2\pi$.

Find the area included between this arc and the x-axis.

5. Find the area enclosed between the curve $y = \sqrt{(x^3)}$ and the straight line $y = 2x$.

   Find the volume generated when this area is rotated through four right angles
   (i) about the x-axis,
   (ii) about the y-axis.
   Find the coordinates of the centroid of the area.

6. A uniform lamina is bounded by the curve $y = x \cos x$ between $x = 0$ and $\frac{1}{2}\pi$ and by the x-axis. Find the area of the lamina and the coordinates of its centre of gravity.

7. Sketch the curve whose parametric equations are

$$x = a \sin t, \qquad y = a \tan t.$$

   Find the area in the positive quadrant enclosed by the curve, the axis of x, and the line $x = \frac{1}{2}a$.
   Prove that the volume generated by rotating this area through four right angles about the axis of x is equal to

$$\tfrac{1}{2}\pi a^3(\ln 3 - 1).$$

8. Show that the area enclosed between the curve $y = \tan x$ and the lines $y = 0$ and $x = \frac{1}{4}\pi$ is $\frac{1}{2} \ln 2$.

   If the above area is rotated through four right angles about the line $y = 0$, prove that the volume generated is $\pi(4 - \pi)/4$.

9. Show that the curve $3ay^2 = x^2(3a - 2x)$ possesses a loop, and draw a rough sketch.

   P, Q are the points on the loop at which the tangents to the curve are parallel to the x-axis. Find the area of the loop, and also the fraction of this area which lies between the lines OP, OQ produced, where O is the origin.

10. Prove that the curve

$$y = xe^{-(3x^2/2)}$$

   is symmetrical about the origin.
   Prove that the curve has turning-points where

$$x = \pm 1/\sqrt{3},$$

   and points of inflexion where $x = 0, \pm 1$.
   Sketch the curve.
   [In sketching the curve you may assume that $e^{-\frac{1}{2}} \approx 0.6$, $e^{-\frac{3}{2}} \approx 0.2$, $1/\sqrt{3} \approx 0.6$, and that $y \to 0$ as $x \to \infty$.]

11. Prove that the volume of a cap of a sphere, of radius r, cut off by a plane distant $(r - h)$ from the centre is $\pi h^2(r - \frac{1}{3}h)$.

   Water is poured into a hemispherical vat at the rate of 11 000 cm³ s⁻¹. If the internal radius of the vat is 80 cm, find the rate at which the depth of water in the vat is increasing when this depth is 50 cm.

12. The area between the axes and the curve $y = \cos x$ from $x = 0$ to $x = \frac{1}{2}\pi$ is divided into two parts by that piece of the curve $y = \sin x$ which crosses it. Each part is rotated through four right angles about the axis that forms its straight edge. Prove that the volumes of the solid figures thus generated are

$$\tfrac{1}{2}\pi(\tfrac{1}{2}\pi - 1) \quad \text{and} \quad 2\pi(\tfrac{1}{4}\pi\sqrt{2} - 1).$$

13. Prove that the area enclosed by the curves $y = \cos x$ and $y = \sin x$ between the lines $x = 0$ and $x = \frac{1}{4}\pi$ is $\sqrt{2} - 1$.

   Find the coordinates of the centre of gravity of this area.

14. A curve is defined by the parametric equations

$$x = a(1 - t^2), \qquad y = a(2 - t)(1 - t),$$

where a is a positive constant. Prove that
   (i) the curve passes through the points A, O, B, whose coordinates are $(0, 6a)$, $(0, 0)$, $(-3a, 0)$,
   (ii) the point t is in the first quadrant when $-1 < t < 1$ and in the third quadrant when $1 < t < 2$.

   Make a rough sketch showing the part of the curve corresponding to values of t between -1 and $+2$.

   Find the equation of the tangent to the curve at O, and prove that the area bounded by the arc AB and the chord AB is $27a^2/2$.

Tangents and Asymptotes*

The equation of the tangent at the point t of the astroid $x = a \cos^3 t$, $y = a \sin^3 t$ was found on p. 99 to be

$$x \sin t + y \cos t = a \cos t \sin t.$$

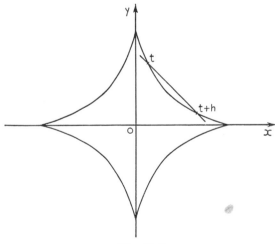

Fig. 11.5

The method employed was to find the gradient at t and then use

$$y - y_1 = m(x - x_1).$$

An alternative method is now illustrated (Fig. 11.5) using the same curve.

---

* This section may be deferred till later.

From coordinate geometry, the equation of the line joining the points (x_1, y_1), (x_2, y_2) is

$$\begin{vmatrix} x & y & 1 \\ x_1 & y_1 & 1 \\ x_2 & y_2 & 1 \end{vmatrix} = 0.$$

Thus the line joining the points t and $t + h$ of the astroid is

$$\begin{vmatrix} x & y & 1 \\ a \cos^3 t & a \sin^3 t & 1 \\ a \cos^3 (t + h) & a \sin^3 (t + h) & 1 \end{vmatrix} = 0.$$

Taking the second row from the third and then dividing the new third row by h, we have

$$\begin{vmatrix} x & y & 1 \\ a \cos^3 t & a \sin^3 t & 1 \\ \dfrac{a\{\cos^3 (t + h) - \cos^3 t\}}{h} & \dfrac{a\{\sin^3 (t + h) - \sin^3 t\}}{h} & 0 \end{vmatrix} = 0.$$

Now as $h \to 0$, this will give the tangent at t. The elements of the third row of the determinant become the gradients of the corresponding elements of the second; we have, as the equation of the tangent,

$$\begin{vmatrix} x & y & 1 \\ a \cos^3 t & a \sin^3 t & 1 \\ -3a \cos^2 t \sin t & 3a \sin^2 t \cos t & 0 \end{vmatrix} = 0$$

or, if $\sin t$, $\cos t \neq 0$,

$$\begin{vmatrix} x & y & 1 \\ a \cos^3 t & a \sin^3 t & 1 \\ -\cos t & \sin t & 0 \end{vmatrix} = 0$$

yielding $x \sin t + y \cos t = a \cos t \sin t$, as before. The reader should now generalize this result, proving that the tangent at t to the curve given by $x:y:a = f(t):g(t):h(t)$ is

$$\begin{vmatrix} x & y & a \\ f(t) & g(t) & h(t) \\ f'(t) & g'(t) & h'(t) \end{vmatrix} = 0.$$

An *asymptote* is a line to which a curve approximates when x and/or y tends to infinity. It may be described as a 'tangent at infinity'; this idea is illustrated by the following example.

EXAMPLE 11.4   *To investigate the curve whose equation is $x^3 + y^3 = 3axy$.*

To plot this curve by taking values of x and solving a succession of cubics for y would be very laborious. Instead, we find parametric expressions for x and y, by putting $y = tx$. This is equivalent to sending out search-rays from the origin; the line $y = tx$ meets the curve where $x^3 + t^3x^3 = 3atx^2$, i.e. (apart from the origin) where

$$x = \frac{3at}{1 + t^3}, \qquad y = \frac{3at^2}{1 + t^3},$$

or $x : y : 3a = t : t^2 : 1 + t^3$. The tangent at t is given by

$$\begin{vmatrix} x & y & 3a \\ t & t^2 & 1 + t^3 \\ 1 & 2t & 3t^2 \end{vmatrix} = 0$$

or $-xt(2 - t^3) + y(1 - 2t^3) + 3at^2 = 0$, $(t \neq -1)$. Now the only value of t for which x or y is not defined is -1. As $t \to -1$, the equation of the tangent at t becomes $x + y + a = 0$, which is accordingly the equation of the only asymptote (Fig. 11.6).

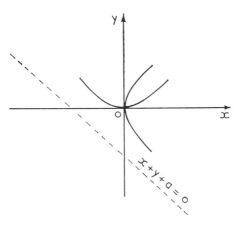

Fig. 11.6

When t is small, $x \approx 3at$, $y \approx 3at^2$ (i.e. $x^2 \approx 3ay$) and when t is large, $x \approx \dfrac{3a}{t^2}$, $y \approx \dfrac{3a}{t}$ (i.e. $y^2 \approx 3ax$). Near the origin, therefore, the curve approximates to parts of the two parabolas $x^2 = 3ay$, $y^2 = 3ax$.

The equation $x^3 + y^3 = 3axy$ shows that the curve is symmetrical about the line $x = y$. It is now possible to sketch the curve (Fig. 11.7), caution perhaps suggesting that a few points be plotted by giving t various values in the parametric equations.

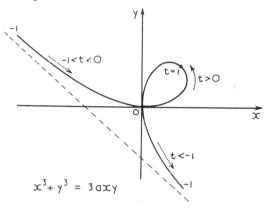

Fig. 11.7   The Folium of Descartes

Another method of finding asymptotes is illustrated by the next two examples.

EXAMPLE 11.5   *To sketch the curve* $(y - x)(x^2 + y^2) = 3x + 4y$.

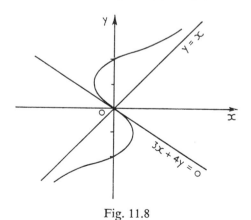

Fig. 11.8

(i) An indication of the direction of the curve when x and/or y is large is given by equating the terms of highest degree to zero,* i.e.

---

* For such values, terms of first degree are small compared with those of third degree. Near the origin it is the other way round, cf. (ii).

$(y - x)(x^2 + y^2) = 0$. This yields only $y = x$, which gives the direction of the only asymptote.

Now $y - x = \dfrac{3x + 4y}{x^2 + y^2}$, and for a second approximation we set $y = x$ in the right-hand side, obtaining $y - x \approx \dfrac{7}{2x}$. Thus the equation of the asymptote is $y = x$, and since $y \approx x + \dfrac{7}{2x}$, the curve is above it when $x > 0$ and below when $x < 0$.

(ii) For the shape near the origin, we equate the terms of lowest degree to zero; thus the tangent at the origin is $3x + 4y = 0$.

(iii) When $y = 0$, $x = 0$; when $x = 0$, $y = 0$ or ± 2.

EXAMPLE 11.6   *To find the asymptotes of the curve*

$$y(y - x)(y + x) = y^2 - 4x^2.$$

The directions of the asymptotes are given by $y = 0$, $y = \pm x$. These are the first approximations to the curve 'at infinity'.

$$\text{(i)} \quad y = \frac{y^2 - 4x^2}{y^2 - x^2} = \frac{\dfrac{y^2}{x^2} - 4}{\dfrac{y^2}{x^2} - 1} \to 4 \qquad \text{as } x \to \infty.$$

The equation of the first asymptote is accordingly $y = 4$.

$$\text{(ii)} \quad y - x = \frac{y^2 - 4x^2}{y(y + x)}.$$

Putting $y = x$ in the right-hand side* yields $\dfrac{-3x^2}{2x^2} = -\tfrac{3}{2}$. The equation of the second asymptote is $y - x = -\tfrac{3}{2}$, or $2x - 2y = 3$.

$$\text{(iii)} \quad y + x = \frac{y^2 - 4x^2}{y(y - x)} \approx \frac{-3x^2}{2x^2} = -\frac{3}{2}.$$

Thus the equations of the asymptotes are $y = 4$, $2x - 2y = 3$, $2x + 2y + 3 = 0$.

EXERCISE LIII

1. Find the equation of the tangent at the point t of the curves (i) $x = t^3$, $y = t^2$; (ii) $x = \dfrac{2t}{1 + t^2}$, $y = \dfrac{1 - t^2}{1 + t^2}$. Sketch the curves.

---

* Substituting the first approximation here leads to the second, which provides the equation of the asymptote. We suppress a justification of this, but it may not be beyond the reader's powers to provide one for himself.

2. Find the asymptotes of the curve $x = \dfrac{a(1 + t^2)}{1 - t^2}$, $y = \dfrac{2bt}{1 - t^2}$.

3. Find the asymptote of the Folium of Descartes, whose equation is
$$x^3 + y^3 = 3axy,$$
using the method of Example 11.5.

4. Find the asymptotes of the curve $y^2(x - 2) = x(x^2 - 1)$. (Cf. Example 6.13, p. 103.)

5. (i) By putting $y = tx$, show that $x^3 + y^3 = y^2$ can be represented by the parametric equations $x = \dfrac{t^2}{1 + t^3}$, $y = \dfrac{t^3}{1 + t^3}$.

   (ii) Find the equation of the tangent at the point t, and deduce the equation of the asymptote.

   (iii) Find where the curve meets the axes and its asymptote.

   (iv) Show that near the origin $y^2 \approx x^3$.

   (v) Sketch the curve.

   (vi) Show that the tangents at three points t_1, t_2, t_3 are concurrent if and only if $t_1 + t_2 + t_3 = 0$. [Find the condition for the tangent at t to pass through a fixed point, say (x_1, y_1). Compare the resulting cubic for t with that whose roots are t_1, t_2, t_3.]

6. Find the tangent at t_1 to the curve $x = t^3$, $y = t + 1$ and show that it meets the curve again at the point whose parameter is $-2t_1$. [$(t^3, t + 1)$ lies on the straight line $lx + my + n = 0$ if $lt^3 + m(t + 1) + n = 0$. This is a cubic in which the coefficient of t^2 is zero.]

7. Sketch the curves (i) $y(y - 2x) = 3x + 4y$, (ii) $(x^2 + y^2)(y + 2x) = x - y$.

8. The equation of a curve is $x^4 - y^4 = x^2 - 4y^2$.
   Find the equations of the asymptotes and of the tangents at the origin.
   Prove that the curve is parallel to the x-axis at the points $(0, \pm 2)$, and also at the eight points at which $x^2 = \frac{1}{2}$, $y^2 = \frac{1}{2}(4 \pm \sqrt{15})$.
   Sketch the curve.

9. (i) Prove that the curve
$$y^2 = x^2 \frac{1 - x}{1 + x}$$
   lies wholly between the lines $x = -1$ and $x = +1$, and that the line $x = -1$ is its only real asymptote.

   (ii) Sketch the curve.

   (iii) Prove that the volume of the solid formed by revolving the loop of the curve through four right angles about its axis of symmetry is equal to
$$\pi(2 \ln 2 - \tfrac{4}{3}).$$

10. Prove that no part of the curve
$$y^2(8 - 4x) = x^2(8 - x)$$
lies between the lines $x = 2$ and $x = 8$, and that the line $x - 2y = 3$ is an asymptote of the curve.
    Find the equations of the other asymptotes and of the tangents to the curve at the origin. Sketch the curve.

11. Show that no part of the curve

$$y^2(x^2 - 4) = x^2(x^2 - 1)$$

lies between the lines $x = 1$ and $x = 2$.
Give the equations of
  (i) any axes of symmetry of the curve,
  (ii) the asymptotes,
  (iii) the tangents at the origin.
Sketch the curve.

Length of Arc

Let AB be an arc of a curve whose gradient is positive and increasing, and let points $A = P_0, P_1, P_2, \ldots, P_n = B$ be in order on the curve (Fig. 11.9).

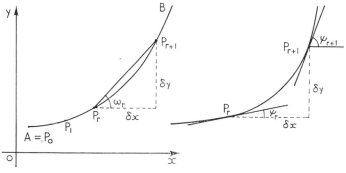

Fig. 11.9

Suppose further that the angle ψ_r which the tangent at any point P_r makes with the x-axis is acute.

If the sum of the lengths of the chords, $P_0P_1 + P_1P_2 + \ldots + P_{n-1}P_n$, tends to a definite limit as $n \to \infty$ and the length of each chord tends to zero, regardless of how the intermediate points are chosen, then this limit is our definition of the *length of arc AB*.

Now $P_rP_{r+1} = \sec \omega_r \, \delta x$, where δx is the difference between the x-coordinates of P_r, P_{r+1} and ω_r is the angle the chord makes with the x-axis; ω_r lies between ψ_r and ψ_{r+1}. Thus the length of arc AB is given by

$$s = \lim_{n \to \infty} \sum_{r=0}^{n-1} P_rP_{r+1}$$

$$= \lim_{\delta x \to 0} \sum_{r=0}^{n-1} \sec \omega_r \, \delta x$$

$$= \int_a^b \sec \psi \, dx,$$

where a, b are the respective x-coordinates of A, B. Now

$$\sec \psi = \sqrt{(1 + \tan^2 \psi)} = \sqrt{\left\{1 + \left(\frac{dy}{dx}\right)^2\right\}}.$$

Thus

$$s = \int_a^b \sqrt{\left\{1 + \left(\frac{dy}{dx}\right)^2\right\}}\, dx \qquad (1)$$

If B is now taken as a variable point and we replace b by x, we have

$$s = \int_a^x \sqrt{\left\{1 + \left(\frac{dy}{dx}\right)^2\right\}}\, dx$$

and, differentiating with respect to x,

$$\frac{ds}{dx} = \sqrt{\left\{1 + \left(\frac{dy}{dx}\right)^2\right\}}$$

(cf. p. 128, result (8)) or

$$\left(\frac{ds}{dx}\right)^2 = 1 + \left(\frac{dy}{dx}\right)^2. \qquad (2)$$

This may be written symbolically as

$$(ds)^2 = (dx)^2 + (dy)^2, \qquad (3)$$

implying that if x, y are functions of a parameter t, then

$$\left(\frac{ds}{dt}\right)^2 = \left(\frac{dx}{dt}\right)^2 + \left(\frac{dy}{dt}\right)^2. \qquad (4)$$

It is easy to remember (3) from the fact that $(\delta s)^2 \approx (\delta x)^2 + (\delta y)^2$ (Fig. 11.10), but of course this does not constitute a proof.

We considered a particular type of arc in obtaining (1) (the gradient being increasing and the angle between tangent and x-axis acute); it can be shown similarly that (1) holds whenever $\frac{dy}{dx}$ is continuous over the range of integration. From (4),

$$\frac{ds}{dt} = \sqrt{\left\{\left(\frac{dx}{dt}\right)^2 + \left(\frac{dy}{dt}\right)^2\right\}}$$

so that

$$s = \int_{t_1}^{t_2} \sqrt{\left\{\left(\frac{dx}{dt}\right)^2 + \left(\frac{dy}{dt}\right)^2\right\}}\, dt$$

provided that $\frac{dx}{dt}$, $\frac{dy}{dt}$ are continuous, t increases steadily from t_1 to t_2 and $\frac{ds}{dt} > 0$, i.e. s increases with t. Care must be taken over the last condition, as is shown in Example 11.8, below.

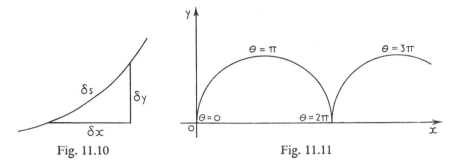

Fig. 11.10   Fig. 11.11

EXAMPLE 11.7   *To find the length of an arch of the cycloid*

$$\begin{cases} x = a(\theta - \sin \theta), \\ y = a(1 - \cos \theta). \end{cases}$$

Refer to Fig. 11.11 for the shape of the curve, cf. Exercise XXIX No. 3, p. 99.
We have

$$\left(\frac{ds}{d\theta}\right)^2 = \left(\frac{dx}{d\theta}\right)^2 + \left(\frac{dy}{d\theta}\right)^2$$

$$= a^2(1 - \cos \theta)^2 + a^2 \sin^2 \theta$$

$$= 2a^2(1 - \cos \theta)$$

$$= 4a^2 \sin^2 \frac{\theta}{2}$$

so $\dfrac{ds}{d\theta} = 2a \sin \dfrac{\theta}{2}$, and the required length is given by

$$s = \int_0^{2\pi} 2a \sin \frac{\theta}{2} \, d\theta$$

$$= \left[-4a \cos \frac{\theta}{2} \right]_0^{2\pi}$$

$$= 8a.$$

Note that we took $+2a \sin \dfrac{\theta}{2}$ as the square root of $4a^2 \sin^2 \dfrac{\theta}{2}$, since $\sin \dfrac{\theta}{2} \geqslant 0$ in the range $0 \leqslant \theta \leqslant 2\pi$.

EXAMPLE 11.8   *A circle C of radius a rolls without slipping round another circle C' of radius 2a. To find the distance travelled by a point on the circumference of C during one complete traverse of its locus.*

Let axes be taken as shown in Fig. 11.12, P having started at A. If K is the centre of C, angle $AOK = \theta$, angle $PKO = \phi$, the coordinates of P are given by

$$x = 3a \sin \theta - a \sin (\theta + \phi) \quad (= OL - PM)$$

$$y = 3a \cos \theta - a \cos (\theta + \phi) \quad (= KL - KM).$$

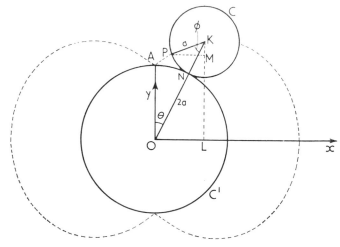

Fig. 11.12

Now the rolling condition is that $2a\theta = a\phi$,* so that $x = 3a \sin \theta - a \sin 3\theta$, $y = 3a \cos \theta - a \cos 3\theta$.

$$\left(\frac{ds}{d\theta}\right)^2 = \left(\frac{dx}{d\theta}\right)^2 + \left(\frac{dy}{d\theta}\right)^2$$

$$= (3a \cos \theta - 3a \cos 3\theta)^2 + (-3a \sin \theta + 3a \sin 3\theta)^2$$

$$= 18a^2 - 18a^2(\cos \theta \cos 3\theta + \sin \theta \sin 3\theta)$$

$$= 18a^2(1 - \cos 2\theta)$$

$$= 36a^2 \sin^2 \theta.$$

Thus $\dfrac{ds}{d\theta} = \pm 6a \sin \theta$. Since θ varies from 0 to 2π for a complete circuit; we must take $\dfrac{ds}{d\theta} = +6a \sin \theta$ for $0 \leqslant \theta \leqslant \pi$ and $-6a \sin \theta$ for $\pi < \theta \leqslant 2\pi$.

In fact, by symmetry, the required length is

$$2 \int_0^{\pi} 6a \sin \theta \, d\theta = 12a[-\cos \theta]_0^{\pi} = 24a.$$

---

* This expresses the fact that the arc PN of C is equal in length to the arc AN of C'.

1. (D) (i) Under what conditions is the arc-length of a curve given by

$$s = \int_{y_1}^{y_2} \sqrt{\left\{1 + \left(\frac{dx}{dy}\right)^2\right\}}\, dy\,?$$

    (ii) Find the length of arc of the parabola $y^2 = 2x$ from the origin to the point $(\frac{9}{50}, \frac{3}{5})$. [Investigate the gradient at the origin to explain why the formula of the first part of the question should be used rather than (1) of the text.]

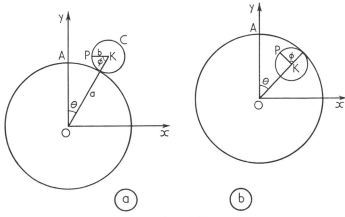

Fig. 11.13

2. (D) (i) Generalize Example 11.8 by taking the radii of C', C as a, b respectively (Fig. 11.13a); obtain the coordinates of P as

$$\begin{aligned} x &= (a + b)\sin\theta - b\sin(\theta + \phi) \\ y &= (a + b)\cos\theta - b\cos(\theta + \phi), \end{aligned} \qquad (1)$$

where $a\theta = b\phi$.
For any given ratio $b:a$ these are the parametric equations of an *epicycloid*. Sketch the locus when $b:a = \frac{1}{3}, \frac{1}{4}, \dfrac{1}{\sqrt{2}}, 2$.

    (ii) Consider the special case when $b = a$. The locus of P is then a *cardioid*; sketch it, and find its total arc-length.

    (iii) Suppose now that C rolls *inside* C' (Fig. 11.13b). P then traces out a *hypocycloid*. Show that parametric equations are given by

$$\begin{aligned} x &= (a - b)\sin\theta - b\sin(\phi - \theta) \\ y &= (a - b)\cos\theta + b\cos(\phi - \theta), \end{aligned} \qquad (2)$$

where $a\theta = b\phi$.

    (iv) Sketch the three-cusped hypocycloid, or *deltoid*, obtained by putting $a = 3b$ in (2), and find its total arc-length.

3. Find the arc-length of $y = x^{\frac{3}{2}}$ between the points (4, 8) and (9, 27).

4. (i) Find the arc-length of the curve $x = a \cos t$, $y = a \sin t$ between $t = 0$ and $t = \dfrac{\pi}{2}$; identify the curve.

   (ii) Find the total length of the astroid (cf. p. 192) $x = a \cos^3 t$, $y = a \sin^3 t$.

5. The parametric equations of an ellipse are $x = a \cos \theta$, $y = b \sin \theta$, the *eccentricity* e being defined by the equation $b^2 = a^2(1 - e^2)$. Show that the perimeter of the ellipse is given by $4a \displaystyle\int_0^{\frac{\pi}{2}} \sqrt{(1 - e^2 \cos^2 \theta)}\, d\theta$. This is an 'elliptic integral', which we cannot evaluate by formula; show, however, that if e is so small that e^3 can be neglected, then the perimeter is given approximately by $2\pi a - \frac{1}{2}\pi a e^2$. Sketch the curve when e is small.

6. Find the length of the curve $y = e^x$ from $y = \frac{5}{12}$ to $y = \frac{3}{4}$.

7. Show that the length of the arc of the curve $y = f(x)$ from $x = a$ to $x = b$ is

$$\int_a^b \sqrt{\left\{ 1 + \left(\frac{dy}{dx}\right)^2 \right\}}\, dx.$$

Find the arc-length of the curve

$$y = \frac{3}{4\sqrt{2}}(x^{\frac{4}{3}} - x^{\frac{2}{3}})$$

from $x = 0$ to $x = X$. Verify that, if S is the arc-length and if $y = Y$ when $x = X$, then

$$8S^2 = 9X^2 + 8Y^2.$$

8. The line $x = 4a$ cuts the parabola $y^2 = 4ax$ at the points A and B. Find
   (i) the gradient of the curve at A and at B;
   (ii) the area enclosed between the chord AB and the arc AB of the parabola;
   (iii) the length of the arc AB.

9. Show that the curve $9y^2 = x(3 - x)^2$ has a loop lying between the lines $x = 0$ and $x = 3$. Prove that at any point on the curve

$$\left(\frac{ds}{dx}\right)^2 = \frac{1}{4}\left(x + 2 + \frac{1}{x}\right),$$

where s is the length of the curve measured from a fixed point on it, and prove that the length of the loop is $4\sqrt{3}$.

The Pappus-Guldin Theorems

THEOREM I. *If a surface is generated by the rotation of an arc of a curve about an axis which it does not cut, the area of the surface so formed is equal to the product of the length of the arc and the distance travelled by its C.M.*

Let the arc, of length s, be rotated through four right angles about the x-axis (Fig. 11.14). We assume that y is a one-to-one function of s; then the surface area, A, is given by $\lim\limits_{\delta s \to 0} \Sigma\, 2\pi y\, \delta s$.* Thus $A = \int 2\pi y\, ds$, over the appropriate range for s. Now if y is the ordinate of the C.M. of the arc, we have

$$\bar{y} \cdot s = \lim_{\delta s \to 0} \sum y\, \delta s$$

$$= \int y\, ds.$$

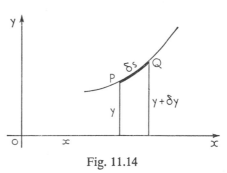

y + δy

Fig. 11.14

Thus $A = \int 2\pi y\, ds = (2\pi\bar{y})s$, which establishes the theorem.

THEOREM II. *If a plane convex closed area is rotated about an axis which it does not cut, then the volume of the solid formed is equal to the product of the area and the distance travelled by its C.M.*

Let the area, R, be rotated about the x-axis, the tangents parallel to the y-axis being $x = a$, $x = b$ with points of contact A, B respectively (Fig. 11.15).

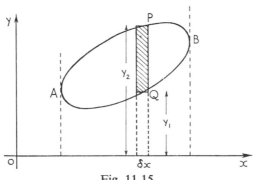

Fig. 11.15

Let P be the point (x, y_2) on the upper arc between A, B and Q the point (x, y_1) on the lower arc; then y_1, y_2 are distinct functions of x in the interval $[a, b]$. The volume formed on rotation is given by

$$V = \lim_{\delta x \to 0} \sum \pi y_2{}^2\, \delta x - \lim_{\delta x \to 0} \sum \pi y_1{}^2\, \delta x$$

$$= \int \pi(y_2{}^2 - y_1{}^2)\, dx$$

$$= 2\pi \int \frac{y_2 + y_1}{2}(y_2 - y_1)\, dx$$

$$= 2\pi\bar{y}R,$$

---

* We shall take this as intuitively 'obvious'. A proof can be given by going back to the definition of s as the limit of the sum of chord-lengths.

where $\bar{y}$ is the y-coordinate of the C.M. of R, since the area of the shaded rectangle is $(y_2 - y_1)\,\delta x$ and the ordinate of its C.M. is $\dfrac{y_2 + y_1}{2}$. This establishes the theorem.

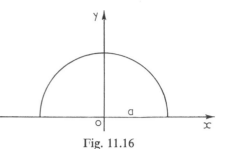

Fig. 11.16

EXAMPLE 11.9 *To find the C.M. of a semicircular arc of radius a.*

Let the C.M. be a distance $\bar{y}$ from the bounding diameter along the radius of symmetry (Fig. 11.16). Rotation about the x-axis yields a spherical shell of surface area $4\pi a^2$, so the first theorem gives $4\pi a^2 = 2\pi\bar{y} \times \pi a$, whence

$$\bar{y} = \frac{2a}{\pi}.$$

EXAMPLE 11.10 *To find the volume of the solid formed by rotating one loop of the curve $a^2y^2 = x^2(a^2 - x^2)$ about the x-axis.*

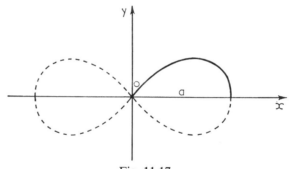

Fig. 11.17

From Example 11.1 (p. 190), the area of the half-loop in the positive quadrant is $\frac{1}{3}a^2$ and the coordinates of its C.M. are $\left(\dfrac{3\pi a}{16}, \dfrac{a}{5}\right)$. Using the second theorem of Pappus-Guldin, the volume V required is given by

$$V = \tfrac{1}{3}a^2 \times 2\pi\,\frac{a}{5} = \frac{2\pi a^3}{15}.$$

EXERCISE LV

1. Find the C.M. of a semicircular disc of radius a.

2. A *torus* is formed by rotating a circle of radius a about an axis in its plane at a distance $b\ (> a)$ from its centre. Find the volume of the torus.

3. An arch of the cycloid $x = a(\theta - \sin\theta)$, $y = a(1 - \cos\theta)$ is rotated through 4 right angles about the x-axis. Show that the area of the surface generated is $\dfrac{64\pi a^2}{3}$.

4. Draw a rough sketch of the curve

$$3ay^2 = x(a - x)^2.$$

Prove that the length of the arc of the loop of the curve is $4\sqrt{3}a/3$.
Find the area of the surface generated by rotating the loop about the x-axis.

5. The portion of the circle $x^2 + y^2 + 2ay\cos\alpha - a^2\sin^2\alpha = 0$ which lies in the positive quadrant (a and $\cos\alpha$ being positive) is rotated through four right angles about the x-axis. Prove that

(i) the volume V of the solid generated is

$$\pi a^3(\sin\alpha - \tfrac{1}{3}\sin^3\alpha - \alpha\cos\alpha),$$

(ii) the area of the curved surface of the solid is

$$2\pi a^2(\sin\alpha - \alpha\cos\alpha),$$

(iii) the distance of the centre of gravity of the solid from its plane face is

$$\pi a^4(1 - \cos\alpha)^3(3 + \cos\alpha)/12V.$$

6. An area is bounded by the arc of the curve

$$y = \sin^2 x,$$

terminated at the points where $x = 0$ and $x = \pi$, and by the segment of the x-axis joining these points. Find the coordinates of its centroid.
Hence find the volume generated by revolving this area about the y-axis.

7. A curve is given parametrically by the equations

$$x = e^t \sin t, \qquad y = e^t \cos t.$$

Prove that the tangent at the point with parameter t is inclined at an angle $(\tfrac{1}{4}\pi - t)$ to the x-axis.
Sketch the curve for values of t between 0 and 2π, indicating clearly on the sketch the correspondence between the curve and the values taken by the parameter.
Find the length of the curve between the points with parameters 0 and t.
The part of the curve for which $0 \leqslant t \leqslant \tfrac{1}{2}\pi$ is revolved through four right angles about the axis of x. Prove that the area of the surface generated is $\tfrac{2}{5}\pi(e^\pi - 2)\sqrt{2}$.

8. (D) (i) The diagram, Fig. 11.18, represents a sphere, centre O and radius a, and a circumscribing cylinder. These are cut by two planes parallel to the base of the cylinder, represented above by AB, CD (the continuous lines show a central section of the figure). Angle $XOA = \alpha$, angle $XOC = \beta$.

(a) Show that the surface area of the zone of the sphere between the parallel planes is given by $\lim\limits_{\delta\theta \to 0} \Sigma_{\theta=\alpha}^{\beta} 2\pi a \cos\theta \cdot a\, \delta\theta$.

(b) Find the surface area of the part of the cylinder contained between the two planes, and deduce:

Archimedes' Theorem. If a zone of a sphere is cut off by two planes parallel to the base of a circumscribing cylinder, then the surface area of the zone is equal to that of the part of the cylinder contained between the two planes.

(ii) Use the above result and the first Pappus-Guldin theorem to find the C.M. of a circular arc (cf. Example 11.3, p. 193).

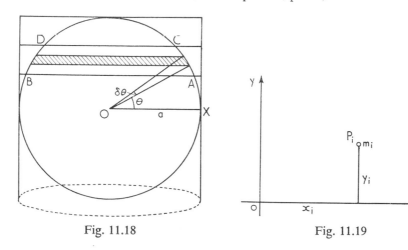

Fig. 11.18   Fig. 11.19

Moments of Inertia

Moments of inertia (M.I.) are needed in the study of the dynamics of a rigid body.

Consider a system of particles of mass m_i at points $P_i(x_i, y_i)$, ($i = 1, 2, \ldots, n$), Fig. 11.19. Their M.I. about the x-axis is defined by

$$I_x = \sum_{i=1}^{n} m_i y_i^2.$$

Similarly the M.I. about the y-axis is given by $I_y = \sum_{i=1}^{n} m_i x_i^2$.

This idea can be extended at once to three dimensions. For example, the M.I. of the mass m at $P(x, y, z)$ about the axis Oz is $m \cdot PN^2 [= m(x^2 + y^2)]$, where N is the foot of the perpendicular from P to Oz (Fig. 11.20).

The M.I. of a rigid body about an axis may be defined by considering it as made up of a number of small elements (cf. the treatment of centres of mass, p. 70).

EXAMPLE 11.11 *To find the M.I. of a uniform rod of length 2a and mass M about an axis through its centre perpendicular to its length.*

Refer to Fig. 11.20. Let the mass per unit length be ρ, so that $2a\rho = M$, and consider the rod as the limit of the sum of small elements mass $\rho \, \delta x$ at a distance x from 0 $(-a \leqslant x \leqslant a)$. Then the M.I. of the rod is given by

$$I_0 = \lim_{\delta x \to 0} \sum_{x=-a}^{a} x^2 \rho \, \delta x$$

$$= \int_{-a}^{a} \rho x^2 \, dx$$

$$= \frac{2\rho a^3}{3}$$

$$= \frac{Ma^2}{3} .$$

If the M.I. of a body is k^2 times its mass, then k is called the *radius of gyration*. Thus, in the above example, $k = \dfrac{a}{\sqrt{3}}$.

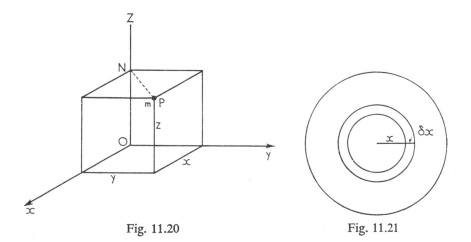

Fig. 11.20   Fig. 11.21

EXAMPLE 11.12   *To find the M.I. of* (i) *a circular ring of mass m and radius r about an axis through its centre perpendicular to its plane;* (ii) *a circular disc of mass M and radius a about an axis through its centre perpendicular to its plane;* (iii) *a cylinder of mass M' and radius a about its axis.*

    (i) Since each element of the ring (say δm) is at a distance r from the axis, the M.I. is $\Sigma \, (\delta m \, r^2) = (\Sigma \, \delta m) r^2 = m r^2$.

    (ii) Let the density be ρ, so that $\pi a^2 \rho = M$, and consider the disc as made up of the limit of the sum of concentric rings (Fig. 11.21) of width δx distant x from the centre $(0 \leqslant x \leqslant a)$. The mass of such a ring may be taken as $\rho \,.\, 2\pi x\ \delta x$, so using (i) the M.I. of the disc is given by

$$\lim \sum x^2 \,.\, \rho 2\pi x\ \delta x = \int_0^a 2\pi \rho x^3\ dx$$

$$= 2\pi \rho \,.\, \frac{a^4}{4}$$

$$= \tfrac{1}{2}Ma^2.$$

(iii) The cylinder can be considered as the limit of the sum of a number of thin discs (Fig. 11.22) of mass δM whose planes are perpendicular to the axis. From (ii), the M.I. is accordingly $\lim \Sigma\ (\tfrac{1}{2}\ \delta M\ a^2) = \lim (\Sigma\ \delta M)\tfrac{1}{2}a^2 = \tfrac{1}{2}M'a^2.$

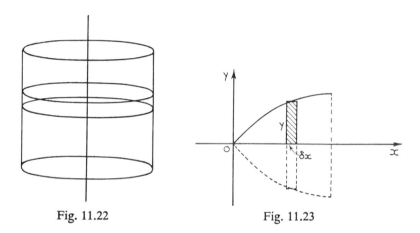

Fig. 11.22   Fig. 11.23

EXAMPLE 11.13 *A solid of mass M is formed by rotating the curve $y = \sin x$ $(0 \leqslant x \leqslant \pi/2)$ through four right angles about the x-axis; to find its M.I. about the x-axis.*

Consider the solid, as usual, to be made up of the limit of the sum of a number of cylinders of radius y and height δx (Fig. 11.23). Let the mass per unit volume be ρ. Then

$$M = \int_0^{\frac{\pi}{2}} \rho\pi y^2\ dx = \rho\pi \int_0^{\frac{\pi}{2}} \sin^2 x\ dx$$

$$= \rho\frac{\pi}{2} \int_0^{\frac{\pi}{2}} (1 - \cos 2x)\ dx$$

$$= \frac{\rho\pi^2}{4}. \tag{1}$$

Now the M.I. about the x-axis, I_x, is given by $\lim \Sigma \frac{1}{2} y^2 \times \rho \pi y^2 \, \delta x$ (the M.I. of a cylinder of mass m and radius r about its axis being $\frac{1}{2} mr^2$).
Thus

$$I_x = \int_0^{\frac{\pi}{2}} \frac{1}{2} \rho \pi y^4 \, dx$$

$$= \frac{1}{2} \rho \pi \int_0^{\frac{\pi}{2}} \sin^4 x \, dx$$

$$= \frac{1}{8} \rho \pi \int_0^{\frac{\pi}{2}} (1 - \cos 2x)^2 \, dx$$

$$= \frac{1}{8} \rho \pi \int_0^{\frac{\pi}{2}} \{1 - 2 \cos 2x + \frac{1}{2}(1 + \cos 4x)\} \, dx$$

$$= \frac{1}{8} \rho \pi \times \frac{3\pi}{4} = \frac{3 \rho \pi^2}{32} = \frac{3}{8} M, \qquad \text{from (1)}.$$

EXAMPLE 11.14   *A frustum of a uniform solid right circular cone is bounded by circles of radii r and R and is of mass M. Show that the moment of inertia of the frustum about its axis is*

$$\tfrac{3}{10} M \frac{R^5 - r^5}{R^3 - r^3}.$$

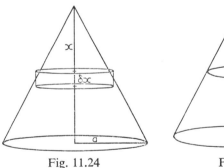

Fig. 11.24   Fig. 11.25

Refer to Fig. 11.24. We first find the M.I. about its axis of a cone of base-radius a and mass m. Let its height be h and density ρ, so that $\frac{1}{3} \pi a^2 h \rho = m$.
Consider this cone as the limit of the sum of cylinders of height δx at distance x from the vertex ($0 \leqslant x \leqslant h$). The radius of such a cylinder is $\dfrac{xa}{h}$ and its M.I. about its axis is

$$\frac{1}{2} \left(\frac{xa}{h}\right)^2 . \pi \left(\frac{xa}{h}\right)^2 \rho \, \delta x = \frac{\rho \pi a^4 x^4}{2h^4} . \delta x.$$

Thus the M.I. of the cone is

$$\int_0^h \frac{\rho\pi a^4 x^4}{2h^4}\, dx = \frac{\rho\pi a^4}{2h^4}\cdot\frac{h^5}{5} = (\tfrac{1}{3}\pi a^2 h\rho)\tfrac{3}{10}a^2 = \tfrac{3}{10}ma^2.$$

Now consider the given frustum together with a conical cap of radius r, the two together forming a cone of radius R (Fig. 11.25). If the mass of the cap is kr^3 and that of the complete cone is kR^3, then that of the frustum is

$$M = k(R^3 - r^3). \tag{1}$$

Now, since moments of inertia are additive,

$$I_R = I_F + I_r$$

where I_R, I_F, I_r are the respective M.I.'s of the complete cone, the frustum and the cap about their axis. Thus

$$I_F = I_R - I_r$$
$$= \tfrac{3}{10}kR^3 \cdot R^2 - \tfrac{3}{10}kr^3 \cdot r^2$$

(from above)

$$= \tfrac{3}{10}k(R^5 - r^5)$$
$$= \tfrac{3}{10}M\frac{R^5 - r^5}{R^3 - r^3}, \qquad\text{from (1).}$$

EXERCISE LVI

1. (D) A rectangle of sides $2a$, $2b$ $(a > b)$ is of mass M. By considering it as the limit of the sum of a number of strips, show that its M.I. about an axis through its centre parallel to the shorter side is $\tfrac{1}{3}Ma^2$.
2. Find the radius of gyration about the x-axis of the lamina bounded by the curve $y = x^3$, the y-axis and the line $y = 8$. [Consider strips parallel to the x-axis.]
3. A solid of mass M is formed by rotating the area under the curve $y^2 = x$ between $x = 2$ and $x = 3$ through four right angles about the x-axis. Find the M.I. of this solid about the x-axis.

The Lamina Theorem

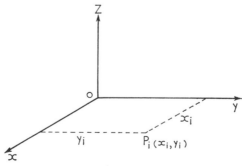

Fig. 11.26

We return to the system of particles of mass m_i at points $P_i(x_i, y_i)$ in the plane xOy (Fig. 11.26). Let Oz be perpendicular to this plane. If I_x, I_y, I_z denote the M.I.'s of the system about the three axes Ox, Oy, Oz respectively, then we have

$$I_z = \sum (m_i \, OP_i{}^2)$$
$$= \sum m_i(x_i{}^2 + y_i{}^2)$$
$$= \sum (m_i x_i{}^2) + \sum (m_i y_i{}^2)$$
$$= I_x + I_y.$$

By considering a rigid body as made up of a number of such particles, we have the theorem:

The M.I. of a lamina about an axis perpendicular to its plane, through a point O in its plane, is equal to the sum of its M.I.'s about any two perpendicular axes through O in its plane.

EXAMPLE 11.15 *To find the M.I. of a circular disc of mass M and radius a about a diameter.*

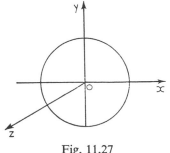

Fig. 11.27

Let Ox, Oy be two perpendicular diameters and let Oz be at right angles to the plane containing the disc (Fig. 11.27).

By symmetry,

$$I_x = I_y.$$

Also

$$I_x + I_y = I_z \text{ (lamina theorem)}$$
$$= \tfrac{1}{2}Ma^2 \text{ (from Example 11.12).}$$

Thus the required M.I., I_x, is $\tfrac{1}{4}Ma$.

The Parallel Axes Theorem

Refer to Fig. 11.28. Let Gx, Gy, Gz be rectangular axes of coordinates, where G is the C.M. of particles of mass m_i at points $P_i(x_i, y_i, z_i)$, $(i = 1, 2, \ldots, n)$. Then, since G is the origin, it follows from the definition of C.M. that

$$\sum m_i x_i = 0. \tag{1}$$

Let A be the point $(h, 0, 0)$ and let AK be a line parallel to Gz, the distance between Gz and AK accordingly being h. Let the M.I. of the system about Gz be I_G and that about AK be I_A. We shall prove that

$$I_A = I_G + (\sum m_i)h^2. \tag{2}$$

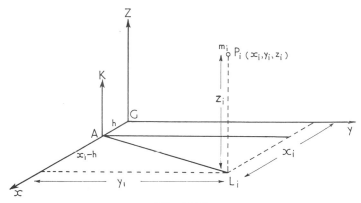

Fig. 11.28

For $I_A = \sum m_i AL_i^2$ (where L_i is the foot of the perpendicular from P_i on to the plane xGy)

$$= \sum m_i\{(x_i - h)^2 + y_i^2\}$$

$$= \sum m_i\{x_i^2 + y_i^2) - 2hx_i + h^2\}$$

$$= \sum m_i(x_i^2 + y_i^2) - 2h \sum m_i x_i + (\sum m_i)h^2$$

$$= I_G + (\sum m_i)h^2, \text{ since } \sum m_i x_i = 0 \text{ from eqn. (1).}$$

Considering a rigid body as made up of a number of such particles, we have the Parallel Axes Theorem:

If the M.I. of a rigid body of mass M about an axis through its C.M. is I_G, and the M.I. about a parallel axis a distance h away is I_A, then

$$I_A = I_G + Mh^2.$$

EXAMPLE 11.16   *To find the M.I. of a uniform rod of length 2a and mass M about an axis through one end perpendicular to its length.*

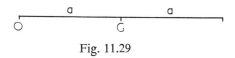

Fig. 11.29

From Example 11.11, the M.I. about a parallel axis through G is $\frac{1}{3}Ma^2$. Thus, the required M.I. is $\frac{1}{3}Ma^2 + Ma^2 = \dfrac{4Ma^2}{3}$, from the parallel axes theorem. The reader should verify this result by taking O as origin and proceeding directly, as in Example 11.11.

EXAMPLE 11.17   *A solid of mass M is formed by rotating the area between* $y^2 = x$, *the x-axis and the line* $x = 2$ *through 4 right angles about the x-axis. To find the M.I. of this solid about the y-axis.*

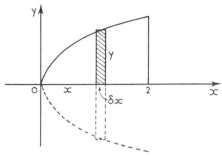

Fig. 11.30

If the density of the solid is ρ, we have

$$M = \rho \int_0^2 \pi y^2 \, dx = \pi \rho \int_0^2 x \, dx = 2\pi\rho.$$

We consider the solid as the limit of the sum of cylindrical discs of radius y and height δx. The M.I. of such a strip about a diameter is $\frac{1}{4}y^2 \cdot \rho\pi y^2 \, \delta x$ (cf. Example 11.16) and so, by the parallel axes theorem, its M.I. about Oy is $(x^2 + \frac{1}{4}y^2)\rho\pi y^2 \, \delta x$. Thus the M.I. of the solid about Oy is

$$\int_0^2 \{(x^2 + \tfrac{1}{4}y^2)\rho\pi y^2 \, dx = \rho\pi \int_0^2 (x^3 + \tfrac{1}{4}x^2) \, dx$$
$$= \rho\pi(4 + \tfrac{2}{3})$$
$$= 2\rho\pi \times \tfrac{7}{3}$$
$$= \frac{7M}{3}.$$

EXERCISE LVII

1. (D) (i) Show that the M.I. of a rectangle of mass M and sides $2a$, $2b$ about a line through its centre perpendicular to its plane is $\frac{1}{3}M(a^2 + b^2)$.
   (ii) Deduce the M.I. of a cuboid of sides $2a$, $2b$, $2c$ about an axis joining the centres of the faces with dimensions $2a \times 2b$.
2. Find the radius of gyration of a circular disc of radius a about a tangent.
3. Find the M.I. of a cone of base-radius r, height h and mass M about a line through the vertex perpendicular to its axis.
4. Find the radius of gyration about the x-axis of the area bounded by the curve $y = x^2$, the x-axis and the line $x = 5$.

5. Show that the M.I. of a body about an axis through its C.M. is less than that about any parallel axis.

6. A uniform circular disc of radius $3a$ has a circular hole of radius a pierced in it, the edge of the hole passing through the centre of the original disc. Show that the moment of inertia about the diameter which is tangent to the edge of the hole is $\frac{19}{8}Ma^2$, where M is the mass of the pierced disc.

7. A uniform rectangular plate is of mass M and its sides are of length a and b. Find its moment of inertia about a diagonal.

8. A uniform lamina is bounded by the chord AB of a circle, of radius a, and the arc ACB which subtends an angle of $90°$ at the centre of the circle. Prove that the radius of gyration of the lamina about the diameter parallel to AB is

$$\tfrac{1}{2}a\sqrt{\left(\frac{\pi}{\pi-2}\right)}.$$

9. A flat circular disc is of radius $\frac{1}{2}a\pi$, and its density at a distance r from its centre is

$$\rho = \rho_0 \cos\left(\frac{r}{a}\right),$$

where ρ_0 is a constant. Find
   (i) the total mass of the disc;
   (ii) the moment of inertia of the disc about an axis through its centre and perpendicular to its plane.

Revision Exercise II

1. Differentiate the following functions with respect to x:

(i) $\dfrac{1 - x^2}{1 + x^2}$;   (ii) $x^2 \ln x$;   (iii) $\sin^2 3x \tan 3x$.

If $y = e^{-2x} \sin 2x$, prove that

$$\frac{d^2y}{dx^2} = -8e^{-2x}\cos 2x.$$

2. (i) Differentiate with respect to x the functions

$$\frac{1 - x}{(2 - x)(1 + x)}, \qquad \tan^{-1}\frac{1}{x}, \qquad \ln(\sec x + \tan x),$$

giving each answer in its simplest form.
   (ii) If x and y are given in terms of the variable θ by the equations

$$x = a(2\theta + \sin 2\theta), \qquad y = a(1 - \cos 2\theta),$$

prove that

$$\frac{dy}{dx} = \tan\theta \quad \text{and} \quad \frac{d^2y}{dx^2} = \frac{1}{4a}\sec^4\theta.$$

3. Differentiate with respect to x:

(i) $\dfrac{1 + x}{x(1 - x)}$;   (ii) xe^{x^2};   (iii) $\ln\left(\dfrac{1 + \tan x}{1 - \tan x}\right)$.

Given that $y = e^{4x} \cos 3x$, prove that

$$\frac{d^2y}{dx^2} - 8\frac{dy}{dx} + 25y = 0.$$

4. (i) Differentiate with respect to x the functions $x(1 + x^2)^{-\frac{1}{2}}$, $e^x \cos 2x$, ln $(\sec x + \tan x)$, giving your answers in their simplest form.

(ii) Given that $x = \tan \theta$, $y = \sin 2\theta$, prove that

$$(1 + x^2)\frac{dy}{dx} = 2 \cos 2\theta.$$

By differentiating with respect to x, prove that

$$(1 + x^2)^2 \frac{d^2y}{dx^2} + 2x(1 + x^2)\frac{dy}{dx} + 4y = 0.$$

5. (i) Differentiate with respect to x the functions

$$x(x^2 + 1)^{-\frac{1}{2}}, \qquad \ln (\sec x + \tan x),$$

giving your answers in their simplest form.

(ii) Given that $(x - 2y)^4 = a(x + y)$, where a is constant, prove that

$$\frac{dy}{dx} = \frac{x + 2y}{3x + 2y} \quad \text{and} \quad \frac{d^2y}{dx^2} = 4\frac{(x + y)(x - 2y)}{(3x + 2y)^3}.$$

6. (i) If $y = \left(\sinh^{-1}\frac{x}{a}\right)^2$, show that

$$(a^2 + x^2)\frac{d^2y}{dx^2} + x\frac{dy}{dx} = 2.$$

(ii) If $\sin y = x \sin (y + \frac{1}{6}\pi)$, show that

$$\frac{dy}{dx} = 2 \sin^2 (y + \frac{1}{6}\pi).$$

7. Differentiate with respect to x

$$\frac{1 - 2x}{(1 + 3x)^2}, \qquad e^{\tan^2 x}, \qquad \sin^{-1}\frac{1}{\sqrt{(1 + x^2)}}.$$

Prove that, when $y = at^2 + 2bt + c$, $t = ax^2 + 2bx + c$, and a, b, c are constants,

$$\frac{d^3y}{dx^3} = 24a^2(ax + b).$$

8. Differentiate with respect to x:

(i) $(1 + x^2)\sqrt{(1 - x^2)}$; (ii) $\ln (1 + x^2)$; (iii) $\tan^{-1} 3x + \cot^{-1} 3x$.

Given that $y = \sin (\ln x)$, prove that

$$x^2\frac{d^2y}{dx^2} + x\frac{dy}{dx} + y = 0.$$

9. Differentiate the following functions with respect to x:

(i) $\dfrac{\sqrt{(1-x^2)}}{x^3}$;   (ii) $\dfrac{\sin x}{1+\cos x}$;   (iii) $e^{-2x^2}\ln 3x$.

If $x = 3t + t^3$ and $y = 3 - t^{\frac{3}{2}}$,

express $\dfrac{dy}{dx}$ in terms of t and prove that, when $\dfrac{d^2y}{dx^2} = 0$, then x has one of the values 0, $\pm 6\sqrt{3}$.

10. Integrate with respect to x:

$$\frac{1}{(3+x)(2-x)} ; \qquad x \ln x.$$

Evaluate $\displaystyle\int_0^{\frac{\pi}{2}} \sin^3 \theta \, d\theta.$

11. Integrate:

$$\frac{x^3}{1-x^2}, \qquad \ln(1-x), \qquad \frac{1}{(a^2+x^2)^{\frac{3}{2}}}, \qquad \operatorname{cosec} x.$$

Evaluate: $\displaystyle\int_0^2 \frac{dx}{x^2+2x+4}.$

12. Integrate:

$$\int \frac{x}{x^4-1}\,dx; \qquad \int \frac{x^2-x+1}{x^2+x+1}\,dx;$$

$$\int \sec^3 x \tan x \, dx; \qquad \int \sin^{-1} x \, dx.$$

13. Integrate the following functions with respect to x:

(i) $x\sqrt{(1-x^2)}$;   (ii) $\dfrac{\sin^3 x}{\cos^2 x}$;   (iii) $\dfrac{5x}{(2x+1)(x-2)}$.

Evaluate $\displaystyle\int_0^{\frac{\pi}{2}} x^2 \sin x \, dx.$

14. Integrate the following functions with respect to x:

(i) $\cos x \sin^3 x$;   (ii) $\dfrac{2}{4x^2-1}$;   (iii) $\{x + \sqrt{(1+x^2)}\}^2$.

If $y = \sin^2(\theta + \alpha) + \sin^2(\theta - \alpha) - 2\cos 2\alpha \sin(\theta + \alpha)\sin(\theta - \alpha)$ where α is a constant, prove that $dy/d\theta = 0$, and hence show that the value of y is $\sin^2 2\alpha$ for all values of θ.

Evaluate $\displaystyle\int_0^1 x^3 e^{x^2} \, dx.$

15. Integrate with respect to x:

$$\int \frac{x+2}{\sqrt{(1-x^2)}}\,dx; \qquad \int \ln x \, dx.$$

Evaluate: $\displaystyle\int_0^2 \frac{dx}{x^2-2x+2} ; \qquad \int_0^{\pi} (1+\cos\theta)^2 \, d\theta.$

16. (a) Prove that $\displaystyle\int_0^{\frac{\pi}{2}} (\sin x + \cos x)^2 \, dx = \frac{\pi}{2} + 1$

(b) Integrate xe^{2x} with respect to x.

17. (i) Integrate with respect to x:

    (a) $x^{a/b} + x^{-a/b}$,   (b) $\cos^2 3x$,   (c) $e^x \sin x$.

    (ii) Evaluate $\displaystyle\int_1^2 \frac{x^3 \, dx}{x^2 + 1}$.

18. Evaluate the indefinite integrals

$$\int \sec^4 x \, dx, \qquad \int x\sqrt{(2x + 3)} \, dx, \qquad \int \frac{\ln x}{x} \, dx.$$

Find the numerical value, correct to 3 places of decimals, of

$$\int_1^2 \frac{x + 2}{x^2 + 4} \, dx.$$

19. (i) Prove that, if m and n are unequal positive integers,

$$\int_0^\pi \cos mx \cos nx \, dx = 0.$$

Find the value of the integral when m and n are *equal* positive integers.

    (ii) Find the value of

$$\int_1^a \left(x^2 + \frac{1}{x^3}\right)^2 dx.$$

20. Integrate with respect to x:

$$\text{(i)} \ \frac{x^2 + 1}{x + 1} \, ; \qquad \text{(ii)} \ \frac{1}{x^2} \ln x.$$

Evaluate $\displaystyle\int_0^1 \sqrt{(1 - x^2)} \, dx.$

21. (i) Evaluate

$$\int_2^3 \frac{dx}{(x - 1)(x + 2)}, \qquad \int_0^1 \frac{x^2 \, dx}{\sqrt{(1 - x^2)}}.$$

    (ii) Determine $\displaystyle\int \frac{dx}{(x^2 - 1)^{\frac{3}{2}}}$.

22. Show that

$$\frac{d}{dx}\left\{\ln (\sec x + \tan x)\right\} = \sec x$$

Find, correct to three significant figures,

$$\text{(i)} \int_0^{\frac{\pi}{4}} (\cos x + \sec x) \, dx \qquad \text{(ii)} \int_0^{\frac{\pi}{4}} (\cos x + \sec x)^2 \, dx$$

23. Prove that, if $f'(x) > 0$ for $a < x < b$, then

$$f(a) < f(x) < f(b)$$

when $a < x < b$.

Hence, or otherwise, prove that, if $0 < x < \frac{1}{2}\pi$,

(i) $0 < x - \sin x \cos x < \frac{1}{2}\pi$,

(ii) $0 < x \sin x - \frac{1}{6}\sin^3 x < \frac{1}{2}\pi - \frac{1}{6}$.

24. A curve, whose equation is $y = f(x)$, is such that $\dfrac{d^2 y}{dx^2} = ax + b$; the curve has a minimum at $(2, -19)$ and a maximum when $x = -3$. Prove that $a = 2b$ and $a > 0$.

If the curve passes through $(3, -10\frac{1}{2})$, find its equation.

25. State and prove any one set of conditions that are sufficient to ensure that a function of x shall have a minimum when $x = a$.

A cone, whose semivertical angle is θ, is circumscribed about a sphere of radius a. Prove that the volume of the cone is

$$\frac{1}{3}\pi a^3 \frac{(1 + \sin\theta)^2}{\sin\theta(1 - \sin\theta)},$$

and deduce that the *minimum* volume of such a cone is twice the volume of the sphere.

26. (i) If

$$y = A \cos(\ln x) + B \sin(\ln x),$$

where A and B are constants, prove that

$$x^2 \frac{d^2 y}{dx^2} + x \frac{dy}{dx} + y = 0.$$

(ii) Find the maximum and minimum values of

$$\frac{\sin x}{2 + \cos x}$$

distinguishing between them.

27. (a) If $y = \dfrac{1}{x}\sin x$, show that

$$\frac{d^2 y}{dx^2} + \frac{2}{x}\frac{dy}{dx} + y = 0.$$

(b) A right circular cone is inscribed in a sphere of radius a. If θ is the semi-vertical angle of the cone, determine the area of its curved surface. Find the value of θ when this area is a maximum.

28. (i) By considering the sign of its derivative, prove that the function $\tan x - x$ is positive when $0 < x < \frac{1}{2}\pi$.

(ii) Verify that the function $ax \sin 2x - (a + b)\sin^2 x$ has a stationary value at $x = 0$ which is a minimum when $a > b$ and a maximum when $a < b$. By using (i) above, or otherwise, show that the function has a maximum at $x = 0$ when $a = b$ and $a > 0$.

29. If $y = e^{-x}\cos x$, determine the three values of x between 0 and 3π for which $dy/dx = 0$. Show that the corresponding values of y form a geometric progression with common ratio $-e^{-\pi}$.

30. By first putting the expression into partial fractions, or otherwise, find the first and second derivatives of the function

$$\frac{3x - 1}{(4x - 1)(x + 5)}.$$

Find the coordinates of any maxima, minima and points of inflexion that the function may have, and draw a rough sketch of its graph.

31. (i) A right circular cylinder is inscribed in a fixed sphere of radius a. Prove that the total area of its surface (including the ends) is $2\pi a^2$ (sin 2θ + cos^2 θ), where $a \cos \theta$ is the radius of an end. Hence prove that the maximum value of the total area is $\pi a^2(\sqrt{5} + 1)$.

(ii) Prove that the function $x^{-1}e^x$ has a minimum value e at $x = 1$. By considering a rough graph of the function, show that the equation $e^x = kx$ has two real roots if $k > e$, one root real if $k = e$ or $k < 0$, no real roots if $0 \leqslant k < e$.

32. The normal to the parabola $y^2 = 4x$ at the point $P(t^2, 2t)$ meets the parabola again at Q. Show that

$$PQ^2 = \frac{16(t^2 + 1)^3}{t^4}$$

Find the minimum length of PQ.

33. Prove that the graph of the function $f(x)$ has a minimum at the point whose abscissa is a if $f'(a) = 0$ and $f''(a)$ is positive.

A variable straight line passing through (4, 1) meets the positive x-axis at A and the positive y-axis at B. Express OB in terms of OA (where O is the origin), and prove that the minimum value of $(OA + OB)$ is 9.

Find also the minimum area of the triangle OAB.

34. The tangent to the curve $y = 1 - x^2$ at $x = \alpha$, where $0 < \alpha < 1$, meets the axes at P and Q. Prove that, as α varies, the minimum value of the area of the triangle OPQ is $\dfrac{2}{\sqrt{3}}$ times the area bounded by the axes and the part of the curve for which $0 < x < 1$.

35. Prove that, if $f'(a) = 0$ and $f''(a)$ is negative, then the graph of the function $f(x)$ has a maximum at the point whose abscissa is a.

A straight line passes through the point $P(0, 1)$ and meets the negative x-axis at A; a point B is taken on $\overrightarrow{AP}$ produced such that the length of AB is 8 units. Prove that, if x is the abscissa of B,

$$x = 8 \cos \theta - \cot \theta,$$

where θ is the angle of slope of AB. Hence prove that x is a maximum (and *not* a minimum) when $\theta = 30°$.

36. (i) Differentiate with respect to x:

(a) $\dfrac{x}{\sqrt{(1 - x^2)}}$,   (b) ln tan $2x$,   (c) 3^{5x}.

(ii) Using the method of small increments, calculate the value of cot $45°$ 1′ correct to five decimal places.

37. The coordinates of three points on the curve

$$y = A + Bx + Cx^2$$

are (x_1, y_1), (x_2, y_2) and (x_3, y_3), where $x_2 = \frac{1}{2}(x_3 + x_1)$. Prove that the area under the curve between the lines $x = x_1$ and $x = x_3$ is equal to

$$\tfrac{1}{6}(x_3 - x_1)(y_1 + 4y_2 + y_3).$$

Deduce Simpson's rule for five ordinates.

Using five ordinates and four-figure tables, apply Simpson's rule to evaluate the integral $4\int_0^1 \dfrac{dx}{1+x^2}$ and thus to find a value for π correct to three places of decimals.

38. Sketch the graph of $y = x + \cos x - 2$. Show that the equation

$$x + \cos x = 2$$

has a root just less than π and use Newton's method to find it correct to two decimal places.

39. Sketch the curves (i) $y = e^x \sin x$, (ii) $y = xe^{-x}$, (iii) $y = \dfrac{\sin x}{x}$.

40. Draw a rough sketch of the curve

$$y = \frac{a^3}{x^2 + a^2} \qquad (a > 0).$$

Find the equation of the tangent to the curve at the point

$$(a \tan \theta, \ a \cos^2 \theta).$$

Prove that there are two tangents other than the line $y = a$ which go through the point $(0, a)$. Show that the area of the triangle contained by these two tangents and the x-axis is $2a^2$.

41. Show that the curve $a^2 y^2 = x^2(a^2 - x^2)$ consists of two loops. Find the area included in one of the loops and prove that the centroid of this area is a distance $\tfrac{3}{16}\pi a$ from the origin.

If this loop is rotated through four right angles about the y-axis, find the volume of the solid thus formed.

42. Sketch the curve $y = e^{2x}$.

(i) P is any point on the curve and the tangent at P meets the x-axis at T. N is the foot of the perpendicular from P on to the x-axis. Prove that TN is of constant length.

(ii) A, B are the points on the curve for which x has the values 0, 1 respectively. Find the area bounded by the arc AB and the chord AB.

43. A curve is given by the parametric equations

$$x = t^3 - t^2 + 2,$$
$$y = t^2 - 1.$$

Prove that the equation of the tangent at the point t_1 is

$$(3t_1 - 2)y = 2x + (t_1 - 2)(t_1 + 1)^2,$$

and show that this tangent meets the curve again at the point $t = -\tfrac{1}{2}t_1$.

44. Sketch the curve given by $x = 2a \sin \theta - a \sin 2\theta$, $y = 2a \cos \theta - a \cos 2\theta$. Show that tangents at points whose parameters differ by π meet at right angles on the circle $x^2 + y^2 = 9a^2$.

45. (i) Find the maximum and minimum values of the function
$$(1 + 2x^2)e^{-x^2}.$$

(ii) The area bounded by the parabola $y^2 = 20x$ and the lines $x = 0$ and $y = 10$ is rotated about the y-axis. Show that the volume of the solid obtained is 50π, and find the volume of the solid obtained by rotating the same area about the x-axis.

46. Draw a rough sketch of that part of the curve given by the parametric equations

$$x = a(2t + \sin t), \qquad y = a \cos t,$$

for which $-\pi < t < \pi$.

Find the value of $\dfrac{d^2y}{dx^2}$ in terms of t, and deduce the coordinates of the points of inflexion in the range $-\pi < t < \pi$.

Find also the area contained between the x-axis and the part of the curve within this range which lies above it.

47. Show that the curve $ay^2 = x^2(a - x)$, $(a > 0)$, contains a loop one half of which lies in the positive quadrant. Obtain, in terms of X, the slope of the curve at any point (X, Y) on this half, and find the point B where the tangent is parallel to the x-axis.

Prove that the normal at any point on the whole loop except the point $(a, 0)$ cuts the x-axis at a point which is not outside the interval from the origin to the foot of the ordinate through B.

48. A solid hemisphere is divided into two parts by a plane parallel to the flat base. Prove that, if the total surface areas of the separated parts of the hemisphere are the same, their volumes are in the ratio 47/81.

49. The line $3y = 2x$ cuts the curve $6xy = x^4 + 3$ in four points. P and Q are the two of these points which lie in the positive quadrant. Prove that the length of the curve between P and Q is $\frac{1}{3}(\sqrt{3} + 1)$.

50. Find the area contained between the x-axis and that part of the curve

$$x = 2t^2 + 1, \qquad y = t^2 - 2t$$

which corresponds to values of t lying between 0 and 2.

Find also the coordinates of the centroid of this area.

51. Two points A, B are taken on a parabola, both points being on the same side of the axis and B being farther from the vertex than A. Lines AC, BC are drawn parallel to the axis and the directrix respectively. If the area ABC is revolved through four right angles about the axis of the parabola, show that the volume of the solid generated is equal to that of a right circular cylinder whose radius is AC and whose length is the semi-latus rectum of the parabola.

Show that the centroid of the solid divides its axis in the ratio $2:1$.

52. Prove that

$$\int (1 - \cos\theta)^2 \cos\theta \, d\theta = \tfrac{1}{12}\sin 3\theta - \tfrac{1}{2}\sin 2\theta + \tfrac{7}{4}\sin\theta - \theta.$$

The area enclosed between the quadrant of a circle of radius a and the tangents at its extremities is rotated through four right angles about one of these tangents. Prove that the volume of the solid so formed is $\pi a^3(\frac{5}{3} - \frac{1}{2}\pi)$.

The solid is now divided into two parts by a cross-section through the mid-point of its axis of symmetry. Prove that the volumes of the parts are in the ratio

$$\frac{23 - 6\sqrt{3} - 4\pi}{17 + 6\sqrt{3} - 8\pi}.$$

53. Find (i) the C.M. of the part of the ellipse $\dfrac{x^2}{a^2} + \dfrac{y^2}{b^2} = 1$ which lies in the first quadrant, and (ii) the volume of the half-ellipsoid formed by rotating this area about the y-axis.

54. State Taylor's theorem for the expansion of $f(x + h)$ in a series of powers of h and deduce that, if $f(a) = 0$ and $g(a) = 0$, but $f'(a)$ and $g'(a)$ are not zero, then

$$\lim_{x \to a} \left\{ \frac{f(x)}{g(x)} \right\} = \frac{f'(a)}{g'(a)}.$$

Evaluate the following limits:

   (i) $\displaystyle\lim_{x \to a} \dfrac{x^3 - a^3}{x^2 - a^2}$,   (ii) $\displaystyle\lim_{x \to 0} \dfrac{\tan x - x}{x - \sin x}$.

55. State Maclaurin's theorem for the expansion of the function $f(x)$ in a series of ascending powers of x.

If $y = (1 + x)^2 \ln (1 + x)$, evaluate $\dfrac{dy}{dx}$ and $\dfrac{d^2y}{dx^2}$, and prove that, if $n > 2$, then

$$\frac{d^ny}{dx^n} = (-1)^{n-1} \frac{2 \cdot (n - 3)!}{(1 + x)^{n-2}}.$$

Prove that the expansion of y as far as the term in x^4 is

$$y = x + \tfrac{3}{2}x^2 + \tfrac{1}{3}x^3 - \tfrac{1}{12}x^4.$$

56. Differentiate with respect to x:

   (i) $\dfrac{x}{1 - x^2}$;   (ii) $\dfrac{x}{\ln x}$;   (iii) $\sin^{-1} \dfrac{1}{\sqrt{(1 + x^2)}}$.

Use Maclaurin's theorem to show that the expansion of $\sec x$ in a series of ascending powers of x as far as the term in x^4 is

$$1 + \frac{x^2}{2} + \frac{5x^4}{24}.$$

57. Differentiate with respect to x:

   (i) $\dfrac{1 + x^2}{1 - x}$;   (ii) $\ln \{x + \sqrt{(1 + x^2)}\}$;   (iii) $x\sqrt{(1 - x^2)} + \sin^{-1} x$.

If $y = e^{4x} \cos 3x$, prove that

$$\frac{dy}{dx} = 5e^{4x} \cos (3x + \alpha),$$

where $\tan \alpha = \tfrac{3}{4}$. Use Maclaurin's theorem to find the expansion of y in ascending powers of x as far as the term in x^3.

58. (i) If

$$y = \frac{1}{1 + x} \ln (1 + x)$$

prove that

$$(1 + x)\frac{dy}{dx} + y = \frac{1}{1 + x}$$

and

$$(1 + x)\frac{d^2y}{dx^2} + 2\frac{dy}{dx} = -\frac{1}{(1 + x)^2}.$$

Prove that, if x^5 and higher powers are neglected in the expansion of y by Maclaurin's theorem, then

$$y = x - \tfrac{3}{2}x^2 + \tfrac{11}{6}x^3 - \tfrac{25}{12}x^4.$$

(ii) Prove by means of Taylor's expansion that, if x is the radian measure of an angle which is so small that x^3 and higher powers of x can be neglected, then

$$\sin\left(\frac{1}{6}\pi + x\right) = \frac{1}{2} + \frac{\sqrt{3}}{2}x - \frac{1}{4}x^2.$$

59. State Maclaurin's theorem on the expansion of a function of x as a series of powers of x.

Prove that, if $y = e^x \sin x$, then

$$\frac{dy}{dx} = \sqrt{2} \, . \, e^x \sin(x + \tfrac{1}{4}\pi),$$

$$\frac{d^2y}{dx^2} = 2e^x \sin(x + \tfrac{1}{2}\pi).$$

Hence obtain the Maclaurin expansion of $e^x \sin x$ as far as the term in x^4. What is the coefficient of x^n in this expansion?

60. Differentiate with respect to x:

(i) $\dfrac{x^4}{(1 + x^2)^2}$;   (ii) $e^{2x} \ln \sec x$;   (iii) $\tan^{-1}\left(\dfrac{1 - x^2}{1 + x^2}\right)$.

Use Maclaurin's theorem to show that, if x^5 and higher powers of x are neglected,

$$\ln\{x + \sqrt{(1 + x^2)}\} = x - \tfrac{1}{6}x^3.$$

61. State Maclaurin's theorem for the expansion of a function $f(x)$ in a series of ascending powers of x, giving the coefficient of x^n.

Prove that, if $y = e^{\sin^2 x}$, then

$$\frac{dy}{dx} = y \sin 2x, \qquad \frac{d^2y}{dx^2} = \tfrac{1}{2}y(1 + 4\cos 2x - \cos 4x).$$

Obtain the expansion of y as far as the term in x^4.

62. (i) Assuming that a function $f(x)$ can be expanded in a series of ascending powers of x, state Maclaurin's theorem for finding the coefficients.

Use Maclaurin's theorem to obtain the first three non-vanishing terms of the series for $\sinh x$, and also the general term.

(ii) Write down the first three terms, and the general term, of the expansion of $\ln(1 + \tfrac{1}{3}x^2)$, and state the range of values of x for which it is valid.

63. Use Maclaurin's theorem to obtain the expansion, in ascending powers of x, of

(i) $(1 + x)^{-3}$ as far as the term in x^4,

(ii) $e^{-x} \sin x$ as far as the term in x^5.

Evaluate

$$\ln\{(1 + x\cosh x)/(1 - x\cosh x)\}^{\frac{1}{2}}$$

to two decimal places when $x = 0\cdot1$.

64. (i) State and prove Leibniz's theorem.

(ii) Find the n-th derivative of (a) $(x^3 - 3x^2 + 1)e^{2x}$, (b) $\dfrac{1}{(x-1)(x+2)}$.

(iii) Show that the n-th derivative of $e^{2x} \sin 2x$ is $2^{3n/2}e^{2x} \sin \left(2x + \dfrac{n\pi}{4}\right)$.

65. The values at $x = 0$ of $\tan x$ and its successive derivatives are denoted by $a_0, a_1, a_2, \ldots$. Prove, by using Leibniz's theorem on $y \cos x = \sin x$, or otherwise, that $a_0 = a_2 = a_4 = 0$ and that

$$a_1 = 1, \qquad a_3 - 3a_1 = -1, \qquad a_5 - 10a_3 + 5a_1 = 1.$$

Hence, by Maclaurin's theorem, show that the expansion of $\tan x$ begins with the terms

$$x + \frac{x^3}{3} + \frac{2}{15}x^5.$$

Prove in the same way that the next term in the expansion is $\frac{17}{315}x^7$.

66. (i) Prove that $\cosh 2x - 1 = 2 \sinh^2 x$.

(ii) Find the minimum value of $17 \cosh x + 8 \sinh x$.

67. Define the functions $\sinh x$, $\cosh x$ and obtain the identity connecting them. Show that the point $A\,(0, a)$ lies on the curve

$$y = a \cosh \frac{x}{a}$$

P is any point on the curve, N is the foot of the ordinate through P, M is the foot of the perpendicular from N on to the tangent at P. Show that the length of NM is constant and that PM and the arc AP are of the same length.

68. (i) Use the substitution $u = \sqrt{(1 + x^2)}$ to evaluate

$$\int_{\frac{4}{3}}^{\frac{12}{5}} \frac{dx}{x(1 + x^2)^{\frac{3}{2}}}.$$

(ii) A uniform right circular cone is of height h, the radius of its base is a, and its mass is M. Prove by the methods of the integral calculus that the volume of the cone is $\frac{1}{3}\pi a^2 h$ and that its moment of inertia about its axis is $\frac{3}{10}Ma^2$.

[You can assume that the moment of inertia of a uniform circular disc of radius r and mass m about an axis through the centre and perpendicular to the disc is $\frac{1}{2}mr^2$.]

69. Axes are chosen for a uniform trapezium so that the y-axis is parallel to and midway between the parallel sides. These sides are a distance $2h$ apart and the length of the chord through the point $(x, 0)$ parallel to the y-axis is $ax + b$. Find

(i) the area of the trapezium;

(ii) the x-coordinate of its centre of mass;

(iii) its radius of gyration about the y-axis.

70. Find the y-coordinate of the centroid of the area in the first quadrant enclosed by the parabola $y^2 = 4ax$, the x-axis, and the line $x = h$.

Prove that the volume generated when this area is rotated through four right angles about the x-axis is $2\pi a h^2$.

Assuming that the radius of gyration of a uniform circular disc of radius r about a perpendicular axis through its centre is $r/\sqrt{2}$, find the radius of gyration of the solid of revolution about the x-axis.

71. Prove that the tangent to the curve

$$\sqrt{x} + \sqrt{y} = \sqrt{a}$$

at the point (x, y) makes intercepts $\sqrt{(ax)}$ and $\sqrt{(ay)}$ on the axes.

A solid is generated by rotating about the x-axis the area whose complete boundary is formed by (i) the arc of this curve joining the points $(a, 0)$ and $(0, a)$, and (ii) the straight lines joining the origin to these two points. Prove that, when this solid is of uniform density ρ, its mass M is $\frac{1}{15}\pi a^3 \rho$. Find the moment of inertia of the solid about Ox in terms of M and a.

72. A function $f(x)$ is defined for $0 \leqslant x \leqslant \frac{3}{2}$ as follows:

$$0 \leqslant x < 1, \quad f(x) = -\tfrac{1}{6}x^3 + x;$$
$$1 \leqslant x \leqslant \tfrac{3}{2}, \quad f(x) = -\tfrac{1}{2}x^2 + \tfrac{3}{2}x - \tfrac{1}{6}.$$

Discuss the continuity of $f(x)$ and its successive derivatives. Sketch the graphs of $f(x)$ and its derivatives throughout the whose interval $(0, \frac{3}{2})$.

Prove that the function $f(x) - \dfrac{3}{\pi} \sin \dfrac{\pi x}{3}$ has a stationary value at $x = 1$, and determine whether it is a maximum or a minimum.

73. (P) Write short notes on the following, placing them in chronological order and describing their contributions to calculus:

Pappus, Guldin, Leibniz, Maclaurin, Newton, Bishop Berkeley, Archimedes, Rolle, Cavalieri, Taylor, Simpson, Napier, L'Hôpital.

Papers I to P

I

I1. Prove that, if u and v are functions of x, then

$$\frac{d}{dx}(uv) = v\frac{du}{dx} + u\frac{dv}{dx}.$$

Differentiate the following with respect to x:

(i) $(1 + x)^3 \tan 3x$;   (ii) $\dfrac{5 + 3\sin x}{3 + 5\sin x}$.

If $x = \dfrac{2 + t}{1 + 2t}$, $y = \dfrac{3 + 2t}{t}$, prove that

$$\frac{dy}{dx} = \frac{(1 + 2t)^2}{t^2}$$

and find the value of d^2y/dx^2 when $x = 0$.

I2. Integrate the following functions with respect to x:

(i) $x(1 + x^2)^{\frac{3}{2}}$;   (ii) $\dfrac{3 + 2x}{1 - 4x^2}$;   (ii) $x^2 \ln x$.

Evaluate $\displaystyle\int_0^{\frac{\pi}{2}} x \cos^2 3x \, dx.$

I3. A right circular cone of variable height h is inscribed in a sphere of fixed radius R. Show that the volume of the cone is at a maximum when $h = 4R/3$ and calculate the volume in this case.

[You may assume that the volume of a right circular cone of height h and base radius r is $\frac{1}{3}\pi r^2 h$.]

Prove that the maximum volume of a right circular cylinder which can be inscribed in the sphere is $\frac{4}{9}\pi R^3 \sqrt{3}$.

I4. Find the smaller of the two areas bounded by the curve $y^2 = 4x$, the line $y = 2$, and the line $x = 4$.

Prove that the volume of the solid of uniform density formed by rotating this area through four right angles about the x-axis is 18π.

Prove also that the radius of gyration of this solid about the x-axis is $2\sqrt{2}$.

[You may assume that the radius of gyration of a circular disc of radius a about an axis through its centre perpendicular to its plane is $a/\sqrt{2}$.]

I5. (i) State Taylor's theorem for the expansion of $f(a + h)$ in a series of ascending powers of h.

Prove that the first four terms in the Taylor expansion of $\tan^{-1} (1 + h)$ are

$$\tfrac{1}{4}\pi + \tfrac{1}{2}h - \tfrac{1}{4}h^2 + \tfrac{1}{12}h^3.$$

(ii) An open tank is to be constructed with a square horizontal base and vertical sides. The capacity of the tank is to be 2000 m³. The cost of the material for the sides is £4 per square metre and for the base is £2 per square metre. Prove that the minimum cost of the material involved is £2400, and give the corresponding dimensions of the tank.

J

J1. Denoting the point $(1, 1)$ by A, prove that the curves $y = x^3$ and $y = 2 - x^2$ intersect at A and at no other point and that the acute angle at which they intersect is $45°$. Sketch the two curves in the same diagram.

If O is the origin and B the point where the second curve meets the positive x-axis, calculate the area of the region bounded by OB and the arcs OA, AB of the two curves.

J2. Integrate with respect to x:

(i) $\dfrac{x}{(x - 1)(x - 2)}$;   (ii) $x^2 \ln x$;   (iii) $\dfrac{x + 2}{x^2 + 4x + 10}$.

Prove that, if $C = \int e^{ax} \cos bx \, dx$ and $S = \int e^{ax} \sin bx \, dx$,

$$aC - bS = e^{ax} \cos bx,$$

$$aS + bC = e^{ax} \sin bx.$$

Evaluate $\displaystyle\int_0^{\frac{\pi}{2}} e^{2x} \sin 3x \, dx$.

J3. Prove that the volume of a right circular cone of semi-angle α and base radius r is $\frac{1}{3}\pi r^3 \cot \alpha$, and that the area of the curved surface is $\pi r^2 \operatorname{cosec} \alpha$.

A rocket consists of a right circular cone, of semi-angle α and base radius r joined at its base to a circular cylinder also of radius r. The length of the cylinder is l

and the total volume of the rocket is fixed. Prove that, if $\alpha = 30°$, the area of the curved surface is a minimum when

$$\frac{l}{r} = 2 - \sqrt{3}.$$

J4. Prove that the length of the arc of the curve $y = f(x)$ from $x = a$ to $x = b$ is

$$\int_a^b \sqrt{\left\{1 + \left(\frac{dy}{dx}\right)^2\right\}}\, dx.$$

Find the arc-length of the curve

$$y = \tfrac{1}{4}(e^x - e^{-x})^2$$

between the origin and the line $x = b$.

A surface of revolution is formed by rotating the given curve through four right angles about the axis $y = 0$. Find the volume contained by the surface and the two planes generated by rotating the lines $x = 0$ and $x = b$.

J5. State Maclaurin's theorem for the expansion of a function of x in a series of ascending powers of x.

Prove that, if

$$y = \ln\left(\frac{1 + \sin x}{1 - \sin x}\right),$$

then   (i) $\dfrac{dy}{dx} = 2 \sec x$,

and (ii) the expansion of y in a series of powers of x as far as the term in x^4 is $2x + \tfrac{1}{3}x^3$.

Find, correct to four significant figures, the value of y when $x = 1° \, 48'$, taking $\pi = 3\cdot142$.

K

K1. (i) Differentiate with respect to x the functions

$$(1 - x)^5 x(1 + x), \qquad \ln\frac{x^2 - 1}{x^2 + 1}, \qquad \frac{\sin x}{\sin x + \cos x}$$

giving your answers in their simplest form.

(ii) Prove that, if α is any constant,

$$\frac{d}{dx}\{e^{3x} \sin (4x + \alpha)\} = 5e^{3x} \sin (4x + \alpha + \beta),$$

where β is the acute angle $\tan^{-1}\tfrac{4}{3}$.

Find $\dfrac{d^3}{dx^3}\{e^{3x} \sin 4x\}$.

K2. A triangular frame ABC is made from three rods BC, CA, AB. After a certain change of temperature it is found that the rods AB, AC have not changed in length while the length of the rod BC has decreased by $\tfrac{1}{3}$ per cent. Prove that the change in the angle B is $\dfrac{3 \cot C}{5\pi}$ degrees approximately.

Find also the approximate changes in the other two angles.

K3. Find the maximum and minimum values of the function $y = (4x + 1)(x - 1)^4$.

A beam of rectangular cross-section is to be cut from a cylindrical log of diameter d. The stiffness of such a beam is proportional to xy^3, where x is the breadth and y the depth of the section. Find the cross-sectional area of the beam (i) of greatest volume, (ii) of greatest stiffness, that can be cut from the log.

K4. Integrate with respect to x:

$$\text{(i)} \ \frac{1}{x(x-2)} \ ; \qquad \text{(ii)} \ x^2 \cos x; \qquad \text{(iii)} \ \tan^3 x.$$

By means of the substitution $\tan \frac{1}{2}\theta = t$, or otherwise, evaluate the integral

$$\int_0^{\frac{\pi}{2}} \frac{d\theta}{2 + \cos \theta}.$$

K5. (i) Find the position of the centre of gravity of a uniform semicircular lamina.

(ii) A lamina of uniform density is in the shape of the area under the curve $y = \cos x$ between the limits $x = 0$ and $x = \frac{1}{2}\pi$. Find the area of the lamina and prove that the radius of gyration of the lamina about the y-axis is $\frac{1}{2}\sqrt{(\pi^2 - 8)}$.

L

L1. (i) If
prove that

$$y = \sin^3 x \cos 3x,$$

$$\frac{dy}{dx} = 3 \sin^2 x \cos 4x.$$

(ii) Find

$$\int \frac{\sec^2 x \, dx}{\sqrt{(1 + \tan x)}}, \qquad \int (e^x + e^{-x})^2 \, dx.$$

L2. Prove that, if the function y is given by

$$y = \sin x + \tfrac{1}{2} \sin 2x + \tfrac{1}{3} \sin 3x,$$

then $dy/dx = \cos 2x(1 + 2 \cos x)$.

Show that the interval $0 \leqslant x \leqslant \pi$ can be divided into four parts in two of which the slope of the graph of the function is positive and in the other two negative.

Make a rough drawing of the graph over this interval, marking in the maximum and minimum points. Deduce from the graph that when $0 < x < \pi$ the function is positive.

L3. Show that the points $(-h, a)$, $(0, b)$, (h, c) all lie on the parabola

$$2h^2y = (a - 2b + c)x^2 + h(c - a)x + 2bh^2,$$

and prove that the area included between the parabola, the x-axis, and the ordinates at $x = -h$, $x = h$ is

$$\tfrac{1}{3}h(a + 4b + c).$$

By dividing the range of integration into four equal intervals and using Simpson's rule, evaluate the integral $\int_0^1 \frac{dx}{1 + x^2}$, giving your answer correct to four places of decimals.

Deduce an approximate value of π.

L4. (i) State Taylor's theorem for the expansion of $f(x + h)$ and prove that, if terms involving powers higher than h^3 are neglected,

$$\sin (x + h) = \left(1 - \frac{h^2}{2}\right) \sin x + h\left(1 - \frac{h^2}{6}\right) \cos x.$$

(ii) Prove that

$$\ln \cos x = -\tfrac{1}{2}x^2 - \tfrac{1}{12}x^4$$

if terms beyond x^4 are neglected.

L5. The coordinates of a moving particle are given by

$$x = 2a \cos t + a \cos 2t, \qquad y = 2a \sin t + a \sin 2t,$$

where a is a constant number of feet, and t is the time, measured in seconds. Find the components, parallel to the axes, of (i) the velocity, (ii) the acceleration, and prove that $y = 0$ whenever the resultant acceleration is perpendicular to the resultant velocity.

Find the magnitude of the resultant velocity at any instant, and prove that the whole length of the path described by the particle in 2π seconds is 16

M

M1. Assuming that $(\sin \phi)/\phi$ tends to 1 as ϕ tends to 0, obtain from first principles the derivative of $\sin x$ with respect to x.

Determine values of the constants A and B such that if

$$y = e^x (A \cos x + B \sin x)$$

then, for all values of x,

$$\frac{d^2y}{dx^2} - 3\frac{dy}{dx} + 4y = e^x \sin x.$$

M2. By using Taylor's theorem obtain the expansion of $\sin^2 (\tfrac{1}{6}\pi + x)$ in powers of x up to the term in x^3.

[You are advised to use the formula

$$\sin^2 \theta = \tfrac{1}{2}(1 - \cos 2\theta)$$

to simplify the calculation of the derivatives that you require.]

Hence calculate the value of $\sin^2 31° 30'$ correct to four places of decimals. [Take $\pi = 3\cdot142$.]

M3. A water-main of circular cross-section is to deliver a given volume of water per second. The cost of the main is proportional to its radius and the cost of the pumping machinery is proportional to the product of the radius of the main and the cube of the velocity of flow. Find the ratio of the cost of the main to the cost of the machinery for a minimum total cost.

M4. Integrate the following functions with respect to x:

(i) $\left(x^2 - \dfrac{2}{x}\right)^2$;   (ii) $\sin 3x \cos 5x$;   (iii) $\dfrac{2}{x(1 + x^2)}$.

Prove by means of the substitution $t = \tan x$ that

$$\int_0^{\frac{\pi}{4}} \frac{dx}{1 + \sin 2x} = \frac{1}{2},$$

and find the value of

$$\int_0^{\frac{\pi}{4}} \frac{dx}{(1 + \sin 2x)^2}.$$

M5. The perpendicular to the x-axis, OX, from the point $P(2, 6)$ on the curve $y = \frac{3}{4}x^3$ meets OX at Q. Find (i) the coordinates of the centroid of the uniform lamina bounded by the arc OP and the lines OQ, QP, and (ii) the radius of gyration of the lamina about PQ.

Find also the volume generated by the lamina when rotated about the y-axis through four right angles.

N

N1. (i) Differentiate with respect to x

(a) $\dfrac{x}{\sqrt{(1 + x^2)}}$,   (b) e^{-x^2}.

(ii) A spherical balloon is being inflated. When the diameter of the balloon is 10 cm its volume is increasing at the rate of 200 cm^2 s^{-1}. Find the rate at which its surface area is then increasing.

N2. (i) Evaluate, correct to three significant figures,

$$\int_0^2 2^x \, dx.$$

(ii) Use the substitution $x = \sin^2 \theta$ to prove that

$$\int_0^{\frac{1}{2}} \sqrt{\left(\frac{x}{1 - x}\right)} \, dx = \frac{1}{4}(\pi - 2)$$

N3. (i) Prove that the least value of $4 \sec \theta - 3 \tan \theta$ for $0 < \theta < \frac{1}{2}\pi$ is $\sqrt{7}$.

(ii) A body is made up of a hemisphere of radius r with its plane face joined to one of the plane faces of a circular cylinder of radius r and length x. Prove that, if the volume of the body is 45π cm^3, the least value of the total surface area is 45π cm^2.

N4. Prove that the coordinates of any point on the curve $x^{\frac{2}{3}} + y^{\frac{2}{3}} = a^{\frac{2}{3}}$ can be represented in the form

$$x = a \cos^3 t, \qquad y = a \sin^3 t.$$

Sketch the curve and prove that the length of its whole perimeter is $6a$.
Find the area enclosed by the curve.

N5. Prove that, if α is approximately equal to a root of an equation $f(x) = 0$, then $\alpha - \dfrac{f(\alpha)}{f'(\alpha)}$ is in general a closer approximation to the root.

Prove that, if η is small, the equation

$$\theta + \sin \theta \cos \theta = 2\eta \cos \theta$$

has a small root approximately equal to

$$\eta - \tfrac{1}{6}\eta^3.$$

[You may assume the power-series expansions for $\sin \theta$ and $\cos \theta$.]

O

O1. A particle moves along a cycloid so that its coordinates at time t are given by

$$x = a(t - \sin t), \qquad y = a(1 - \cos t).$$

Obtain the speed v of the particle at time t, (i) expressed in terms of t, and (ii) expressed in terms of y.

Sketch the graphs (i) of x against y, (ii) of v against t, and (iii) of v against y, all as t varies in the range 0 to 4π.

O2. (i) Find the following integrals:

$$\int \frac{(x + 1)^2}{x^2 + 4}\, dx; \quad \int \sec^3 \theta \tan \theta\, d\theta; \quad \int x^2 \ln x\, dx.$$

(ii) Evaluate $\quad \displaystyle\int_0^1 \frac{dx}{(2 - x)(1 + x)} \quad$ and $\quad \displaystyle\int_0^{\frac{\pi}{4}} \cos^3 \theta\, d\theta.$

O3. Prove that the area enclosed by the curve $y = \tan x$, the x-axis, and $x = \tfrac{1}{3}\pi$ is $\ln 2$.

If the point $(\bar{x}, \bar{y})$ is the centroid of this area, prove that

$$\bar{y} = (3\sqrt{3} - \pi)/6 \ln 2$$

and calculate $\bar{x}$ by means of Simpson's rule with ordinates at intervals of $\tfrac{1}{12}\pi$.

$$[\tan \tfrac{1}{12}\pi = 2 - \sqrt{3}.]$$

O4. Prove that a function $f(x)$ has a maximum value when $x = a$ if $f'(a) = 0$ and $f''(a)$ is negative.

A cone of semi-vertical angle θ is inscribed in a sphere of radius a. Prove that the volume of the cone is

$$\tfrac{8}{3}\pi a^3 \sin^2 \theta \cos^4 \theta,$$

and find the area of its curved surface.

Prove that if a is fixed and θ allowed to vary, the maximum volume of the cone is $\tfrac{8}{27}$ of the volume of the sphere.

Prove further that if the volume of the cone is a maximum, the area of its curved surface is also a maximum.

O5. (i) A sphere, of radius a, is cut into two unequal parts by a plane at a distance d from the centre of the sphere. Prove that the volume of the smaller part is

$$\tfrac{1}{3}\pi(2a + d)(a - d)^2.$$

(ii) AB is a non-uniform straight rod of length l, the density of which at a point at a distance x from A is $a + bx^2$, where a and b are constants. Prove that the mass of the rod is

$$\tfrac{1}{3}l(3a + bl^2).$$

Find also the distance of the centre of mass of the rod from A and the radius of gyration about an axis through A perpendicular to AB.

P

P1. Find from first principles the derivative of $1/x^3$ with respect to x.

What is the derivative of $1/x^3$ with respect to x^2?

Differentiate with respect to x:

(i) $\ln \cos x$;   (ii) $x\sqrt{(1 - x^2)}$;   (iii) $\dfrac{e^x}{(1 + x^2)^2}$.

P2. Prove that a plane at distance $\tfrac{1}{2}a$ from the centre of a solid sphere of radius a divides the sphere into two segments whose volumes are in the ratio $27:5$.

If the sphere is uniform, find the distance of the centre of gravity of the larger segment from the plane.

P3. Prove that the function

$$\frac{1 - \cos 2x}{\sqrt{(4 + 3\cos 2x)}}$$

has turning points when $x = 0$ and $x = \tfrac{1}{2}\pi$. Find the maximum and minimum values of the function.

Find the equation of the tangent at the point whose abscissa is $\tfrac{1}{4}\pi$.

Sketch the curve for values of x from 0 to $\tfrac{1}{2}\pi$.

P4. Given that $y = e^{ax} \sin x$, where a is a constant, prove that

$$\frac{d^2y}{dx^2} - 2a\frac{dy}{dx} + (a^2 + 1)y = 0.$$

Calculate the first five derivatives of y for the value $x = 0$.

Show, by Maclaurin's theorem or otherwise, that the expression

$$\tfrac{1}{2}(e^{x/\sqrt{3}} + e^{-x/\sqrt{3}}) \sin x$$

differs from x by a power series beginning with the term $-\tfrac{2}{135}x^5$.

P5. Prove that the area of the surface generated by rotating the ellipse $\dfrac{x^2}{a^2} + \dfrac{y^2}{b^2} = 1$ through four right angles about the major axis $(y = 0)$ is

$$2\pi ab\left(\sqrt{(1 - e^2)} + \frac{\sin^{-1} e}{e}\right),$$

where e is the eccentricity.

12
Curves, Surfaces and Solids—(II)

Everyone knows what a curve is, until he has studied enough
mathematics to become confused through the countless number
of possible exceptions. KLEIN

Curvature

Let the arc-length s be measured from a point O on a curve. The *positive* direction of the tangent at a point P is that in which s is increasing; let this direction make an angle ψ with a fixed line (normally the x-axis).

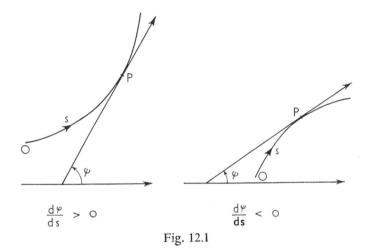

$$\frac{d\psi}{ds} > 0 \qquad\qquad \frac{d\psi}{ds} < 0$$

Fig. 12.1

We suppose that s is a one-to-one function of ψ over a certain length of the curve, so that ψ is also a function of s. The relationship between s, ψ is called the *intrinsic equation* of the curve. $\dfrac{d\psi}{ds}$ is called the *curvature* of the arc at a given point P. It is a measure of the 'rate at which the curve curves'.

The reciprocal of the curvature is denoted by ρ and called the *radius of curvature*. Thus $\rho = \dfrac{ds}{d\psi}$, which may be positive or negative as shown in the diagrams, Fig. 12.1. $\left(\text{Some writers define the radius of curvature as }\left|\dfrac{ds}{d\psi}\right|.\right)$

239

EXAMPLE 12.1 *To find the intrinsic equation of the 'catenary' formed by a uniform chain hanging freely under gravity, and hence find an expression for its radius of curvature.*

Let the mass per unit length be μ, the tension at the lowest point O be T_0 and that at a point P, an arc-length s from O, be T (Fig. 12.2). Further, let the tangent at P make an angle ψ with that at O (the horizontal).

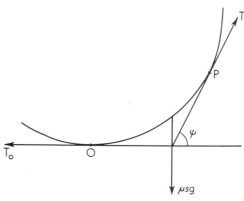

Fig. 12.2

Considering the forces shown in Fig. 12.2 we have:

resolving vertically, $T \sin \psi = \mu s g$

resolving horizontally, $T \cos \psi = T_0$.
Thus

$$s = c \tan \psi, \qquad \text{where } c = \frac{T_0}{\mu g}.$$

This is the intrinsic equation of the catenary. The radius of curvature at P is given by $\rho = \dfrac{ds}{d\psi} = c \sec^2 \psi$.

A formula for ρ in terms of cartesian coordinates (x, y) can be obtained by using the relations

$$\tan \psi = \frac{dy}{dx} \qquad (1)$$

$$(ds)^2 = (dx)^2 + (dy)^2. \qquad (2)$$

From (1),

$$\sec^2 \psi \frac{d\psi}{dx} = \frac{d^2y}{dx^2}$$

so that

$$\frac{d\psi}{dx} = \frac{\dfrac{d^2y}{dx^2}}{1 + \tan^2 \psi} = \frac{\dfrac{d^2y}{dx^2}}{1 + \left(\dfrac{dy}{dx}\right)^2}.$$

Thus

$$\rho = \frac{ds}{d\psi} = \frac{\dfrac{ds}{dx}}{\dfrac{d\psi}{dx}}$$

$$= \sqrt{\left\{1 + \left(\frac{dy}{dx}\right)^2\right\}} \times \frac{1 + \left(\dfrac{dy}{dx}\right)^2}{\dfrac{d^2y}{dx^2}}$$

$$= \frac{\left\{1 + \left(\dfrac{dy}{dx}\right)^2\right\}^{\frac{3}{2}}}{\dfrac{d^2y}{dx^2}}.$$

If x, y are given in terms of a parameter t, using dots to denote differentiation with respect to t we have

$$(ds)^2 = (dx)^2 + (dy)^2,$$

so

$$\dot{s} = \sqrt{(\dot{x}^2 + \dot{y}^2)}. \qquad (1)$$

Also

$$\tan \psi = \frac{dy}{dx} = \frac{\dfrac{dy}{dt}}{\dfrac{dx}{dt}} = \frac{\dot{y}}{\dot{x}},$$

whence

$$\sec^2 \psi \frac{d\psi}{dt} = \frac{d}{dt}\left(\frac{\dot{y}}{\dot{x}}\right) = \frac{\ddot{y}\dot{x} - \ddot{x}\dot{y}}{\dot{x}^2},$$

so that

$$\dot{\psi} = \frac{\ddot{y}\dot{x} - \ddot{x}\dot{y}}{\dot{x}^2\left(1 + \dfrac{\dot{y}^2}{\dot{x}^2}\right)} = \frac{\ddot{y}\dot{x} - \ddot{x}\dot{y}}{\dot{x}^2 + \dot{y}^2}. \qquad (2)$$

Now

$$\rho = \frac{ds}{d\psi}$$

$$= \frac{\dot{s}}{\dot{\psi}}$$

$$= \frac{(\dot{x}^2 + \dot{y}^2)^{\frac{3}{2}}}{\ddot{y}\dot{x} - \ddot{x}\dot{y}} \, .$$

Circle of Curvature

The *circle of curvature* at a point P on a curve is defined as the circle which touches the curve at P and has the same curvature as the curve at that point. The radius of this circle is accordingly $|\rho|$; its centre is called the *centre of curvature*.

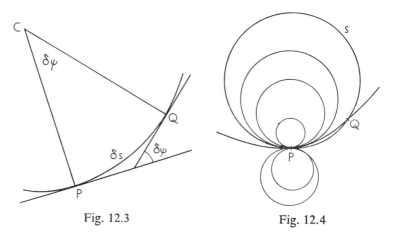

Fig. 12.3   Fig. 12.4

In this section and the next, our approach will be intuitive and *no pretence is made of giving formal justifications*. To simplify matters, we suppose for the moment that $\frac{ds}{d\psi} > 0$. Let Q be a neighbouring point on the curve, the arc-length PQ being δs, and let the normals at P, Q meet at C (Fig. 12.3). Then angle $PCQ = \delta\psi$.

As $Q \to P$, $\frac{CQ}{CP} \to 1$ and we will assume that $CP \to \rho$ as $\delta\psi \to 0$. This is plausible since $\rho = \frac{ds}{d\psi} = \lim_{\delta\psi \to 0} \frac{\delta s}{\delta\psi}$. The centre of curvature is thus the limit of the point of intersection of the normals at P, Q as $Q \to P$.

The circle of curvature may also be thought of as the 'best-fitting' circle at the point P. There is an infinity of circles which touch the curve at P;

consider one such circle, S, which also meets the curve again at a neighbouring point Q (Fig. 12.4). Then if we vary this circle by letting $Q \to P$, in the limit we shall have the circle of curvature at P.

Let the tangent and normal to the curve at P be taken as axes of coordinates (Fig. 12.5) and let the circle S meet the curve again at $Q(x, y)$.

The chord QR of S is drawn parallel to the x-axis, to meet the diameter OP at N. If the radius of S is r, by the 'intersecting chords theorem' we have

$$NP \cdot NO = NQ \cdot NR$$

or

$$y(2r - y) = x^2.$$

Thus $r = \dfrac{x^2}{2y} + \tfrac{1}{2}y$, and so

$$|\rho| = \lim_{x, y \to 0} r = \lim_{x, y \to 0} \frac{x^2}{2y}.$$

This is known as *Newton's formula.*

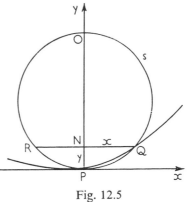

Fig. 12.5

EXAMPLE 12.2 *To find the radius of curvature of the parabola* $y^2 = 4ax$ *at the point* $(at^2, 2at)$.

Using dots to denote differentiation with respect to t, we have $\dot{x} = 2at$, $\ddot{x} = 2a$, $\dot{y} = 2a$, $\ddot{y} = 0$, so that

$$\rho = \frac{(\dot{x}^2 + \dot{y}^2)^{\frac{3}{2}}}{\ddot{y}\dot{x} - \ddot{x}\dot{y}} = \frac{(4a^2t^2 + 4a^2)^{\frac{3}{2}}}{-4a^2} = -2a(1 + t^2)^{\frac{3}{2}}.$$

EXAMPLE 12.3 *To find the radius of curvature at the origin of the curve* $y = 3x^2 - 4x^3$.

The equation of the curve (Fig. 12.6) may be written as

$$\frac{1}{2} = \frac{3x^2}{2y} - \frac{2x^3}{y} = \frac{x^2}{2y}(3 - 4x)$$

so that $\dfrac{x^2}{2y} = \dfrac{1}{2(3 - 4x)}$. Since Ox, Oy are tangent and normal respectively at O, we may use Newton's formula, obtaining

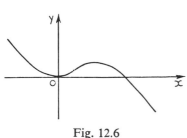

Fig. 12.6

$$|\rho| = \lim_{x, y \to 0} \frac{x}{2y} = \frac{1}{6}.$$

Exercise LVIII

1. Find ρ in terms of x for the curves (i) $y = x^3 + x$; (ii) $y = \sin x$; (iii) $y = \ln \sin x$. Draw the graph of the last function.

2. Find the curvature at the origin of the parabolas (i) $y = x^2$; (ii) $y^2 = 2x$.

3. Find the radius of curvature at the point t of the astroid $x = a \cos^3 t$, $y = a \sin^3 t$.

4. A particle moves in a plane so that, at time t, its coordinates referred to rectangular axes in the plane are given by

$$x = a \cos 2t + 2a \cos t, \qquad y = a \sin 2t + 2a \sin t.$$

Find the components of the velocity parallel to the axes and the resultant speed of the particle.

Use the formula

$$\rho = \left\{ 1 + \left(\frac{dy}{dx}\right)^2 \right\}^{\frac{3}{2}} \Big/ \frac{d^2y}{dx^2},$$

or any other formula, to show that ρ, the radius of curvature at a point of the path, is proportional to the speed of the particle at that point.

5. Show that the function

$$y = \frac{9}{x} - \frac{1}{x-2}$$

has a maximum at $x = 3$ and find the value of x for which y has a minimum. Sketch the graph of the function. Calculate the radius of curvature at the maximum point.

6. The coordinates of the points of the curve

$$4y^3 = 27x^2$$

are expressed parametrically in the form $(2t^3, 3t^2)$. By using this parametric representation or otherwise, prove that the length of the curve between the origin and the point P with parameter t_1 is

$$2(1 + t_1^2)^{\frac{3}{2}} - 2,$$

and that the radius of curvature at P is numerically equal to

$$6t_1(1 + t_1^2)^{\frac{3}{2}}.$$

7. The coordinates of a point of a curve are given in terms of a parameter t by the equations $x = te^t$, $y = t^2 e^t$. Find $\dfrac{dy}{dx}$ in terms of t, and prove that

$$\frac{d^2y}{dx^2} = \frac{t^2 + 2t + 2}{(t+1)^3} e^{-t}.$$

Prove also that the radii of the circles of curvature at the two points at which the curve is parallel to the x-axis are in the ratio $e^2:1$.

8. Define the radius of curvature at a point of a plane curve, and from your definition deduce the formula

$$\left\{ 1 + \left(\frac{dy}{dx}\right)^2 \right\}^{\frac{3}{2}} \Big/ \frac{d^2y}{dx^2}.$$

Prove that the radius of curvature of the curve

$$y = \tfrac{1}{4}x^2 - \tfrac{1}{2}\ln x \qquad (x > 0)$$

at the point (x, y) is $(1 + x^2)^2/4x$.
Find the point at which this curve is parallel to the x-axis, and prove that the circle of curvature at this point touches the y-axis.

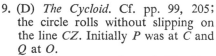

9. (D) *The Cycloid.* Cf. pp. 99, 205; the circle rolls without slipping on the line CZ. Initially P was at C and Q at O.

Fig. 12.7

(i) Taking axes as shown in Fig. 12.7, verify that the coordinates of Q are given by

$$x = a\theta + a\sin\theta, \quad y = a - a\cos\theta.$$

(ii) Deduce the intrinsic equation of the cycloid in the form $s = 4a\sin\psi$.

10. (D) *The Catenary.* Starting from the intrinsic equation $s = c\tan\psi$ (cf. Example 12.1), show that, with suitable choice of origin,

(i) $\dfrac{ds}{dx} = c\dfrac{d^2y}{dx^2}$;

(ii) $\sqrt{(1 + p^2)} = c\dfrac{dp}{dx}$, where $p = \dfrac{dy}{dx}$;

(iii) $x = c\sinh^{-1}p$;

(iv) $y = c\cosh\dfrac{x}{c}$.

Centres of Curvature and Evolutes

We recall that the centre of curvature can be considered in two ways, (i) as the centre of the best-fitting circle at P, i.e. at a distance $|\rho|$ along the inward normal, (ii) as the limit of the point of intersection of neighbouring normals.

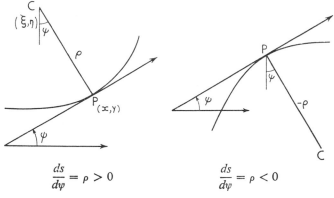

$$\frac{ds}{d\psi} = \rho > 0 \qquad\qquad \frac{ds}{d\psi} = \rho < 0$$

Fig. 12.8

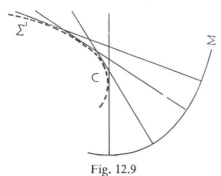

Fig. 12.9

(i) If the coordinates of the centre of curvature C (Fig. 12.8) are (ξ, η), we have

$$\xi = x - \rho \sin \psi$$
$$\eta = y + \rho \cos \psi.$$

The reader should satisfy himself that these equations hold for both positive and negative ρ.

(ii) By drawing a number of neighbouring normals of a curve Σ (Fig. 12.9), it can be seen that these *envelop* a new curve Σ', called the *evolute* of Σ. This 'envelope' of the normal is also the locus of the centre of curvature at points of Σ.

EXAMPLE 12.4   *To find the evolute of the ellipse $x = a \cos \theta$, $y = b \sin \theta$.*

Method 1. $\rho = \dfrac{(\dot{x}^2 + \dot{y}^2)^{\frac{3}{2}}}{\dot{y}\ddot{x} - \ddot{x}\dot{y}}$ where dots denote differentiation with respect to θ. We have $\dot{x} = -a \sin \theta$, $\ddot{x} = -a \cos \theta$; $\dot{y} = b \cos \theta$, $\ddot{y} = -b \sin \theta$, whence

$$\rho = \frac{(a^2 \sin^2 \theta + b^2 \cos^2 \theta)^{\frac{3}{2}}}{ab}.$$

Also

$$\tan \psi = \frac{dy}{dx} = \frac{dy}{d\theta} \Big/ \frac{dx}{d\theta} = \frac{b \cos \theta}{-a \sin \theta} = -\frac{b}{a} \cot \theta.$$

We measure s counter-clockwise from A. The right-hand diagram (Fig. 12.10) shows that in the fourth quadrant $\sin \psi = \dfrac{b \cos \theta}{(a^2 \sin^2 \theta + b^2 \cos^2 \theta)^{\frac{1}{2}}}$,

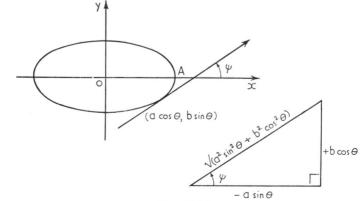

Fig. 12.10

$\cos \psi = \dfrac{-a \sin \theta}{(a^2 \sin^2 \theta + b^2 \cos^2 \theta)^{\frac{1}{2}}}$. The reader should verify that these equations hold for every point on the ellipse.

Thus the centre of curvature at the point θ is given by

$$\xi = x - \rho \sin \psi$$
$$= a \cos \theta - \frac{b \cos \theta(a^2 \sin^2 \theta + b^2 \cos^2 \theta)}{ab}$$
$$= \frac{a^2 - b^2}{a} \cos^3 \theta. \tag{1}$$

$$\eta = y + \rho \cos \psi$$
$$= b \sin \theta - \frac{a \sin \theta(a^2 \sin^2 \theta + b^2 \cos^2 \theta)}{ab}$$
$$= \frac{a^2 - b^2}{b} \sin^3 \theta. \tag{2}$$

Elimination of θ between eqns. (1) and (2) gives

$$(a\xi)^{\frac{2}{3}} + (b\eta)^{\frac{2}{3}} = (a^2 - b^2)^{\frac{2}{3}},$$

so that the equation of the evolute is

$$(ax)^{\frac{2}{3}} + (by)^{\frac{2}{3}} = (a^2 - b^2)^{\frac{2}{3}}.$$

Method 2. The gradient at θ is $\dfrac{b \cos \theta}{-a \sin \theta} = -\dfrac{b}{a} \cot \theta$, so the equation of the normal is

$$y - b \sin \theta = \frac{a}{b} \tan \theta(x - a \cos \theta),$$

or

$$ax \sin \theta - by \cos \theta = (a^2 - b^2) \sin \theta \cos \theta$$
$$= \tfrac{1}{2}(a^2 - b^2) \sin 2\theta. \quad (1)$$

The normal at the neighbouring point $\theta + \delta\theta$ is

$$ax \sin (\theta + \delta\theta) - by \cos (\theta + \delta\theta)$$
$$= \tfrac{1}{2}(a^2 - b^2) \sin \{2(\theta + \delta\theta)\}. \quad (2)$$

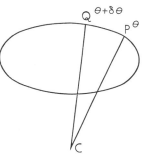

Fig. 12.11

The centre of curvature is obtained by solving eqns. (1) and (2) together and letting $\delta\theta \to 0$. Taking (1) from (2) and dividing by $\delta\theta$, we have

$$\frac{ax\{\sin (\theta + \delta\theta) - \sin \theta\}}{\delta\theta} - \frac{by\{\cos (\theta + \delta\theta) - \cos \theta\}}{\delta\theta}$$
$$= \tfrac{1}{2}(a^2 - b^2)\frac{\{\sin 2(\theta + \delta\theta) - \sin 2\theta\}}{\delta\theta}. \tag{3}$$

Now $f'(\theta) = \lim\limits_{\delta\theta\to 0} \dfrac{f(\theta + \delta\theta) - f(\theta)}{\delta\theta}$, so letting $\delta\theta \to 0$, (3) becomes

$$ax \cos\theta + by \sin\theta = (a^2 - b^2)\cos 2\theta. \tag{4}$$

The coordinates of the centre of curvature satisfy both (1) and (4), and eliminating θ between them gives the locus of the centre of curvature, which is also the 'envelope' of the normal to the ellipse. Multiplying (1) by $\sin\theta$, (4) by $\cos\theta$ and adding yields

$$\begin{aligned} ax &= (a^2 - b^2)\{\tfrac{1}{2}\sin 2\theta \sin\theta + \cos\theta\cos 2\theta\} \\ &= (a^2 - b^2)\cos\theta(\sin^2\theta + \cos^2\theta - \sin^2\theta) \\ &= (a^2 - b^2)\cos^3\theta. \end{aligned}$$

Similarly, $by = (a^2 - b^2)\sin^3\theta$, whence

$$(ax)^{\frac{2}{3}} + (by)^{\frac{2}{3}} = (a^2 - b^2)^{\frac{2}{3}}, \text{ as before.}$$

This last method can be generalized to find the envelope of a system of curves; we restrict ourselves here to straight lines (Fig. 12.12).

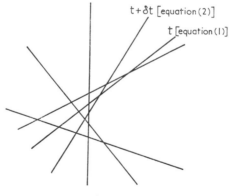

t+δt [equation(2)]

t [equation(1)]

Fig. 12.12

If a system of lines is given by

$$xf(t) + yg(t) + h(t) = 0, \tag{1}$$

for different values of t, the line '$t + \delta t$' is

$$xf(t + \delta t) + yg(t + \delta t) + h(t + \delta t) = 0. \tag{2}$$

Where these meet,

$$x\,\frac{f(t + \delta t) - f(t)}{\delta t} + y\,\frac{g(t + \delta t) - g(t)}{\delta t} + \frac{h(t + \delta t) - h(t)}{\delta t} = 0;$$

or, letting $\delta t \to 0$,

$$xf'(t) + yg'(t) + h'(t) = 0. \tag{3}$$

Elimination of t between eqns. (1) and (3) gives the envelope of the system of lines. Equation (3) can be obtained from eqn. (1) by differentiating with respect to t treating x, y as constants.

EXAMPLE 12.5 *To find the envelope of $x - yt + at^2 = 0$.*

Using the above procedure, we have

$$x - yt + at^2 = 0 \tag{1}$$

and, differentiating with respect to t (treating x, y as constants),

$$-y + 2at = 0. \tag{2}$$

Elimination of t between eqns. (1) and (2) leads to the equation of the envelope, namely $y^2 = 4ax$. [The given equation is, in fact, that of the tangent at the point $(at^2, 2at)$.]

EXERCISE LIX

1. Find the centre of curvature at the point t of the rectangular hyperbola

$$x = ct, \quad y = \frac{c}{t}.$$

2. Find the envelopes of the following systems of lines, (i) $x \cos t + y \sin t = a$; (ii) $bx \sec \varphi - ay \tan \varphi = ab$.

3. Find the equation of the normal at the point t to the parabola $x = at^2, y = 2at$. Hence find the equation of the evolute of the parabola, and sketch the two curves in the same diagram.

4. Sketch the ellipse $\dfrac{x^2}{a^2} + \dfrac{y^2}{b^2} = 1$ and its evolute in the same diagram (cf. Example 12.4).

5. The coordinates x and y of a point on a plane curve are given in terms of a parameter. Prove that the centre of curvature at the point (x, y) is at the point (x_1, y_1), where

$$x_1 = x - \frac{\dot{y}(\dot{x}^2 + \dot{y}^2)}{\dot{x}\ddot{y} - \dot{y}\ddot{x}}, \qquad y_1 = y + \frac{\dot{x}(\dot{x}^2 + \dot{y}^2)}{\dot{x}\ddot{y} - \dot{y}\ddot{x}}.$$

A cycloid is given by the equations $x = a\theta - a\sin\theta$, $y = a - a\cos\theta$. Prove that the radius of curvature at any point is $|4a\sin\frac{1}{2}\theta|$.
Find the locus of the centre of curvature and prove that this is also a cycloid.

6. Prove that the radius of curvature at the point $\left(\dfrac{a}{2}, \dfrac{\sqrt{3}a}{4}\right)$ on the curve

$$a^2y^2 = x^2(a^2 - x^2)$$

has magnitude $4a/5$, and that the corresponding centre of curvature is $(9a/10, -3\sqrt{3}a/20)$.

7. Find an expression for the radius of curvature at any point on the curve $8ay^2 = x^3$.

   Prove that the radius of curvature at the point $(2a, a)$ is $125a/12$ and find the coordinates of the centre of curvature at this point.

   Prove that the length of the arc of the curve measured from the origin to the point $(2a, a)$ is $61a/27$.

8. A curve is given by the equations

$$x = a(2 \cos \theta + \cos 2\theta), \qquad y = a(2 \sin \theta - \sin 2\theta).$$

   Prove that the radius of curvature at any point has magnitude $8a \sin (3\theta/2)$.

   Prove further that the locus of the centre of curvature is

$$x = 3a(2 \cos \theta - \cos 2\theta), \qquad y = 3a(2 \sin \theta + \sin 2\theta).$$

9. Find the equation of the circle of curvature at the origin on the parabola $y = x^2$.

   Show that the locus of the centre of curvature for a variable point on the parabola is

$$27x^2 = 2(2y - 1)^3.$$

10. The centre of curvature Q corresponding to a point $P(\xi, \eta)$ on the curve $y = \ln \sec x$ has coordinates (X, Y). Prove that

$$X - \xi = -\tan \xi, \qquad Y - \eta = 1.$$

   Using these relations as parametric equations for the locus of Q, prove that the radius of curvature of the locus of Q is of magnitude $\sec \xi \tan \xi$.

11. The tangent to a curve at the point (x, y) makes an angle ψ with the x-axis. Prove that the centre of curvature at (x, y) is

$$\left(x - \frac{dy}{d\psi}, \, y + \frac{dx}{d\psi} \right).$$

   A curve is given parametrically by the equations

$$x = 2a \cos t + a \cos 2t, \qquad y = 2a \sin t - a \sin 2t.$$

   Prove that $\psi = -\tfrac{1}{2}t$.

   P is a variable point on the curve, and Q is the centre of curvature at P; the point R divides PQ internally so that $PR = \tfrac{1}{4}PQ$. Prove that the locus of R is the circle

$$x^2 + y^2 = 9a^2.$$

   Prove also that no point of the curve lies outside this circle.

   Draw a rough sketch of the curve.

Polar Coordinates

Instead of specifying the position of a point P by its cartesian coordinates (x, y) it is sometimes convenient to use 'polars' (r, θ) instead (Fig. 12.13).

r is the distance of P from a fixed point O and θ the angle OP makes with a fixed 'initial line' through O, namely the positive direction of the x-axis, Fig. 12.13. Transformation from one set of coordinates to the other can be effected by means of the relationships

$$x = r \cos \theta, \qquad y = r \sin \theta;$$

and

$$r^2 = x^2 + y^2, \qquad \tan \theta = \frac{y}{x}.$$

The reader should check for himself that (r, θ), $(r, \theta + 2\pi)$, $(-r, \theta + \pi)$ represent the same point. Thus, when we refer to r as being a function of θ, it must be understood that their ranges of values are suitably restricted.

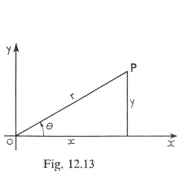

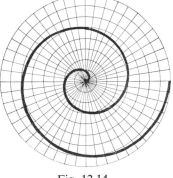

Fig. 12.13   Fig. 12.14

EXAMPLE 12.6   *To plot the curve* $r = \theta$, $(\theta > 0)$.

When $\theta = 0$, $r = 0$ so that the curve passes through the origin. When $\theta = \frac{\pi}{2}$, $r = \frac{\pi}{2}$ and so on, r increasing uniformly with θ. The resulting curve is *Archimedes' spiral* (Fig. 12.14); in cartesians, its equation would be $\sqrt{(x^2 + y^2)} = \tan^{-1}\left(\frac{y}{x}\right)$.

If we also took negative values for θ, we would have for example $\theta = -\frac{\pi}{4}$, $r = -\frac{\pi}{4}$, which would be plotted by turning through *clockwise* from the initial line and then proceeding backwards $\left(\text{i.e. along the ray } \theta = \frac{3\pi}{4}\right)$ a distance $\frac{\pi}{4}$. The reader should sketch this part of the curve for himself.

EXAMPLE 12.7 *To sketch the curve $r = \cos \theta$.*

Method 1. Multiplying by r, we have $r^2 = r \cos \theta$, or

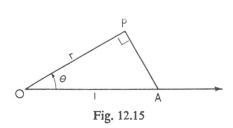

Fig. 12.15

$$x^2 + y^2 = x,$$

that is

$$(x - \tfrac{1}{2})^2 + y^2 = (\tfrac{1}{2})^2,$$

which represents a circle, centre $(\tfrac{1}{2}, 0)$, radius $\tfrac{1}{2}$.

Method 2. 'By inspection', if A is the point $(1, 0)$ in polar coordinates, then OA subtends a right angle at P and so the locus of P is the circle on OA as diameter (Fig. 12.15).

EXERCISE LX

Investigate the curves given by the following polar equations. Some at least should be plotted accurately, using polar graph paper.

1. (i) $r = \dfrac{1}{\theta}$; (ii) $r = \theta + \dfrac{1}{\theta}$.

2. $r = \sin n\theta$ $(n = 1, 2, 3, 4, 5)$.

3. $r = 3 \sec \theta + k$ $(k = 3, 1, 0, -3, -5)$.

4. $r = 3 + k \cos \theta$ $(k = 1, 3, 5)$.

5. $\dfrac{1}{r} = 3 + k \cos \theta$ $(k = 1, 3, 5)$.

6. $r = 4 \sin \theta \tan \theta$ (the 'cissoid of Diocles'; it may be simplest to find the cartesian equation).

Area of Sector

A sector is formed by an arc of the curve $r = f(\theta)$ and the lines joining O to the points $A(\theta = \alpha)$ and $B(\theta = \beta)$. P, Q are neighbouring points on the arc, with coordinates (r, θ), $(r + \delta r, \theta + \delta \theta)$ respectively (Fig. 12.16).

Let the greatest value of the radius vector between P and Q be $r + \lambda \, \delta r$ and the least $r + \mu \, \delta r$. Then the area of the sector OPQ lies between those of circular sectors with radii $r + \lambda \, \delta r$, $r + \mu \, \delta r$, namely

$$\tfrac{1}{2}(r + \lambda \, \delta r)^2 \, \delta \theta \quad \text{and} \quad \tfrac{1}{2}(r + \mu \, \delta r)^2 \, \delta \theta.$$

Thus by the general theory of integration the area of the whole sector is given by the common limit of $\Sigma_{\theta = \alpha}^{\beta} \tfrac{1}{2}(r + \lambda \, \delta r)^2 \, \delta \theta$ and $\Sigma_{\theta = \alpha}^{\beta} \tfrac{1}{2}(r + \mu \, \delta r)^2 \, \delta \theta$, that is

$$\frac{1}{2} \int_{\alpha}^{\beta} r^2 \, d\theta.$$

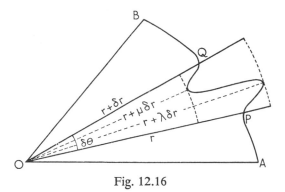

Fig. 12.16

EXAMPLE 12.8   *To find the area enclosed by one of the loops of the curve*
$r = \sin 3\theta$ (Fig. 12.17).

The required area is given by

$$\frac{1}{2}\int_0^{\frac{\pi}{3}} r^2\, d\theta = \frac{1}{2}\int_0^{\frac{\pi}{3}} \sin^2 3\theta\, d\theta$$

$$= \frac{1}{4}\int_0^{\frac{\pi}{3}} (1 - \cos 6\theta)\, d\theta$$

$$= \frac{1}{4}\left[\theta - \frac{1}{6}\sin 6\theta\right]_0^{\frac{\pi}{3}}$$

$$= \frac{\pi}{12}.$$

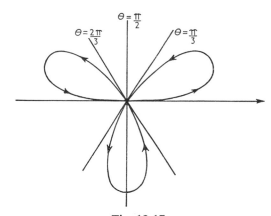

Fig. 12.17

Length of Arc

From $x = r \cos \theta$, $y = r \sin \theta$ we have

$$\frac{dx}{d\theta} = \frac{dr}{d\theta} \cos \theta - r \sin \theta, \qquad \frac{dy}{d\theta} = \frac{dr}{d\theta} \sin \theta + r \cos \theta.$$

Thus

$$\left(\frac{ds}{d\theta}\right)^2 = \left(\frac{dx}{d\theta}\right)^2 + \left(\frac{dy}{d\theta}\right)^2$$

$$= \left(\frac{dr}{d\theta} \cos \theta - r \sin \theta\right)^2 + \left(\frac{dr}{d\theta} \sin \theta + r \cos \theta\right)^2$$

$$= r^2 + \left(\frac{dr}{d\theta}\right)^2.$$

This result can be written symbolically as

$$(ds)^2 = (dr)^2 + r^2(d\theta)^2 \tag{1}$$

meaning that if s, r, θ are functions of a variable t, then

$$\left(\frac{ds}{dt}\right)^2 = \left(\frac{dr}{dt}\right)^2 + r^2\left(\frac{d\theta}{dt}\right)^2$$

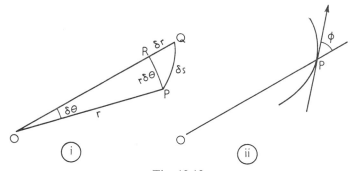

Fig. 12.18

(1) can be remembered by drawing a 'triangle' (Fig. 12.18(i)) consisting of an arc of curve δs, a circular arc $r\,\delta\theta$ and a side δr, so that

$$(\delta s)^2 \approx r^2(\delta \theta)^2 + (\delta r)^2,$$

but this does not, of course, form a basis for a proof of (1).

The angle between the tangent at P (drawn in the direction of s increasing) and OP is denoted by ϕ, Fig. 12.18(ii). We shall accept without proof that

$$\cot \phi = \frac{1}{r}\frac{dr}{d\theta} ;$$

this is again plausible from the 'triangle' PQR, since angle PQR is approximately equal to ϕ.

EXAMPLE 12.9 *To find the class of curves for which the angle ϕ is constant.*

Let $\phi = \alpha$; then $\dfrac{1}{r}\dfrac{dr}{d\theta} = \cot\alpha$, whence

$$\cot\alpha\,\frac{d\theta}{dr} = \frac{1}{r},$$

so

$$\theta\cot\alpha = \int \frac{1}{r}\,dr = \log r + c = \log\frac{r}{a}.$$

Thus the required curves (*equiangular spirals*) are given by $r = a\,e^{\theta\cot\alpha}$, a being an arbitrary constant.*

EXAMPLE 12.10 *To find* (i) *the total arc-length of the 'cardioid'*

$$r = a(1 + \cos\theta),$$

(ii) *the area enclosed by the curve,* (iii) *the C.M. of the lamina bounded by its upper half and the initial line,* (iv) *the volume obtained by revolving this upper half through 4 right angles about the initial line,* (v) *the surface area of the solid thus formed.*

(i)
$$\left(\frac{ds}{d\theta}\right)^2 = r^2 + \left(\frac{dr}{d\theta}\right)^2$$

$$= a^2(1 + \cos\theta)^2 + a^2\sin^2\theta$$

$$= 2a^2(1 + \cos\theta)$$

$$= 4a^2\cos^2\frac{\theta}{2}.$$

Thus $\dfrac{ds}{d\theta} = \pm 2a\cos\dfrac{\theta}{2}$. *For $0 < \theta < \pi$, the positive sign must be taken, for $\pi < \theta < 2\pi$ the negative one.*

By symmetry, the total arc-length is

$$2\int_0^\pi 2a\cos\frac{\theta}{2}\,d\theta = 8a\left[\sin\frac{\theta}{2}\right]_0^\pi = 8a.$$

---

* The interested reader is referred to *On Growth and Form* by D'Arcy Wentworth Thompson, (C.U.P.).

(ii) The area is given by

$$A = \frac{1}{2}\int_0^{2\pi} r^2 \, d\theta$$

$$= \frac{1}{2}\int_0^{2\pi} a^2(1 + \cos\theta)^2 \, d\theta$$

$$= \frac{a^2}{2}\int_0^{2\pi} \{1 + 2\cos\theta + \tfrac{1}{2}(1 + \cos 2\theta)\} \, d\theta$$

$$= \frac{a^2}{2}\left[\frac{3}{2}\theta + 2\sin\theta + \frac{1}{4}\sin 2\theta\right]_0^{2\pi}$$

$$= \frac{3\pi a^2}{2}.$$

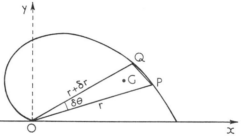

Fig. 12.19

(iii) We assume that the area can be considered as the limit of the sum of triangles such as OPQ, when $\delta\theta \to 0$ (Fig. 12.19). The area of this triangle is $\frac{1}{2}r(r + \delta r)\sin\delta\theta$; if G is its C.M., then $OG = \frac{2}{3}(r + \lambda\,\delta r)$ where $0 < \lambda < 1$, and so the x-coordinate of its C.M. is

$$\tfrac{2}{3}(r + \lambda\,\delta r)\cos\theta.$$

The area of the lamina is $\dfrac{3\pi a^2}{4}$ and so, using $\Sigma\,mx = M\bar{x}$, we have

$$\lim_{\delta\theta\to 0}\Sigma_{\theta=0}^{\pi}\tfrac{1}{2}r(r + \delta r)\sin\delta\theta \cdot \tfrac{2}{3}(r + \lambda\,\delta r)\cos\theta = \frac{3\pi a^2}{4}\bar{x}.$$

We may choose the angles $\delta\theta$ to be equal, so the left-hand side can be written

$$\lim_{\delta\theta\to 0}\left\{\frac{\sin\delta\theta}{\delta\theta}\Sigma_{\theta=0}^{\pi}\tfrac{1}{3}r(r + \delta r)(r + \lambda\,\delta r)\cos\theta\,\delta\theta\right\}$$

and (assuming the limit of a product is equal to the product of the two limits), we thus have

$$\frac{3\pi a^2}{4}\bar{x} = \int_0^\pi \tfrac{1}{3}r^3 \cos\theta \, d\theta$$

$$= \frac{a^3}{3}\int_0^\pi (1 + \cos\theta)^3 \cos\theta \, d\theta$$

$$= \frac{a^3}{3}\int_0^\pi \{\cos\theta + 3\cos^2\theta + 3\cos^3\theta + \cos^4\theta\} \, d\theta$$

$$= \frac{a^3}{3}\int_0^\pi \{\tfrac{3}{2}(1 + \cos 2\theta) + \tfrac{1}{4}(1 + \cos 2\theta)^2\} \, d\theta$$

$$\left(\int_0^\pi \cos^{2n+1}\theta \, d\theta = 0, \text{ cf. p. 130}\right)$$

$$= \frac{a^3}{3}\int_0^\pi \{\tfrac{3}{2} + \tfrac{3}{2}\cos 2\theta + \tfrac{1}{4} + \tfrac{1}{2}\cos 2\theta + \tfrac{1}{8}(1 + \cos 4\theta)\} \, d\theta$$

$$= \frac{a^3}{3} \times \frac{15\pi}{8} = \frac{5\pi a^3}{8}.$$

Thus

$$\bar{x} = \frac{4}{3\pi a^2} \times \frac{5\pi a^3}{8} = \frac{5a}{6}.$$

Similarly,

$$\frac{3\pi a^2}{4}\bar{y} = \int_0^\pi \tfrac{1}{3}r^3 \sin\theta \, d\theta$$

$$= \frac{a^3}{3}\int_0^\pi (1 + \cos\theta)^3 \sin\theta \, d\theta$$

$$= \frac{a^3}{3}\int_{-1}^1 (1 + c)^3 \, dc,$$

putting $\cos\theta = c$,

$$= \frac{4a^3}{3},$$

so

$$\bar{y} = \frac{4}{3\pi a^2} \times \frac{4a^3}{3} = \frac{16a}{9\pi}.$$

Thus the coordinates of the C.M. of the lamina are $\left(\dfrac{5a}{6}, \dfrac{16a}{9\pi}\right)$.

(iv) By the second Pappus-Guldin theorem, the volume is equal to the area of the lamina times the distance travelled by its C.M., i.e.

$$V = \frac{3\pi a^2}{4} \times 2\pi\left(\frac{16a}{9\pi}\right) = \frac{8\pi a^3}{3}.$$

(v) The surface area is given by

$$S = \int 2\pi y \, ds$$

$$= \int_0^\pi 2\pi y \, \frac{ds}{d\theta} \, d\theta,$$

on expressing y, s as functions of θ,

$$S = \int_0^\pi 2\pi r \sin\theta \,.\, 2a \cos\frac{\theta}{2} \, d\theta$$

$$= 4\pi a^2 \int_0^\pi (1 + \cos\theta) \sin\theta \cos\frac{\theta}{2} \, d\theta$$

$$= 16\pi a^2 \int_0^\pi \cos^4\frac{\theta}{2} \sin\frac{\theta}{2} \, d\theta$$

$$= 32\pi a^2 \int_0^1 c^4 \, dc, \qquad \text{putting } \cos\frac{\theta}{2} = c, \; -\tfrac{1}{2}\sin\frac{\theta}{2} \, d\theta = dc, \,.$$

$$= \frac{32\pi a^2}{5} \,.$$

EXERCISE LXI

1. Find the area of one loop of the curve $r = a \cos 4\theta$.
2. Find the length of arc of the equiangular spiral $r = ae^{\theta \cot\alpha}$ from $\theta = 0$ to 2π.
3. Sketch the lemniscate $r^2 = \cos 2\theta$ and find the area of one of its loops.
4. Find the area contained between the two loops of the 'limaçon' $r = 1 + 2\cos\theta$.
5. Find the area enclosed by both the cardioid $r = a(1 - \cos\theta)$ and the circle $r = 2a \sin\theta$.
6. Express the total arc-length of the curve $r = 5 + 4\cos\theta$ as an integral. Can you evaluate this exactly?
7. Find the area of the surface of revolution generated by rotating one loop of the curve $r^2 = a^2 \cos 2\theta$ about the line $\theta = 0$.
8. By transforming to polar coordinates, prove that the area common to the two ellipses

$$\frac{x^2}{a^2} + \frac{y^2}{b^2} = 1 \quad \text{and} \quad \frac{x^2}{b^2} + \frac{y^2}{a^2} = 1 \quad (\text{where } a > b)$$

is

$$4 \int_0^{\frac{\pi}{4}} \frac{a^2 b^2}{a^2 \cos^2\theta + b^2 \sin^2\theta} \, d\theta.$$

By changing the variable to $t = \tan\theta$ prove that the value of this is

$$4ab \tan^{-1}\left(\frac{b}{a}\right).$$

9. The equation of a curve referred to polar coordinates is $r = f(\theta)$, and the curve is symmetrical about the initial line $\theta = 0$. Prove that the formula for the distance between the origin and the centre of gravity of the sector bounded by the radii $\theta = \alpha$, $\theta = -\alpha$ is expressible in the form

$$\frac{2 \displaystyle\int_0^\alpha r^3 \cos \theta \, d\theta}{3 \displaystyle\int_0^\alpha r^2 \, d\theta}.$$

Prove that the distance between the origin and the centre of gravity of the sector bounded by the radii $\theta = \frac{1}{4}\pi$, $\theta = -\frac{1}{4}\pi$, and the portion of the curve $r = a \sec^2 \frac{1}{2}\theta$ which lies between them, is $\frac{12}{35}(2\sqrt{2} - 1)a$.

[Use the result $\tan \frac{1}{8}\pi = \sqrt{2} - 1$, if required.]

EXERCISE LXII

Miscellaneous questions on Chapters 11 *and* 12.

1. (P) Write a treatise on either (i) the cycloid, or (ii) the cardioid.

2. (i) Sketch together the graphs of the circle $r = 3a \cos \theta$ and the cardioid $r = a(1 + \cos \theta)$.

    Show that the area of the part of the circle outside the cardioid is πa^2.

   (ii) The arc of the curve $y = +2x^{\frac{3}{2}}$ between $x = 0$ and $x = 3$ is rotated through four right angles about the x-axis. Prove that the area of the surface traced out is $56\pi/3$.

3. Find the centre of gravity of a uniform solid hemisphere of radius a.

    Show that the centre of gravity of a thin uniform hemispherical shell of mass m and radius a is at a distance $\frac{1}{2}a$ from the plane of the rim, and find the moment of inertia of the shell about the radius perpendicular to the base.

4. By putting $y = tx$, obtain a parametric representation of the curve

$$x^3 + y^3 = 3axy.$$

    Sketch this curve, and obtain the radii of curvature of the two branches at the origin, proving any formula that you use for radius of curvature.

5. A given closed curve is cut by any line parallel to the y-axis in, at most, two points whose ordinates are y_1 and y_2. Prove that the area enclosed by the curve is given by

$$\int (y_2 - y_1) \, dx$$

taken between certain limits.

    Hence, or otherwise, find the area enclosed by the curve whose equation is

$$2x^2 - 4xy + 12y^2 = 1.$$

6. A solid of uniform density is formed by revolving about the x-axis the area enclosed by the curve $y = \cos x$ from $x = 0$ to $x = \frac{1}{2}\pi$ and the two axes. Find the position of the centre of gravity of the solid, and the moment of inertia of the solid about the x-axis in terms of its mass M.

    [You may assume without proof that the moment of inertia of a cylinder of mass m and radius r about its axis is $\frac{1}{2}mr^2$.]

7. Prove that the intrinsic form of the equation of the parabola $y^2 = 4ax$ is

$$s = a \operatorname{cosec} \psi \cot \psi - a \ln \tan \tfrac{1}{2}\psi,$$

where s is the length of arc measured from the vertex and ψ is the inclination of the tangent to the x-axis. Find the length of arc from the origin to the point $(a, 2a)$.

8. Find the mass of a hemisphere of radius a and variable density σ when (i) $\sigma = kx^2$, where x is the distance from the base of the hemisphere; (ii) $\sigma = kr^2$, where r is the distance from the axis of symmetry of the hemisphere.

9. Find the area bounded by the curve

$$x = 2a(t^3 - 1), \qquad y = 3at^2,$$

and the straight lines $x = 0$, $y = 0$.
Prove that the tangent to the curve at the point $t = 1$ meets the curve again at the point $t = -\tfrac{1}{2}$, and find the area bounded by the parts of the tangent and of the curve that lie between these points.

10. Find the centre of gravity of a uniform arc AB of a circle, of radius a, subtending an angle 2α radians at the centre O of the circle.
Hence show that the centre of gravity of the uniform sector OAB is at a distance $\dfrac{2a \sin \alpha}{3\alpha}$ from O.
Prove that the radius of gyration of a uniform semicircular lamina about the straight edge is $a/2$.

11. If $P(x, y)$, $Q(x + \delta x, y + \delta y)$ are two points referred to rectangular axes, show that the area of the triangle OPQ, where O is the origin, is

$$\tfrac{1}{2}(x\,\delta y - y\,\delta x).$$

Hence show that, if t_1, t_2 are the parameters of two points A, B on a curve, whose parametric equations are $x = f(t)$ $y = F(t)$, the area enclosed by the arc AB and the radius vectors OA, OB is

$$\frac{1}{2}\int_{t_1}^{t_2} \left(x\frac{dy}{dt} - y\frac{dx}{dt} \right) dt.$$

Use this result to obtain the area of the ellipse $x = a \cos t$, $y = b \sin t$, putting $t_1 = 0$, $t_2 = 2\pi$. Verify by direct integration.

12. A curve is given by the parametric equations

$$x = c \ln (\sec t + \tan t), \qquad y = c \sec t,$$

where c is a constant and t the parameter. Find y as a function of x.
Show that, measured from the point $t = 0$, $s = c \tan t$. Show also that the area enclosed between the curve, the axes of coordinates, and the ordinate of the point t, is equal to cs.

13. A uniform solid right circular cone of mass M has a base of radius a. Calculate its moment of inertia about its axis of symmetry.
The solid cone is cut into two parts by a plane parallel to its circular base and bisecting the axis of symmetry. Obtain the moment of inertia of each of the two parts about the axis of symmetry.

13
Further Integration

The steps were higher that they took.
DRYDEN

The reader should first revise his knowledge of integration up to now, in particular re-reading the sections 'Integral as limit of a sum' (p. 59) and 'Definite integrals' (p. 127).

The following example is 'worse' than the reader might reasonably be asked to do himself at this stage, but it will be instructive to work through it, bearing this firmly in mind.

EXAMPLE 13.1   (i) *Prove the identity*

$$(1 - 2\lambda \cos x + \lambda^2)(1 + 2\lambda \cos x + \lambda^2) = 1 - 2\lambda^2 \cos 2x + \lambda^4.$$

(ii) *Given that*

$$\phi(\lambda) = \int_0^\pi \ln (1 - 2\lambda \cos x + \lambda^2) \, dx,$$

prove that

$$\phi(\lambda) = \phi(-\lambda) = \tfrac{1}{2}\phi(\lambda^2).$$

Deduce that, if $|\lambda| < 1$, $\phi(\lambda) = 0$, *and hence prove that if* $|\lambda| > 1$, $\phi(\lambda) = 2\pi \ln \lambda$.

(i) Left-hand side $= \{(4\lambda^2) - 2\lambda \cos x\}\{(1 + \lambda^2) + 2\lambda \cos x\}$

$\qquad\qquad\qquad = (1 + \lambda^2)^2 - (2\lambda \cos x)^2$

$\qquad\qquad\qquad = 1 - 2\lambda^2(2 \cos^2 x - 1) + \lambda^4$

$\qquad\qquad\qquad = 1 - 2\lambda^2 \cos 2x + \lambda^4$

$\qquad\qquad\qquad = $ right-hand side.

(ii)
$$\phi(\lambda) = \int_0^\pi \ln (1 - 2\lambda \cos x + \lambda^2) \, dx$$

$$= -\int_\pi^0 \ln \{1 - 2\lambda \cos (\pi - y) + \lambda^2\} \, dy,$$

$$\text{(putting } x = \pi - y)$$

$$= \int_0^\pi \ln (1 + 2\lambda \cos y + \lambda^2) \, dy$$

$$= \phi(-\lambda).$$

261

[$\phi(\lambda)$ is, of course, not a function of x, since the integral is definite; replacing x by y throughout does not affect its value.]

Now

$$\phi(\lambda) + \phi(-\lambda) = \int_0^\pi \{\ln(1 - 2\lambda \cos x + \lambda^2) + \ln(1 + 2\lambda \cos x + \lambda^2)\}\, dx$$

$$= \int_0^\pi \ln(1 - 2\lambda^2 \cos 2x + \lambda^4)\, dx \qquad \text{(from (i))}$$

$$= \frac{1}{2}\int_0^{2\pi} \ln(1 - 2\lambda^2 \cos z + \lambda^4)\, dz \qquad \text{(putting } 2x = z) \quad (1)$$

Also

$$\phi(\lambda^2) = \int_0^\pi \ln(1 - 2\lambda^2 \cos z + \lambda^4)\, dz$$

$$= \int_0^\pi \ln(1 + 2\lambda^2 \cos z + \lambda^4)\, dz, \qquad \text{since } \phi(\lambda^2) = \phi(-\lambda^2)$$

$$= \int_\pi^{2\pi} \ln(1 - 2\lambda^2 \cos w + \lambda^4)\, dw, \qquad \text{putting } z = w - \pi;$$

thus

$$\phi(\lambda^2) = \frac{1}{2}\int_0^{2\pi} \ln(1 - 2\lambda^2 \cos z + \lambda^4)\, dz. \qquad (2)$$

Since $\phi(\lambda) = \phi(-\lambda)$, it follows from (1) and (2) that

$$\phi(\lambda) = \tfrac{1}{2}\phi(\lambda^2). \qquad (3)$$

Repeated use of (3) gives

$$\phi(\lambda) = \frac{1}{2}\,\phi(\lambda^2) = \frac{1}{2^2}\,\phi(\lambda^4) = \ldots = \frac{1}{2^n}\,\phi(\lambda^{2n}).$$

Now

$$\phi(\lambda^{2n}) = \int_0^\pi \log(1 - 2\lambda^{2n} \cos x + \lambda^{4n})\, dx$$

$$= \pi \log(1 - 2\lambda^{2n} \cos \theta + \lambda^{4n})$$

where $0 < \theta < \pi$, by result (3), p. 60,

$$\to 0 \text{ as } n \to \infty \text{ if } |\lambda| < 1.$$

Since $\phi(\lambda) = \dfrac{1}{2^n}\, \phi(\lambda^{2n})$ for all n, and the right-hand side tends to zero as $n \to \infty$ ($|\lambda| < 1$), it follows that

$$\phi(\lambda) = 0 \quad \text{if} \quad |\lambda| < 1.$$

If $|\lambda| > 1$, writing $\lambda = \dfrac{1}{\mu}$ so that $|\mu| < 1$, we have

$$\phi(\lambda) = \int_0^{\pi} \ln\left(1 - \frac{2}{\mu}\cos x + \frac{1}{\mu^2}\right) dx$$

$$= \int_0^{\pi} \{\ln(\mu^2 - 2\mu\cos x + 1) - \ln \mu^2\}\, dx$$

$$= \varphi(\mu) - \pi \ln \mu^2$$

$$= -2\pi \ln \mu, \quad \text{since } \varphi(\mu) = 0 \text{ from above,}$$

$$= 2\pi \ln \lambda.$$

EXERCISE LXIII

1. Without attempting to evaluate them, determine whether the following integrals are positive, negative, or zero:

$$\int_0^1 x^3(1-x)^3\, dx; \qquad \int_0^{\pi} \sin^2 x \cos^3 x\, dx; \qquad \int_0^{\pi} e^{-x} \sin x\, dx.$$

2. $f(x)$ is a function of x which is positive but diminishes steadily as x increases. $I(n)$ denotes the integral $\displaystyle\int_1^n f(x)\, dx$. Show from graphical considerations that the sum of the series

$$f(1) + f(2) + \ldots + f(n)$$

lies between $I(n + 1)$ and $I(n) + f(1)$.

3. Let $I(z) = \displaystyle\int_1^z \dfrac{(x-1)^p(2-x)^p}{x^{p+1}}\, dx$, $(p > 0)$. By writing $x = 2/y$, prove that $2I(\sqrt{2}) = I(2)$.

4. If $I = \displaystyle\int_0^{\frac{\pi}{2}} \ln \sin x\, dx$, show by means of the substitution $x = \tfrac{1}{2}\pi - y$ that

$$I = \int_0^{\frac{\pi}{2}} \ln \cos y\, dy \text{ and deduce that}$$

$$2I = \int_0^{\frac{\pi}{2}} \ln \sin 2x\, dx + \tfrac{1}{2}\pi \ln (\tfrac{1}{2}).$$

Use the further substitution $2x = z$ to prove that

$$I = \tfrac{1}{2}\pi \ln (\tfrac{1}{2}).$$

5. Prove, when $b > a > 0$, that

$$\int_0^\pi \frac{\sin \theta \, d\theta}{\sqrt{(a^2 + b^2 - 2ab \cos \theta)}} = \frac{2}{b},$$

$$\int_0^\pi \frac{\sin \theta \cos \theta \, d\theta}{\sqrt{(a^2 + b^2 - 2ab \cos \theta)}} = \frac{2a}{3b^2}.$$

Make it clear at what points of your proofs you use the condition $b > a > 0$.

6. By using the substitution $x = \dfrac{1 - u}{1 + u}$ or otherwise, show that

$$F(\alpha) = \int_0^1 \frac{\sin \alpha}{x^2 + 2x \cos \alpha + 1} \, dx$$

is equal to $\frac{1}{2}\alpha$ when $-\pi < \alpha < \pi$.

Hence show that $\quad G(\alpha) = \displaystyle\int_0^1 \frac{\sin \alpha}{x^2 - 2x \cos \alpha + 1} \, dx$

is equal to $\frac{1}{2}(\pi - \alpha)$ when $0 < \alpha < 2\pi$.

What is the value of $F(\alpha)$ when $\pi < \alpha < 2\pi$ and of $G(\alpha)$ when $-\pi < \alpha < 0$?

Infinite Integrals

There are two types of integral involving 'infinity'.

(i) Suppose $\displaystyle\int_a^X f(x) \, dx$ tends to a limit as $X \to \infty$; then we write this limit

as $\displaystyle\int_a^\infty f(x) \, dx$. Similarly $\displaystyle\int_{-\infty}^b f(x) \, dx$ may be defined as $\displaystyle\lim_{X \to \infty} \int_{-X}^b f(x) \, dx$.

EXAMPLE 13.2 *To evaluate* $I = \displaystyle\int_0^\infty \frac{1}{1 + x^2} \, dx$.

$$\int_0^X \frac{dx}{1 + x^2} = \left[\tan^{-1} x \right]_0^X = \tan^{-1} X \to \frac{\pi}{2} \quad \text{as } X \to \infty.$$

Thus $\qquad\qquad\qquad\qquad I = \dfrac{\pi}{2}.$

(ii) The integrand may become infinite for some value(s) within the range of integration.

EXAMPLE 13.3 *To evaluate* $I = \displaystyle\int_0^1 \frac{dx}{\sqrt{(1 - x^2)}}$.

$\dfrac{1}{\sqrt{(1 - x^2)}}$ is undefined ('infinite') when $x = 1$. We can, however, give a meaning to the integral, namely as $\displaystyle\lim_{k \nearrow 1} \int_0^k \frac{dx}{\sqrt{(1 - x^2)}}$, provided this limit exists.

Now $\displaystyle\int_0^k \frac{dx}{\sqrt{(1 - x^2)}} = \sin^{-1} k \to \frac{\pi}{2}$ as $k \nearrow 1$. Thus $I = \dfrac{\pi}{2}$.

EXAMPLE 13.4 *To evaluate* $I = \int_0^1 \dfrac{x^2}{\sqrt{(1 - x^2)}}\, dx.$

$$I = \int_0^1 \frac{x^2\, dx}{\sqrt{(1 - x^2)}}$$

$$= \int_0^{\frac{\pi}{2}} \frac{\sin^2 \theta \cos \theta\, d\theta}{\cos \theta}, \qquad \text{putting } x = \sin \theta,$$

$$= \frac{1}{2} \int_0^{\frac{\pi}{2}} (1 - \cos 2\theta)\, d\theta$$

$$= \frac{\pi}{4}.$$

This procedure can be justified by considering first the range of integration for x to be 0 to $(1 - \varepsilon)$, which for θ becomes 0 to $\left(\dfrac{\pi}{2} - \eta\right)$, η tending to zero as $\varepsilon \nearrow 0$; thus

$$I = \lim_{\eta \to 0} \int_0^{\frac{\pi}{2} - \eta} \sin^2 \theta\, d\theta = \int_0^{\frac{\pi}{2}} \sin^2 \theta\, d\theta \quad \left(= \frac{\pi}{4} \right).$$

EXAMPLE 13.5 $I = \int_{\frac{\pi}{4}}^{\frac{\pi}{2}} \dfrac{dx}{(\tan^{-1} x)^2 (1 + x^2)}.$

Putting $x = \tan y$, $I = \int_1^\infty \dfrac{\sec^2 y}{y^2 \sec^2 y}\, dy = \left[-\dfrac{1}{y} \right]_1^\infty = 1.$

It is unnecessary to give a formal justification in each case; here one would first consider the upper limit as $\dfrac{\pi}{2} - \varepsilon$, which on substitution would become X, where $X \to \infty$ as $\varepsilon \to 0$.

EXAMPLE 13.6 $I = \int_a^b \dfrac{dx}{\sqrt{\{(x - a)(b - x)\}}}.$

Put $\qquad x = a \cos^2 \theta + b \sin^2 \theta,$

$$dx = (-2a \cos \theta \sin \theta + 2b \cos \theta \sin \theta)\, d\theta$$

$$= 2(b - a) \sin \theta \cos \theta\, d\theta.$$

Then $\qquad x - a = (b - a) \sin^2 \theta,\ b - x = (b - a) \cos^2 \theta.$

When $x = a$, $a(1 - \cos^2 \theta) = b \sin^2 \theta$, so $\sin \theta = 0$; when $x = b$,

$$b(1 - \sin^2 \theta) = a \cos^2 \theta, \qquad \text{so } \cos \theta = 0.$$

We can accordingly take the limits for θ as 0 to $\frac{\pi}{2}$, and we have

$$I = \int_0^{\frac{\pi}{2}} \frac{2(b-a)\sin\theta\cos\theta}{(b-a)\sin\theta\cos\theta}\,d\theta$$

$$= \pi.$$

EXAMPLE 13.7   *To investigate* $\int_1^\infty \frac{1}{x^2}\,dx.$

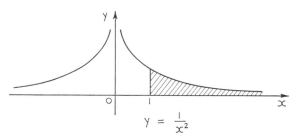

$$y = \frac{1}{x^2}$$

Fig. 13.1

$$\int_1^X \frac{1}{x^2}\,dx = \left[-\frac{1}{x}\right]_1^X = 1 - \frac{1}{X} \to 1 \text{ as } X \to \infty. \text{ Thus } \int_1^\infty \frac{1}{x^2}\,dx = 1. \text{ This, in}$$

Fig. 13.1, is the shaded 'area' under $y = \frac{1}{x^2}$ from 1 to ∞ (i.e. the limit of the area from 1 to x as $x \to \infty$).

EXAMPLE 13.8   *To investigate* $I = \int_0^\infty \frac{1}{x^3}\,dx.$

$$\int_a^b \frac{1}{x^3}\,dx = \left[-\frac{1}{2x^2}\right]_a^b = -\frac{1}{2b^2} + \frac{1}{2a^2}.$$

As $b \to \infty$, $\frac{1}{b^2} \to 0$, but as $a \to 0$, $\frac{1}{a^2} \to \infty$, so that the integral is meaningless.

EXERCISE LXIV

Investigate the following integrals, evaluating those that exist.

1. $\displaystyle\int_0^1 \frac{dx}{(1-x)^{\frac{3}{2}}}.$

2. $\displaystyle\int_0^\infty \frac{dx}{9x^2+4}.$

3. $\displaystyle\int_1^\infty \frac{dx}{9x^2-4}.$

4. $\displaystyle\int_0^\infty \sin x\,dx.$

5. $\displaystyle\int_{-1}^{1} \frac{1}{x^2}\, dx.$

6. $\displaystyle\int_{0}^{\infty} x e^{-x}\, dx.$ $[x e^{-x} \to 0$ as $x \to \infty$, see Appendix 2].

7. $\displaystyle\int_{0}^{1} x \ln x\, dx.$ 8. $\displaystyle\int_{a}^{b} \sqrt{\{(x-a)(b-x)\}}\, dx.$

9. (D) Show that $\displaystyle\int_{1}^{\infty} \frac{1}{x}\, dx$ does not exist, but that $\displaystyle\int_{1}^{\infty} \frac{\pi}{x^2}\, dx$ does. (Thus, in Fig.

13.2, although the area under $y = \dfrac{1}{x}$ from 1 to X tends to infinity as $X \to \infty$,

the volume obtained by rotating this area about the x-axis does not.)

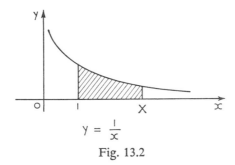

$$y = \frac{1}{x}$$

Fig. 13.2

Reduction Formulae

It is sometimes possible to evaluate an integral by finding a formula (which may be used repeatedly) connecting it with a related but simpler one.

EXAMPLE 13.9   *To find an expression for* $I_n = \displaystyle\int_{0}^{\frac{\pi}{2}} \cos^n \theta\, d\theta$; *in particular, to evaluate* I_8, I_9.

$$I_n = \int_{0}^{\frac{\pi}{2}} \cos^{n-1} \theta \cos \theta\, d\theta$$

$$= \left[\sin \theta \cos^{n-1} \theta \right]_{0}^{\frac{\pi}{2}} + \int_{0}^{\frac{\pi}{2}} \sin^2 \theta (n-1) \cos^{n-2} \theta\, d\theta \qquad \text{(by 'parts')}$$

$$= (n-1) \int_{0}^{\frac{\pi}{2}} \cos^{n-2} \theta (1 - \cos^2 \theta)\, d\theta$$

$$= (n-1)(I_{n-2} - I_n).$$

Thus $n I_n = (n-1) I_{n-2}$

or

$$I_n = \frac{n-1}{n} I_{n-2}.$$

Such a relationship is called a *reduction formula*.

By repeated application, we have $I_{n-2} = \frac{n-3}{n-2} I_{n-4}$ and so on, so that

$$I = \frac{n-1}{n} \cdot \frac{n-3}{n-2} \cdot \frac{n-5}{n-4} \cdot \ldots \cdot \frac{1}{2} I_0 \qquad \text{if } n \text{ is even, and}$$

$$I = \frac{n-1}{n} \cdot \frac{n-3}{n-2} \cdot \ldots \cdot \frac{2}{3} I_1 \qquad \text{if } n \text{ is odd.}$$

Now $I_1 = \int_0^{\frac{\pi}{2}} \cos\theta \, d\theta = 1$, $I_0 = \int_0^{\frac{\pi}{2}} d\theta = \frac{\pi}{2}$. Thus for *n even*,

$$I_n = \frac{n-1}{n} \cdot \frac{n-3}{n-2} \cdot \ldots \cdot \frac{1}{2} \cdot \frac{\pi}{2},$$

and for *n odd*,

$$I_n = \frac{n-1}{n} \cdot \frac{n-3}{n-2} \cdot \ldots \cdot \frac{2}{3}.$$

In particular,

$$I_8 = \int_0^{\frac{\pi}{2}} \cos^8\theta \, d\theta = \frac{7 \cdot 5 \cdot 3 \cdot 1}{8 \cdot 6 \cdot 4 \cdot 2} \cdot \frac{\pi}{2} = \frac{35\pi}{256} \ ;$$

$$I_9 = \int_0^{\frac{\pi}{2}} \cos^9\theta \, d\theta = \frac{8 \cdot 6 \cdot 4 \cdot 2}{9 \cdot 7 \cdot 5 \cdot 3} = \frac{128}{315}.$$

EXAMPLE 13.10   *To find a reduction formula for* $I_n = \int x^n e^{-x} \, dx$.

$$I_n = \int x^n e^{-x} \, dx$$

$$= -e^{-x} x^n + n \int e^{-x} x^{n-1} \, dx \qquad \text{(by parts)}$$

$$= -e^{-x} x^n + n I_{n-1}.$$

EXERCISE LXV

1. Evaluate (i) $\int_0^{\frac{\pi}{2}} \sin^{12}\theta \, d\theta$; (ii) $\int_0^{\frac{\pi}{2}} \sin^{13}\theta \, d\theta$.

2. Find a reduction formula for (i) $\int \tan^n x \, dx$; (ii) $\int (\ln x)^n \, dx$.

3. (i) Prove that, if

$$u_n = \int_0^{\frac{\pi}{2}} \sin^{2n+1} \theta \, d\theta$$

and $n \geqslant 1$, then $(2n + 1)u_n = 2nu_{n-1}$.
   Prove that, if n is a positive integer,

$$u_n = \frac{2n(2n - 2) \ldots 2}{(2n + 1)(2n - 1) \ldots 3}$$

(ii) Prove that, if

$$I_{m,n} = \int_0^{\frac{\pi}{2}} \cos^m \theta \sin^n \theta \, d\theta,$$

and $m \geqslant 2$, $n \geqslant 0$,

$$(m + n)I_{m,n} = (m - 1)I_{m-2,n}.$$

(iii) Using the results of (i) and (ii) prove that

$$\int_0^{\frac{\pi}{2}} \cos^4 \theta \sin^9 \theta \, d\theta = \frac{2 \cdot 4 \cdot 6 \cdot 8}{5 \cdot 7 \cdot 9 \cdot 11 \cdot 13}.$$

4. Integrate $\sin^4 x \cos^3 x$, with respect to x. Prove that, if

$$I_{m,n} = \int_0^1 \frac{x^m}{(1 + x^2)^n} \, dx \qquad (m \geqslant 1),$$

then

$$I_{m,n} = \frac{m - 1}{2(n - 1)} I_{m-2,n-1} - \frac{1}{2^n(n - 1)}.$$

Hence evaluate

$$\int_0^1 \frac{x^4}{(1 + x^2)^3} \, dx.$$

5. If $u_n = \int_0^{\frac{\pi}{2}} x^n \sin x \, dx$, prove that, for $n \geqslant 2$,

$$u_n + n(n - 1)u_{n-2} = n(\tfrac{1}{2}\pi)^{n-1}.$$

Calculate u_1 and deduce that $u_3 = \tfrac{3}{4}\pi^2 - 6$.

6. If

$$I_{2n} = \int_{-\frac{\pi}{2}}^{\frac{\pi}{2}} e^x \sin^{2n}x \, dx \qquad (n \geqslant 1),$$

prove that

$$(4n^2 + 1)I_{2n} = 2n(2n - 1)I_{2n-2} + 2 \sinh \tfrac{1}{2}\pi.$$

7. (i) Prove that, if

$$u_n = \int_0^{\frac{\pi}{2}} \frac{\sin 2n\theta}{\sin \theta} \, d\theta,$$

where n is a positive integer,

then

$$u_n - u_{n-1} = \frac{2(-1)^{n-1}}{2n - 1}.$$

Hence prove that

$$u_n = 2\left\{1 - \frac{1}{3} + \frac{1}{5} - \ldots + \frac{(-1)^{n-1}}{2n-1}\right\}.$$

(ii) Prove that, if $I_n = \int \sec^{2n} \theta \, d\theta$, and $n > 1$,

$$(2n-1)I_n = 2(n-1)I_{n-1} + \sec^{2n-2} \theta \tan \theta.$$

(iii) Using the result of (ii) prove that

$$\int_0^{\frac{\pi}{4}} \sec^{10} \theta \, d\theta = \frac{1328}{315}.$$

8. If $u_n = \int_0^a \frac{\sin (2n-1)x}{\sin x} \, dx$ and $v_n = \int_0^a \frac{\sin^2 nx}{\sin^2 x} \, dx$, prove that

(i) $n(u_{n+1} - u_n) = \sin 2na$,

(ii) $v_{n+1} - v_n = u_{n+1}$.

Deduce that, if n is a positive integer,

$$\int_0^{\frac{\pi}{2}} \frac{\sin^2 nx \, dx}{\sin^2 x} = \frac{n\pi}{2}.$$

9. Show that if $I_n = \int_0^{\frac{\pi}{2}} \cos^n \theta \, d\theta$, where n is a positive integer,

(i) $I_{2n+1} < I_{2n} < I_{2n-1}$.

(ii) $\dfrac{I_{2n-1}}{I_{2n+1}} \to 1$ as $n \to \infty$.

(iii) Deduce that $\dfrac{\pi}{2} = \lim_{n \to \infty} \dfrac{1}{2n+1} \cdot \dfrac{2^2 \cdot 4^2 \ldots \ldots (2n)^2}{1^2 \cdot 3^2 \ldots \ldots (2n-1)^2}$.

EXERCISE LXVI

Miscellaneous examination questions.

1. Integrate the following functions with respect to x:

(i) $\dfrac{x^2}{x^3 + 4}$;   (ii) $\dfrac{6x}{x^2 - 4x + 3}$;   (iii) $x \sin 2x$.

Evaluate $\displaystyle\int_1^2 \dfrac{dx}{x^2 \sqrt{(x-1)}}$

by means of the substitution $x = \sec^2 \phi$.

2. (i) Use the rule for integration by parts to evaluate

$$\int (\ln x)^3 \, dx.$$

(ii) Prove that, if

$$I_n = \int_0^a (a^2 - x^2)^n \, dx,$$

then $\qquad (2n + 1)I_n = 2na^2I_{n-1} \qquad (n > 0);$

and evaluate I_n.

(iii) Use the substitution $\tan x = t$ to evaluate

$$\int_0^{\frac{\pi}{4}} \frac{dx}{9 \sin^2 x + 4 \cos^2 x}.$$

3. If $I_n = \int \cos^n \theta \, d\theta$, prove that

$$nI_n = \cos^{n-1} \theta \sin \theta + (n - 1)I_{n-2}.$$

Prove that $\qquad \int_0^{\frac{\pi}{2}} \cos^4 \theta \, d\theta = \frac{3}{16}\pi.$

By putting $x = a \sin \theta$, or otherwise, prove that

$$\int_0^a x^2 (a^2 - x^2)^{\frac{1}{2}} \, dx = \frac{1}{16}\pi a^4.$$

4. Integrate with respect to x:

(i) $\dfrac{x^2}{x^2 - 4}$, (ii) $x^3 e^{-x^2/2}$, (iii) $\dfrac{1}{\sqrt{(5 + 4x - x^2)}}$.

By means of the substitution $u = \tan x$, or otherwise, evaluate the integral

$$\int_0^{\frac{\pi}{2}} \frac{dx}{2 + \cos^2 x}.$$

5. (i) Prove that, if

$$I_{m,n} = \int x^m (\ln x)^n \, dx,$$

then $\qquad (m + 1)I_{m,n} = x^{m+1} (\ln x)^n - nI_{m,n-1}.$

Hence evaluate $\qquad \int_1^e x^3 (\ln x)^2 \, dx.$

(ii) Show, graphically or otherwise, that

$$\int_1^{1+h} e^{-x^2} \, dx < \int_1^{1+h} xe^{-x^2} \, dx \qquad (h > 0),$$

and deduce that

$$\int_1^{1+h} e^{-x^2} \, dx < \frac{e^{-1} - e^{-(1+h)^2}}{2}.$$

6. Prove that

$$\int_0^{\frac{\pi}{2}} \cos^{2n} \theta \, d\theta = \frac{2n - 1}{2n} \int_0^{\frac{\pi}{2}} \cos^{2n-2} \theta \, d\theta,$$

and obtain a formula for $\int_0^{\frac{\pi}{2}} \cos^{2n} \theta \, d\theta$, where n is a positive integer.

An ellipse is given by the equations

$$x = 3 \cos \theta, \qquad y = 2 \sin \theta.$$

Show that the total length of the ellipse is given by

$$12 \int_0^{\frac{\pi}{2}} [1 - \tfrac{5}{9} \cos^2 \theta]^{\frac{1}{2}} \, d\theta.$$

Find an approximation to this length by expanding the integrand and retaining only three terms in the expansion. Give your answer correct to three significant figures, taking π to be 3·142.

7. Given that a and b are both positive, use the substitution $\tan \theta = t$ to evaluate (in any order) the integrals

$$\int_0^{\frac{\pi}{2}} \frac{\cos^2 \theta \, d\theta}{a^2 \cos^2 \theta + b^2 \sin^2 \theta} \, ; \quad \int_0^{\frac{\pi}{2}} \frac{\sin^2 \theta \, d\theta}{a^2 \cos^2 \theta + b^2 \sin^2 \theta} \, ;$$

$$\int_0^{\frac{\pi}{2}} \frac{d\theta}{a^2 \cos^2 \theta + b^2 \sin^2 \theta} \, .$$

8. (i) Prove that, if m and n are positive integers,

$$\int_0^{\frac{\pi}{2}} \cos^m x \cos nx \, dx = \frac{m}{m + n} \int_0^{\frac{\pi}{2}} \cos^{m-1} x \cos (n - 1)x \, dx.$$

Show that $\int_0^{\frac{\pi}{2}} \cos^4 x \cos 8x \, dx = 0$.

(ii) By means of the substitution $y^2 = (1 - x^2)/(1 + x^2)$, or otherwise, evaluate the integral

$$\int_0^1 \frac{dx}{(1 + x^2)\sqrt{(1 - x^2)}} \, .$$

9. Integrate with respect to x:

(i) $\int (2x - 3)^{-\frac{3}{2}} \, dx$,   (ii) $\int x \sin x \, dx$,   (iii) $\int x\sqrt{(x - 4)} \, dx$.

By means of the substitution $x = 3 \cos^2 \theta + 6 \sin^2 \theta$, or otherwise, evaluate

$$\int_3^6 \frac{dx}{\sqrt{\{(x - 3)(6 - x)\}}} \, .$$

10. (i) Prove that, if n is an integer greater than unity,

(a) $(n - 1) \int_0^{\frac{\pi}{4}} \tan^n \theta \, d\theta = 1 - (n - 1) \int_0^{\frac{\pi}{4}} \tan^{n-2} \theta \, d\theta,$

(b) $(n - 1) \int_0^{\frac{\pi}{4}} \sec^n \theta \, d\theta = 2^{\frac{1}{2}n-1} + (n - 2) \int_0^{\frac{\pi}{4}} \sec^{n-2} \theta \, d\theta.$

(ii) Prove that

(a) $\int_0^1 \dfrac{x^5 \, dx}{x^2 + 1} = \frac{1}{2} \ln 2 - \frac{1}{4},$

(b) $\int_0^1 (10x^2 + 1)(x^2 + 1)^{\frac{3}{2}} \, dx = 16\sqrt{2}.$

11. A curve has parametric equations

$$x = \frac{3t}{1 + t^3}, \qquad y = \frac{3t^2}{1 + t^3};$$

obtain its polar equation in the form

$$r = \frac{3 \sin \theta \cos \theta}{\sin^3 \theta + \cos^3 \theta}.$$

Sketch the curve and find the area of its loop.

12. Prove that

$$\int_1^\infty \frac{dx}{(x + \lambda)\sqrt{(x^2 - 1)}} = 2 \int_1^\infty \frac{du}{u^2 + 2\lambda u + 1}.$$

Hence, or otherwise, prove that, if $1 > \lambda > -1$, the value of the integral is $(\cos^{-1} \lambda)/\sqrt{(1 - \lambda^2)}$, and evaluate the integral for $\lambda < 1$.

Draw a rough sketch to show how the value of the integral varies with λ over these ranges.

13. Prove that, when $a > 0$,

$$\int e^{-x} \cos (ax + b) \, dx = -e^{-x} \cos (ax + b + \alpha) \cos \alpha,$$

where α is the acute angle such that $\tan \alpha = a$.

Prove that, when n is a positive integer,

$$\int_0^\infty x^n e^{-x} \cos ax \, dx = n! \cos^{n+1} \alpha \cos (n + 1)\alpha.$$

14
Differential Equations

Among all the mathematical disciplines the theory of differential equations is the most important. It furnishes the explanation of all those elementary manifestations of nature which involve time.

SOPHUS LIE

An equation involving derivatives is called a *differential equation* (D.E.), e.g. (i) $\dfrac{dy}{dx} = 3x + 1$, (ii) $\dfrac{d^2y}{dx^2} - 5\dfrac{dy}{dx} + 6y = \cos x$. Equation (i) is *first order*, that is contains only the 'first derivative' $\dfrac{dy}{dx}$; (ii) is *second order*, containing a term in $\dfrac{d^2y}{dx^2}$, the 'second derivative' of y with respect to x.

Differential equations have already cropped up. For example, on p. 56 it was shown that the D.E. $\dfrac{dy}{dx} = 2x$ is satisfied by $y = x^2 + c$, where c is an arbitrary constant. Again, from p. 146, Ae^{kx} is a solution of the D.E. $\dfrac{dy}{dx} = ky$.

EXAMPLE 14.1 *To find the D.E. satisfied by the parabolas $y^2 = 4ax$ (for different values of a).*

If
$$y^2 = 4ax, \tag{1}$$

then
$$2y\frac{dy}{dx} = 4a. \tag{2}$$

Eliminating a between eqns. (1) and (2) leads to a D.E. satisfied by all such parabolas, namely

$$\frac{dy}{dx} = \frac{y}{2x}.$$

EXAMPLE 14.2 *To show that* ALL *solutions of the D.E.* $\dfrac{dy}{dx} = ky$ *are of the form $y = Ae^{kx}$.*

Let the D.E. be satisfied by $y = Ae^{kx}f(x)$. We shall show that $f(x)$ must be a constant; for we have

$$\frac{dy}{dx} = Ae^{kx}f'(x) + Ake^{kx}f(x),$$

274

and substitution in the D.E. gives

$$Ae^{kx}f'(x) + Ake^{kx}f(x) = kAe^{kx}f(x)$$

i.e. $f'(x) = 0$, so that $f(x) = c$. Thus *all* solutions are of the given form.

EXERCISE LXVII

1. (D) Give the general solution of the D.E. $\dfrac{dy}{dx} = 3y$ and also the particular one for which $y = 4$ when $x = 0$.
2. Find the D.E. satisfied by (i) the straight lines $y = mx$ (for all m); (ii) the circles $x^2 + y^2 = ax$ (for all a).
3. Verify that the following D.E.'s are satisfied by the given 'solutions'.

   (i) $\dfrac{dy}{dx} = 2y$; $y = e^{2x}$.

   (ii) $\dfrac{dy}{dx} = -\dfrac{y}{x}$; $y = \dfrac{c^2}{x}$.

   (iii) $\dfrac{d^2y}{dx^2} = -9y$; $y = 7\sin 3x + 5\cos 3x$.

   (iv) $\dfrac{d^2y}{dx^2} - 5\dfrac{dy}{dx} + 6y = 0$; $y = 5e^{2x} - 2e^{3x}$.

   (v) $x^2\dfrac{d^2y}{dx^2} - 2x\dfrac{dy}{dx} + (x^2 + 2)y = 0$; $y = x\sin x$.

   Can you find any other solutions of (iii) or (iv)?
4–6. Re-work Exercise XLVI (p. 171), questions 3, 9, 12.

The formulation in mathematical terms of problems in science or engineering often leads to a set of D.E.'s. They are thus of fundamental importance in applications of calculus. There are many which cannot be solved simply or even exactly and which can best be dealt with numerically by computers. The remainder of this chapter is devoted to systematic study of some of the very simplest types.

First-Order Equations

(a) Solutions 'by Inspection'

EXAMPLE 14.3 (i) $\dfrac{dy}{dx} = \dfrac{1}{1 + x^2}$, (ii) $\dfrac{dy}{dx} = \dfrac{1}{1 + y^2}$.

(i) We have at once $y = \tan^{-1} x + c$.

(ii) Assuming the solution to give a one-to-one function, we have $\dfrac{dx}{dy} = 1 + y^2$, so that $x = y + \tfrac{1}{3}y^3 + c$.

(b) Separable Variables

EXAMPLE 14.4 $\dfrac{dy}{dx} = \dfrac{3x^2 + 1}{4y + 2}$

We may write

$$\lim_{\delta x, \delta y \to 0} \frac{\delta y}{\delta x} = \frac{3x^2 + 1}{4y + 2}$$

or

$$\lim_{\delta y \to 0} (4y + 2)\, \delta y = \lim_{\delta x \to 0} (3x^2 + 1)\, \delta x.$$

Summing over a certain range for x, y gives

$$\lim_{\delta y \to 0} \sum (4y + 2)\, \delta y = \lim_{\delta x \to 0} \sum (3x^2 + 1)\, \delta x$$

or

$$\int (4y + 2)\, dy = \int (3x^2 + 1)\, dx.$$

Thus $2y^2 + 2y = x^3 + x + c$.

We may alternatively set out the solution as follows.

$$\frac{dy}{dx} = \frac{3x^2 + 1}{4y + 2}$$

may be written as

$$(4y + 2)\, dy = (3x^2 + 1)\, dx,$$

whence

$$\int (4y + 2)\, dy = \int (3x^2 + 1)\, dx$$

or

$$2y^2 + 2y = x^3 + x + c.$$

It must be understood that this quicker symbolic way of writing the solution is simply a shorthand for the previous one, which provides its justification.

EXAMPLE 14.5   *To find the solution of the equation* $\tan x \dfrac{dy}{dx} = 1 + y^2$ *for which* $y = 1$ *when* $x = \dfrac{\pi}{2}$.

Using the notation explained in the previous example, we write

$$\frac{dy}{1 + y^2} = \cot x \, dx$$

whence

$$\int \frac{dy}{1 + y^2} = \int \cot x \, dx,$$

or

$$\tan^{-1} y = \ln \sin x + c.$$

This is the general solution of the D.E., with an arbitrary constant. If $y = 1$ when $x = \dfrac{\pi}{2}$, we must have

$$\frac{\pi}{4} = \ln 1 + c = c,$$

so that the required solution subject to this condition is

$$\tan^{-1} y = \ln \sin x + \frac{\pi}{4}.$$

(c) Integrating Factor

For equations of the type $\dfrac{dy}{dx} + yf(x) = g(x)$, we seek the solution with the help of an 'integrating factor', which turns the left-hand side into the derivative of a product. We give an example side-by-side with the general method.

General Method	Example

General Method

$$\frac{dy}{dx} + yf(x) = g(x)$$

Example

$$\frac{dy}{dx} + \frac{y}{x + 1} = \sin x.$$

$$f(x) = \frac{1}{x + 1}, \qquad g(x) = \sin x.$$

Multiply through by the 'integrating factor' $e^{\int f(x)\,dx}$, giving

$$e^{\int f(x)dx}\frac{dy}{dx} + yf(x)e^{\int f(x)dx}$$
$$= g(x)e^{\int f(x)dx}$$
$$= h(x), \text{ say.}$$

Now

$$\frac{d}{dx}\{e^{\int f(x)dx}\} = e^{\int f(x)dx}f(x),$$

so the left-hand side is

$$\frac{d}{dx}\{ye^{\int f(x)dx}\}$$

and so we have

$$\frac{d}{dx}\{ye^{\int f(x)dx}\} = h(x)$$

The integrating factor is

$$e^{\int 1/(x+1)dx} = e^{\ln (x+1)} = x + 1.$$

We have

$$(x + 1)\frac{dy}{dx} + y = (x + 1) \sin x.$$

Thus

$$\frac{d}{dx}\{y(x + 1)\} = (x + 1) \sin x.$$

Integrating,   Integrating,

$$ye^{\int f(x)dx} = \int h(x)\, dx + c$$

$$= k(x) + c,$$

$$y(x + 1) = \int (x + 1)\sin x\, dx$$

$$= -\cos x\,(x + 1)$$

$$+ \int \cos x\, dx + c$$

$$= -\cos x\,(x + 1)$$

$$+ \sin x + c.$$

which gives the required solution.   Thus

$$y = -\cos x + \frac{c + \sin x}{x + 1}.$$

It is most important to *remember the arbitrary constant* on integrating.

(d) Homogeneous Equations

A D.E. of the form $(a_0x^n + a_1x^{n-1}y + a_2x^{n-2}y^2 + \ldots + a_rx^{n-r}y^r + \ldots + a_ny^n)\dfrac{dy}{dx} = b_0x^n + b_1x^{n-1}y + \ldots + b_ny^n$ is *homogeneous*, degree n, in x, y. For such an equation, the substitution $y = vx$ is indicated, v being another function of x.

EXAMPLE 14.6 $x\dfrac{dy}{dx} = 2x - y$:

Put $y = vx$, $\dfrac{dy}{dx} = v + x\dfrac{dv}{dx}$. Then

$$x\left(v + x\frac{dv}{dx}\right) = 2x - vx$$

or

$$v + x\frac{dv}{dx} = 2 - v$$

$$x\frac{dv}{dx} = 2(1 - v)$$

Thus

$$\frac{dv}{1 - v} = 2\frac{dx}{x},$$

whence

$$\int \frac{dv}{1 - v} = 2\int \frac{dx}{x},$$

$$\ln x^2 + \ln |1 - v| = c$$

or $\left(\text{remembering that } v = \dfrac{y}{x}\right)$, $\quad x^2\left(1 - \dfrac{y}{x}\right) = A,$

i.e $\qquad\qquad\qquad\qquad y = x + \dfrac{A}{x}.$

It should be noted that *the general solution of each of the first order D.E.'s considered contains one arbitrary constant.*

EXERCISE LXVIII

In questions 1 to 12, solve the given D.E.

1. $\dfrac{dy}{dx} = \tan x.$

2. $\dfrac{dy}{dx} = \tan y.$

3. $(1 + x^2)\dfrac{dy}{dx} = 1 - y^2.$

4. $\tan x \dfrac{dy}{dx} = \tan y.$

5. $\dfrac{dy}{dx} + y(x - 1) = 0.$

6. $\dfrac{dy}{dx} + \dfrac{y}{x} = \sin x.$

7. $x^2 \dfrac{dy}{dx} = y^2 - 5y + 6.$

8. $\dfrac{dy}{dx} = x(y^2 + y + 1).$

9. $x(x + 2)\dfrac{dy}{dx} + 2y = x + 1.$

10. $(1 - x^2)\dfrac{dy}{dx} + y = 1 - x.$

11. $(x - y)\dfrac{dy}{dx} = 2x + y.$

12. $\dfrac{dy}{dx} = \dfrac{x^2 + y^2}{x(x + y)}.$

13. Solve the differential equation
$$(1 - x^2)\frac{dy}{dx} - xy = x,$$
given that $y = 1$ when $x = 0.$

14. Find the solution of the differential equation
$$e^x \frac{dy}{dx} = xy^2$$
for which $y = 1$ when $x = 0.$

15. Solve the differential equation
$$x\frac{dy}{dx} - y = x$$
by means of the substitution $y = vx$, or otherwise.

16. Solve the differential equation
$$\sqrt{(1 - x^2)}\frac{dy}{dx} = xe^{2y}.$$

17. By means of the substitution $y = vx$ reduce the differential equation
$$xy\frac{dy}{dx} = y^2 + \sqrt{(x^2 + y^2)}$$
to an equation in v and x. Find the solution, given that $y = 1$ when $x = 1.$

Second-Order Equations

We shall consider only 'linear' D.E.'s, that is, equations of the type

$$a\frac{d^2y}{dx^2} + b\frac{dy}{dx} + cy = f(x),$$

where a, b, c are constants.*

EXAMPLE 14.7   *To solve the D.E.* $\dfrac{d^2y}{dx^2} - 5\dfrac{dy}{dx} + 6y = 0.$

Try $y = Ae^{kx}$, where A, k are constants,† then $\dfrac{dy}{dx} = Ake^{kx}$, $\dfrac{d^2y}{dx^2} = Ak^2e^{kx}$

and substituting these into the D.E. gives

$$Ak^2e^{kx} - 5Ake^{kx} + 6Ae^{kx} = 0$$

which is satisfied if $k^2 - 5k + 6 = 0$, i.e. if $k = 2$ or 3.

Thus $y = Ae^{2x}$ is a solution and so is Be^{3x}, A, B being arbitrary constants. We now verify that $Ae^{2x} + Be^{3x}$ *is also a solution.* For if

$$y = Ae^{2x} + Be^{3x},$$

$$\frac{dy}{dx} = 2Ae^{2x} + 3Be^{3x},$$

$$\frac{d^2y}{dx^2} = 4Ae^{2x} + 9Be^{3x},$$

and so

$$\frac{d^2y}{dx^2} - 5\frac{dy}{dx} + 6y = (4Ae^{2x} + 9Be^{3x}) - 5(2Ae^{2x} + 3Be^{3x}) + 6(Ae^{2x} + Be^{3x})$$
$$= 0.$$

We assert generally (without proof) that *the general solution of every D.E. of the form* $a\dfrac{d^2y}{dx^2} + b\dfrac{dy}{dx} + cy = 0$ *has two arbitrary constants. Also, the sum of any two solutions is itself a solution.* Thus in the above example the general solution is given by $y = Ae^{2x} + Be^{3x}$.

EXAMPLE 14.8   *To solve the D.E.* $\dfrac{d^2y}{dx^2} + \dfrac{dy}{dx} - y = 0.$

Try $y = Ae^{kx}$; then

$$\frac{dy}{dx} = Ake^{kx}, \qquad \frac{d^2y}{dx^2} = Ak^2e^{kx}.$$

---

* 'Linear' implies that there are no powers of terms such as $\left(\dfrac{dy}{dx}\right)^2$.

† This is a reasonable idea, since then y, $\dfrac{dy}{dx}$ and $\dfrac{d^2y}{dx^2}$ differ only by constant factors.

so that

$$0 = \frac{d^2y}{dx^2} + \frac{dy}{dx} - y = Ae^{kx}(k^2 + k - 1),$$

whence $k^2 + k - 1 = 0, k = \dfrac{-1 \pm \sqrt{5}}{2}$.

Thus the general solution is given by

$$y = Ae^{(-1+\sqrt{5})x/2} + Be^{(-1-\sqrt{5})x/2}$$
$$= e^{-x/2}(Ae^{(\sqrt{5}/2)x} + Be^{(-\sqrt{5}/2)x}).$$

EXAMPLE 14.9 *To solve the D.E.* $\dfrac{d^2y}{dx^2} - 3\dfrac{dy}{dx} + 2y = x + 1.$

(i) We first solve the D.E. $\dfrac{d^2y}{dx^2} - 3\dfrac{dy}{dx} + 2y = 0$, which by the above method yields

$$y = Ae^x + Be^{2x}. \tag{1}$$

(ii) We now seek some value of y which satisfies the given equation,

$$\frac{d^2y}{dx^2} - 3\frac{dy}{dx} + 2y = x + 1.$$

We try $y = a + bx; \dfrac{dy}{dx} = b, \dfrac{d^2y}{dx^2} = 0$. Substituting in the D.E. gives

$$0 - 3b + 2(a + bx) = x + 1,$$

which is satisfied for all x if $a = 1\frac{1}{4}, b = \frac{1}{2}$. Thus

$$y = 1\frac{1}{4} + \tfrac{1}{2}x \tag{2}$$

is a *particular integral* (P.I.), that is, one solution.

(iii) Now consider $y = Ae^x + Be^{2x} + 1\frac{1}{4} + \frac{1}{2}x$. Substituting in the left-hand side of the D.E., namely $\dfrac{d^2y}{dx^2} - 3\dfrac{dy}{dx} + 2y$, gives

$$\frac{d^2}{dx^2}\{Ae^x + Be^{2x} + 1\tfrac{1}{4} + \tfrac{1}{2}x\} - 3\frac{d}{dx}\{Ae^x + Be^{2x} + 1\tfrac{1}{4} + \tfrac{1}{2}x\}$$

$$+ 2\{Ae^x + Be^{2x} + 1\tfrac{1}{4} + \tfrac{1}{2}x\}$$

$$= \left[\frac{d^2}{dx^2}(Ae^x + Be^{2x}) - 3\frac{d}{dx}(Ae^x + Be^{2x}) + 2(Ae^x + Be^{2x})\right]$$

$$+ \frac{d^2}{dx^2}(1\tfrac{1}{4} + \tfrac{1}{2}x) - 3\frac{d}{dx}(1\tfrac{1}{4} + \tfrac{1}{2}x) + 2(1\tfrac{1}{4} + \tfrac{1}{2}x)$$

$$= x + 1, \quad \text{for}$$

(*a*) the contents of the square bracket are equal to zero, since

$Ae^x + Be^{2x}$ satisfies $\dfrac{d^2y}{dx^2} - 3\dfrac{dy}{dx} + 2y = 0$; and

(b) the remainder of the left-hand side equals $x + 1$ since $1\frac{1}{4} + \frac{1}{2}x$ is a P.I. of the D.E.

Thus the general solution is given by

$$y = Ae^x + Be^{2x} + 1\tfrac{1}{4} + \tfrac{1}{2}x.$$

The last example illustrates the method of solution of $a\dfrac{d^2y}{dx^2} + b\dfrac{dy}{dx} + cy = f(x)$; the general solution to $a\dfrac{d^2y}{dx^2} + b\dfrac{dy}{dx} + cy = 0$ is found (the *complementary function*, or C.F.) and is added to one particular solution (the P.I.) of the given complete equation. This sum is the required general solution of the given D.E.

EXAMPLE 14.10   *To find the general solution of the D.E.*
$\dfrac{d^2y}{dx^2} + 2\dfrac{dy}{dx} - 3y = 5 \sin x$, *and also that for which* $y = 0$, $\dfrac{dy}{dx} = 1$ *when* $x = 0$.

(i) The C.F. is found by solving $\dfrac{d^2y}{dx^2} + 2\dfrac{dy}{dx} - 3y = 0$, namely

$$y = Ae^x + Be^{-3x}.$$

(ii) For the P.I., we try $y = C \sin x + D \cos x$. Then

$$\frac{dy}{dx} = C \cos x - D \sin x, \qquad \frac{d^2y}{dx^2} = -C \sin x - D \cos x,$$

and substitution in the D.E. gives

$$-C \sin x - D \cos x + 2(C \cos x - D \sin x) - 3(C \sin x + D \cos x)$$
$$= 5 \sin x.$$

Comparing coefficients of $\sin x$, $\cos x$, this is satisfied if

$$-C - 2D - 3C = 5$$

and

$$-D + 2C - 3D = 0,$$

whence $C = -1$, $D = -\frac{1}{2}$. Thus $-\sin x - \frac{1}{2} \cos x$ is a P.I., and so the general solution is given by

$$y = Ae^x + Be^{-3x} - \sin x - \tfrac{1}{2} \cos x.$$

Now $\dfrac{dy}{dx} = Ae^x - 3Be^{-3x} - \cos x + \tfrac{1}{2}\sin x$, so

if $y = 0$ when $x = 0$, $\qquad 0 = A + B - \tfrac{1}{2}$

and if $\dfrac{dy}{dx} = 1$ when $x = 0$, $\quad 1 = A - 3B - 1$,

whence $A = \tfrac{7}{8}, B = -\tfrac{3}{8}$, and so the solution subject to the given conditions is

$$y = \tfrac{7}{8}e^x - \tfrac{3}{8}e^{-3x} - \sin x - \tfrac{1}{2}\cos x.$$

The form of the P.I. must be guessed intelligently from the form of the right-hand side of the equation, cf. Exercise LXIX, questions 5 to 10.

When some of the terms in the D.E. $a\dfrac{d^2y}{dx^2} + b\dfrac{dy}{dx} + cy = f(x)$ are missing, it is sometimes useful to put $\dfrac{dy}{dx} = p$, as illustrated in the next example.

EXAMPLE 14.11 $\quad \dfrac{d^2y}{dx^2} + 3\dfrac{dy}{dx} = 2x.$

Putting $\dfrac{dy}{dx} = p$ reduces the D.E. to

$$\frac{dp}{dx} + 3p = 2x.$$

Multiplying by the integrating factor $e^{\int 3\,dx}, = e^{3x}$, gives

$$e^{3x}\frac{dp}{dx} + 3e^{3x}p = 2xe^{3x}$$

or

$$\frac{d}{dx}(e^{3x}p) = 2xe^{3x},$$

whence

$$e^{3x}p = 2\int xe^{3x}\,dx = 2\left(\tfrac{1}{3}e^{3x}x - \tfrac{1}{3}\int e^{3x}\,dx\right)$$

$$= \tfrac{2}{3}xe^{3x} - \tfrac{2}{9}e^{3x} + A.$$

Thus

$$\frac{dy}{dx} = \tfrac{2}{3}x - \tfrac{2}{9} + Ae^{-3x},$$

so

$$y = \tfrac{1}{3}x^2 - \tfrac{2}{9}x - \tfrac{1}{3}Ae^{-3x} + B$$

$$= \tfrac{1}{3}x^2 - \tfrac{2}{9}x + Ce^{-3x} + B.$$

Solve the following D.E.'s.

1. $\dfrac{d^2y}{dx^2} - \dfrac{dy}{dx} - 6y = 0.$

2. $\dfrac{d^2y}{dx^2} + 5\dfrac{dy}{dx} - 2y = 0.$

3. $\dfrac{d^2y}{dx^2} + 3\dfrac{dy}{dx} - 10y = 0$, given that $y = 7$, $\dfrac{dy}{dx} = -1$ when $x = 0$.

4. $\dfrac{d^2y}{dx^2} = 7x.$ [Don't make this difficult.]

5. $\dfrac{d^2y}{dx^2} - 2\dfrac{dy}{dx} = 3x + 1.$

6. $\dfrac{d^2y}{dx^2} + 3\dfrac{dy}{dx} - 4y = 3x + 2.$

7. $\dfrac{d^2y}{dx^2} + 5\dfrac{dy}{dx} - 6y = x^2 - 3x + 1.$

8. $\dfrac{d^2y}{dx^2} - 2\dfrac{dy}{dx} - 3y = e^{2x}.$ [Try $y = Ae^{2x}$ for the P.I.]

9. $\dfrac{d^2y}{dx^2} - \dfrac{dy}{dx} - 6y = 3 \cos x.$

10. $\dfrac{d^2y}{dx^2} - 4\dfrac{dy}{dx} - 10y = 4 \cos x - 3 \sin x.$

EXAMPLE 14.12   *To solve the D.E.'s* (i) $\dfrac{d^2y}{dx^2} = -n^2y$; (ii) $\dfrac{d^2y}{dx^2} = -n^2y + k.$

(i) This is the simple harmonic motion (S.H.M.) equation.

It is surprisingly difficult to find the general solution of this equation by elementary means. (It is possible using complex numbers, as the reader should verify after reading the next section.) Here we simply note that

if $$y = \sin nx,$$

then $$\frac{dy}{dx} = n \cos nx,$$

$$\frac{d^2y}{dx^2} = -n^2 \sin nx = -n^2y,$$

and so $\sin nx$ is a solution. Similarly $\cos nx$ is also a solution and (cf. remarks on p. 280) the general solution is given by

$$y = A \sin nx + B \cos nx.$$

This can also be written in the form $a \sin (nx + \varepsilon) = a \cos \varepsilon \sin nx + a \sin \varepsilon \cos nx$, where $a \cos \varepsilon = A$, $a \sin \varepsilon = B$, ε being chosen in such a way that $a > 0$. a is called the *amplitude* of the motion, which has a period of $\dfrac{2\pi}{n}$.

(ii) $\dfrac{d^2y}{dx^2} = -n^2y + k = -n^2\left(y - \dfrac{k}{n^2}\right)$.

Writing $y - \dfrac{k}{n^2} = z$, the equation becomes $\dfrac{d^2z}{dx^2} = -n^2z$,

whence

$$y = \frac{k}{n^2} + z = \frac{k}{n^2} + A \sin nx + B \cos nx.$$

Our method for solving the D.E. $a\dfrac{d^2y}{dx^2} + b\dfrac{dy}{ax} + cy = 0$ has been to try $y = Ae^{kx}$, which satisfies the D.E. if

$$ak^2 + bk + c = 0.$$

If $b^2 > 4ac$, this leads to two real values of k. We now examine what happens if $b^2 < 4ac$.

The reader is assumed to have some acquaintance with complex numbers, and he is asked to accept that $e^{jx} = \cos x + j \sin x$. This is plausible from consideration of the power series, but we make no pretence at proper justification. ($\cos x = 1 - \dfrac{x^2}{2!} + \ldots$, $\sin x = x - \dfrac{x^3}{3!} + \ldots$ suggest that $\cos x + j \sin x = 1 + jx + \dfrac{(jx)^2}{2!} + \ldots$, which is the series for e^x when x is replaced by jx.)

Solution of the equation $ak^2 + bk + c = 0$ gives

$$k = \frac{-b \pm j\sqrt{(4ac - b^2)}}{2a} = \frac{-b \pm jp}{2a}$$

where $p^2 = 4ac - b^2$.

Thus the solution of the D.E. is given by

$$y = Ae^{(-b+jp)x/2a} + Be^{(-b-jp)x/2a}$$
$$= e^{(-bx/2a)}(Ae^{(jpx/2a)} + Be^{(-jpx/2a)})$$
$$= e^{(-bx/2a)}\left\{A\left(\cos\frac{px}{2a} + j\sin\frac{px}{2a}\right) + B\left(\cos\frac{px}{2a} - j\sin\frac{px}{2a}\right)\right\}$$
$$= e^{(-bx/2a)}\left(D\sin\frac{px}{2a} + E\cos\frac{px}{2a}\right), \qquad \begin{array}{l}\text{writing } D = j(A - B), \\ E = A + B.\end{array}$$

EXAMPLE 14.13   *To solve the D.E.* $\dfrac{d^2y}{dx^2} + \dfrac{dy}{dx} + y = 0.$

Putting $y = Ae^{kx}$ yields $k^2 + k + 1 = 0$, or $k = \dfrac{-1 \pm j\sqrt{3}}{2}$. Thus the general solution is given by

$$y = Ae^{(-1+j\sqrt{3})x/2} + Be^{(-1-j\sqrt{3})x/2}$$

$$= e^{(-x/2)}(Ae^{(j\sqrt{3}/2)x} + Be^{(-j\sqrt{3}/2)x})$$

$$= e^{(-x/2)}\left(D \sin \frac{\sqrt{3}}{2}x + E \cos \frac{\sqrt{3}}{2}x \right).$$

Finally, we consider the D.E.

$$\frac{d^2y}{dx^2} - 2a\frac{dy}{dx} + a^2y = 0$$

for which the substitution $y = Ae^{kx}$ gives a perfect square for k (the case '$b^2 = 4ac$', cf. p. 285). We have $k^2 - 2ak + a^2 = 0$; $(k - a)^2 = 0$; $k = a$. Thus $y = Ae^{ax}$ is a solution, but we must suspect the existence of a further one, since there is only one arbitrary constant here.

We start again, writing the D.E. in the form

$$\frac{d^2y}{dx^2} - a\frac{dy}{dx} = a\left(\frac{dy}{dx} - ay\right).$$

Thus

$$\frac{\dfrac{d^2y}{dx^2} - a\dfrac{dy}{dx}}{\dfrac{dy}{dx} - ay} = a.$$

Since the numerator of the left-hand side is the gradient of the denominator, we can integrate with respect to x, obtaining

$$\ln \left(\frac{dy}{dx} - ay\right) = ax + B,$$

so that

$$\frac{dy}{dx} - ay = e^{ax+B} = e^{ax} \cdot e^B = Ce^{ax}.$$

Multiplying through by the integrating factor $e^{-\int a\,dx} = e^{-ax}$,

$$e^{-ax}\frac{dy}{dx} - ae^{-ax}y = C,$$

$$\frac{d}{dx}(ye^{-ax}) = C$$

$$ye^{-ax} = Cx + D.$$

Thus the general solution is $y = e^{ax}(Cx + D).$

EXERCISE LXX

In questions 1 to 6, solve the given D.E.

1. $\dfrac{d^2y}{dx^2} = -9y$, given that $y = 0$, $\dfrac{dy}{dx} = 4$ when $x = 0$.

2. $\dfrac{d^2y}{dx^2} = -5y + 6$.

3. $\dfrac{d^2y}{dx^2} = 9y$. [Try $y = Ae^{kx}$]

4. $\dfrac{d^2y}{dx^2} + \dfrac{dy}{dx} + 2y = 0$.

5. $\dfrac{d^2y}{dx^2} + 2\dfrac{dy}{dx} + 4y = \cos x$.

6. (i) $\dfrac{d^2y}{dx^2} - 4\dfrac{dy}{dx} + 4y = 0$; (ii) $\dfrac{d^2y}{dx^2} - 4\dfrac{dy}{dx} + 4y = 1$.

The remainder of this exercise consists of miscellaneous examination questions.

7. (i) Form a differential equation in x and t by eliminating the constants A and λ from the equation $x = Ae^{\lambda t}$.
   (ii) Find the solution of the equation
   $$x^2 \frac{dy}{dx} - 3x - 1 = 0$$
   for which $y = 0$ when $x = 1$.

8. Solve the differential equations:
   (i) $\dfrac{dy}{dx} - (1 + \cot x)y = 0$;
   (ii) $\dfrac{d^2y}{dx^2} + 9y = 18$.

9. (i) Find values of A and B for which
   $$y = A \sin x + B \cos x$$
   satisfies the differential equation
   $$\frac{d^2y}{dx^2} + 4\frac{dy}{dx} + 3y = \sin x.$$
   Give the complete solution of this equation.
   (ii) Solve the differential equation
   $$(x + 1)\frac{dy}{dx} - xy = e^{2x}.$$

10. Integrate the differential equations:
    (i) $\dfrac{dy}{dx} + \dfrac{2y}{x} = e^x$;
    (ii) $(x - y)\dfrac{dy}{dx} = x + y$,
    using the substitution $y = vx$ in (ii).

11. (i) Find that solution of the differential equation

$$\cos \theta \frac{dr}{d\theta} + a \sin \theta = 0,$$

for which $r = a$ when $\theta = 0$.

(ii) Find the amplitude of the simple harmonic motion defined by the differential equation

$$\frac{d^2x}{dt^2} + 9x = 0,$$

coupled with the conditions

$$x = 5 \quad \text{when} \quad t = 0, \qquad x = 12 \quad \text{when} \quad t = \pi/6.$$

Find also the smallest positive value of t for which $x = 0$.

12. (i) Find values of A and B such that

$$y = (A \cos x + B \sin x)e^{2x}$$

satisfies the differential equation

$$\frac{d^2y}{dx^2} + 3\frac{dy}{dx} - 4y = (12 \cos x - 2 \sin x)e^{2x}.$$

Hence obtain the general solution of the differential equation and the particular solution which satisfies the conditions

$$y = 0, \quad \frac{dy}{dx} = 2 \quad \text{at} \quad x = 0.$$

(ii) Solve the differential equation

$$\frac{dy}{dx} + 4\frac{y}{x} = \frac{3}{x^2}.$$

13. (i) Solve the differential equation

$$\cos x \frac{dy}{dx} - 2y \sin x = 2x + 1.$$

(ii) Find the complete solution of the differential equation

$$\frac{d^2y}{dx^2} + n^2y = cn^2(l - x),$$

c, l, and n being constants, given that $y = dy/dx = 0$ when $x = 0$. Prove that if, in addition, $y = 0$ when $x = l$, then

$$nl = \tan nl.$$

14. (a) The normal at any point of a curve passes through the point $(1, 1)$. Express this condition in the form of a differential equation, and hence find the equation of the family of curves which satisfy the condition.

(b) Given that y satisfies the differential equation

$$\frac{dy}{dx} + 2y \tan x = \sin x,$$

and that $y = 1$ when $x = \pi/3$, express dy/dx in terms of x.

15. (i) By means of the substitution $y = z - x$, or otherwise, solve the differential equation

$$(x + y)\frac{dy}{dx} = x^2 + xy + x + 1$$

(ii) Find the solution of the differential equation

$$\frac{d^2y}{dx^2} + 9y = 18x + 20 \cos 2x,$$

given that $y = 2$ and $\frac{dy}{dx} = -1$ when $x = 0$.

16. If $p = \frac{dy}{dx}$, prove that $\frac{d^2y}{dx^2} = p\frac{dp}{dy}$.

Hence solve the differential equation

$$y\frac{d^2y}{dx^2} = 2\frac{dy}{dx} + \left(\frac{dy}{dx}\right)^2.$$

17. (i) Integrate the differential equation

$$\frac{dy}{dx} = xy(y - 2).$$

(ii) If $\frac{d^2u}{d\theta^2} + u = 1$ and if $\frac{du}{d\theta} = 0$ when $u = 2$, prove that

$$\left(\frac{du}{d\theta}\right)^2 = 2u - u^2.$$

If $u = 0$ when $\theta = 0$, prove also that

$$u = 1 - \cos \theta.$$

18. Find the complete solution of the differential equation

$$\frac{d^2y}{dt^2} + 6\frac{dy}{dt} + 18y = 17 \cos 2t,$$

given that $y = 1$ and $dy/dt = 3$ when $t = 0$.

Prove that when t is large y oscillates between the values $\pm \sqrt{(0 \cdot 85)}$.

EXERCISE LXXI

This exercise illustrates a few of the applications of D.E.'s. Most of the questions need some acquaintance with Newton's second law of motion, 'F = ma'.

1. A particle of unit mass moves on the x-axis, in such a way that its coordinate x at time t satisfies the differential equation

$$\ddot{x} + 2k\dot{x} + p^2x = A \cos qt, \qquad p^2 < k^2.$$

Write down the magnitudes and directions of a system of forces which will produce this motion.

Find values of the constants B and α (in terms of p, q and A) which make

$$x = B \cos (qt - \alpha)$$

a particular solution of the equation.

Determine the general solution of the differential equation and hence find the particular solution for which

$$x = \frac{B}{k}(q \sin \alpha + k \cos \alpha), \qquad \dot{x} = 0,$$

when $t = 0$, giving your answer in terms of B, α, q, k, p, t.

2. [The radial and transverse components of the acceleration of a particle moving in a plane are respectively

$$\ddot{r} - r\dot{\theta}^2 \quad \text{and} \quad \frac{1}{r}\frac{d}{dt}(r^2\dot{\theta}),$$

r, θ being the polar coordinates of the particle at time t.]

A particle moves inside a smooth straight tube of small bore. The tube moves upon a smooth horizontal table, being made to rotate with constant angular velocity about one end which is fixed. At time $t = 0$ the particle is at rest relative to the tube and at a distance l from the fixed end; prove that the trace of the path of the particle on the table is the curve

$$r = l \cosh \theta,$$

where θ is measured from the initial position of the tube.

3. A curve has the property that its centre of curvature at any point lies on the y-axis. Prove that, if dy/dx is written as p, the equation of the curve satisfies the differential equation

$$x \, dp/dx = p + p^3.$$

Show that the general solution of this equation is

$$x^2(1 + p^2) = k^2p^2,$$

where k is an arbitrary constant.

If p is now written as dy/dx, solve the resulting differential equation, and hence show that the only curves with the given property are circles with their centres on the y-axis.

4. A particle of mass m is at a point A of a smooth horizontal plane. It is attracted to a point O in the plane by a force kx, where x is the distance from O. The particle is projected with velocity v in the direction OA. Show that it returns to A after a time $2\sqrt{\dfrac{m}{k}}\tan^{-1}\left(\dfrac{v}{l}\sqrt{\dfrac{m}{k}}\right)$, where $l = OA$.

5. Solve the equation $\dfrac{d^2y}{dt^2} + n^2y = A \cos pt$. What phenomenon is illustrated when $p \approx n$? Show that a particular integral of $\dfrac{d^2y}{dt^2} + n^2y = k \cos nt$ is $y = \dfrac{k}{n}t \sin nt$.

6. A particle of mass m and charge e moves in the plane xOy under the action of constant electric and magnetic fields of strengths E and H respectively. The fields exert forces eE in the direction of Ox and eHv along the normal to the path of the particle, v being the velocity of the particle at any instant. Write down the equations of motion of the particle.

Show that the equations are satisfied by

$$x = A + B \cos \omega t, \qquad y = Ct + B \sin \omega t,$$

where the constants C and ω are given by the formulae

$$C = E/H, \qquad \omega = eH/m.$$

Evaluate the constants A and B if the particle is initially at rest at O; and prove that in these circumstances the particle comes momentarily to rest when $t = 2n\pi/\omega$, where n is any positive integer.

7. A particle of unit mass moves in a straight line under a force directed to a fixed point of the line of such magnitude that, if unrestricted, the particle would execute S.H.M. oscillations of period $\dfrac{2\pi}{n}$. If the resistance is $2k$ times the velocity, find the condition that the motion may be oscillatory. The condition being satisfied, and the particle being at rest at the fixed point, a force proportional to sin nt begins to act upon it at time $t = 0$. Prove that the particle comes to rest whenever s sin $nt = ne^{-kt}$ sin $s\,t$, where $s^2 = n^2 - k^2$.

8. A particle of unit mass is projected vertically upwards in a medium whose resistance is λv^2 newtons, where v is the velocity at any instant. Find the greatest height attained when the initial velocity is V. Show that, when λ is small compared with $\dfrac{g}{V^2}$, the greatest height is approximately $\dfrac{V^2}{2g}\left(1 - \dfrac{\lambda V^2}{2g}\right)$.

15
Partial Differentiation

In this final chapter we give a brief introduction to the idea of a function of several variables. For example, in the formula for the volume of a cylinder

$$V = \pi r^2 h,$$

V depends on both r and h. These may be varied independently of each other, but as long as each is assigned some definite positive value, then there is a *unique* value of V corresponding to them. V is a *function* of the two independent variables r, h, defined for all $r \geqslant 0$, $h \geqslant 0$. We may write

$$V = f(r, h) = \pi r^2 h.$$

For example, $f(1, 2) = \pi \cdot 1^2 \cdot 2 = 2\pi; f(2, 1) = \pi \cdot 2^2 \cdot 1 = 4\pi;$

$$f(r, 2) = 2\pi r^2; \qquad f(2, h) = 4\pi h.$$

This idea can be extended to several independent variables, e.g. with the formula for the area of a triangle

$$\Delta = \tfrac{1}{2}bc \sin A,$$

we may write $\Delta = f(b, c, A)$, since if $b \geqslant 0$, $c \geqslant 0$, $0 \leqslant A \leqslant \pi$, there is a unique value of Δ corresponding to any given set of values for b, c and A.

Setting up rectangular axes, the locus of the point (x, y, z) such that $z = f(x, y)$ is in general a surface. We will illustrate this by investigating the surface $z = x^2 + y^2$.

z is a function of x and y, but it is simplest to consider first what is represented for different fixed values of z (Fig. 15.1).

If $z = 1$, we have $x^2 + y^2 = 1$. Now $z = 1$ represents a plane parallel to xOy (whose equation is $z = 0$). The section of the surface by this plane is thus a circle, centre $(0, 0, 1)$ and radius 1. Similarly, the section by $z = 2$ is a circle radius $\sqrt{2}$, and so on. When $z = 0$, $x = y = 0$ (i.e. we have the origin only) and z cannot be negative, for then $x^2 + y^2 = z$ would yield no real points.

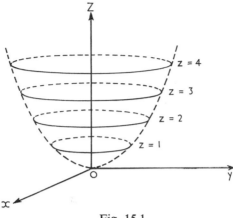

Fig. 15.1

The sections of the surface by planes $z = k$ are thus circular, and in fact we can draw a contour map (Fig. 15.2).

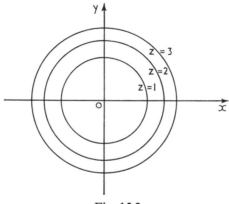

Fig. 15.2

Now if $y = 0$, the equation reduces to $z = x^2$, so that the section by the plane zOx (i.e. $y = 0$) is a parabola. In fact, it can now be seen that the surface can be generated by taking this parabola, given by $z = x^2$ and $y = 0$, (Fig. 15.3), and rotating it about the z-axis.

If we assign to y any fixed value, say $y = k$, we have the section of the surface by the plane $y = k$ (parallel to xOz). This section is given by $z = x^2 + k^2$, $y = k$, i.e. a parabola (shaded in Fig. 15.4).

To find the slope at any point on this parabola, we require the gradient of z ($= x^2 + y^2$) with respect to x, *keeping y constant*, i.e. remaining in the

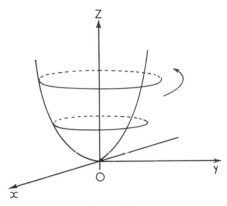

Fig. 15.3

plane $y = k$. This gradient function or *partial derivative* is denoted by $\dfrac{\partial z}{\partial x}$ (spoken as 'partial dz by dx'). Thus, in this case, $\dfrac{\partial z}{\partial x} = 2x$.

To generalize, if $z = f(x, y)$, then the partial derivative $\dfrac{\partial z}{\partial x}$ *is obtained by differentiating z with respect to x, treating y as a constant*; $\dfrac{\partial z}{\partial y}$ is the gradient of z with respect to y, keeping x constant. Similarly for a function of several variables; e.g. if $z = f(x_1, x_2, \ldots, x_n)$, then the partial derivative $\dfrac{\partial z}{\partial x_1}$ will describe the gradient of z with respect to x_1, treating $x_2, x_3, \ldots, x_n$ as constants.

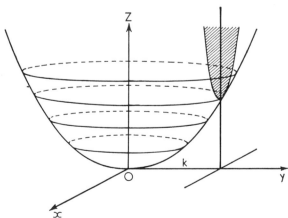

Fig. 15.4

EXAMPLE 15.1   *Prove that, if $V = (1 - 2xy + y^2)^{-\frac{1}{2}}$*

$$(a) \quad \frac{\partial V}{\partial x} + \frac{\partial V}{\partial y} = xV^3,$$

$$(b) \quad x\frac{\partial V}{\partial x} - y\frac{\partial V}{\partial y} = y^2V^3.$$

If

$$V = (1 - 2xy + y^2)^{-\frac{1}{2}},$$

$$\frac{\partial V}{\partial x} = -\tfrac{1}{2}(1 - 2xy + y^2)^{-\frac{3}{2}}(-2y) = y(1 - 2xy + y^2)^{-\frac{3}{2}}$$

and

$$\frac{\partial V}{\partial y} = -\tfrac{1}{2}(1 - 2xy + y^2)^{-\frac{3}{2}}(-2x + 2y) = (x - y)(1 - 2xy + y^2)^{-\frac{3}{2}}.$$

Thus (a)

$$\frac{\partial V}{\partial x} + \frac{\partial V}{\partial y} = (1 - 2xy + y^2)^{-\frac{3}{2}}\{y + (x - y)\} = xV^3,$$

and (b)

$$x\frac{\partial V}{\partial x} - y\frac{\partial V}{\partial y} = (1 - 2xy + y^2)^{-\frac{3}{2}}\{xy - y(x - y)\}$$
$$= y^2V^3.$$

EXAMPLE 15.2   *If $\sin u = 3x^2 - 4xy + 2y^2$, show that*

$$x\frac{\partial u}{\partial x} + y\frac{\partial u}{\partial y} = 2\tan u.$$

There is no need to make u the subject, writing $u = \sin^{-1}(3x^2 - 4xy + 2y^2)$, (though we assume this, so that u is a *function* of x and y). We have

$$\frac{\partial}{\partial x}(\sin u) = \cos u\,\frac{\partial u}{\partial x},$$

for on differentiating $\sin u$ with respect to x we can first differentiate with respect to u and then multiply by the gradient of u with respect to x (keeping y constant throughout the process).

Similarly,

$$\frac{\partial}{\partial y}(\sin u) = \cos u\,\frac{\partial u}{\partial y}.$$

Thus, differentiating partially with respect to x, we have

$$\cos u\,\frac{\partial u}{\partial x} = 6x - 4y,$$

and, with respect to y,

$$\cos u \frac{\partial u}{\partial y} = -4x + 4y.$$

Thus

$$\cos u \left(x \frac{\partial u}{\partial x} + y \frac{\partial u}{\partial y} \right) = x(6x - 4y) + y(-4x + 4y)$$

$$= 6x^2 - 8xy + 4y^2$$

$$= 2 \sin u,$$

so that

$$x \frac{\partial u}{\partial x} + y \frac{\partial u}{\partial y} = 2 \tan u.$$

EXERCISE LXXII

1. (P) Make a cardboard model of the surface $z = x^2 y$, by cutting out sections by the planes $x = -5$ to $+5$ and $y = -5$ to $+5$ and interleaving them.* (Better still, make up your own equation and see what turns out.)

2. Draw contour lines for the surfaces (i) $z = xy$, (ii) $z = x^2 + 2y^2$.

3. If $\Delta = \frac{1}{2}bc \sin A$, write down the values of $\frac{\partial \Delta}{\partial b}$, $\frac{\partial \Delta}{\partial A}$.

4. What do you understand by the partial derivative

$$\frac{\partial}{\partial x} f(x, y)?$$

   (i) Find the value of

$$\frac{\partial}{\partial x} (xe^y + ye^x)$$

   when $x = 0, y = 1$.

   (ii) The variables r, θ, x, y are connected by the equations

$$r = \surd(x^2 + y^2), \qquad \theta = \tan^{-1}(y/x).$$

   Determine the partial derivatives $\dfrac{\partial r}{\partial x}$, $\dfrac{\partial r}{\partial y}$, $\dfrac{\partial \theta}{\partial x}$, $\dfrac{\partial \theta}{\partial y}$ and verify that

$$\frac{\partial r}{\partial y} \Big/ \frac{\partial r}{\partial x} = - \frac{\partial \theta}{\partial x} \Big/ \frac{\partial \theta}{\partial y}.$$

5. (i) Given that $f(x, y) = \sin(x^2 + y^2)$, prove that

$$y \frac{\partial f}{\partial x} = x \frac{\partial f}{\partial y}.$$

   (ii) Given that $v(r, \theta) = r^2 + 2ar \cos \theta + a^2$, evaluate the partial derivatives $\dfrac{\partial v}{\partial r}$, $\dfrac{\partial v}{\partial \theta}$, and find the values of r and θ which make them simultaneously zero.

6. State what you understand by the partial derivative

$$\frac{\partial}{\partial x} f(x, y).$$

---

* Cf. Cundy and Rollett, *Mathematical Models*, O.U.P., 2nd edition, p. 198.

If $S_n = x^n + x^{n-1}y + x^{n-2}y^2 + \ldots + y^n$, prove that

$$x \frac{\partial S_4}{\partial x} + y \frac{\partial S_4}{\partial y} = 4S_4$$

and that

$$\frac{\partial S_4}{\partial x} + \frac{\partial S_4}{\partial y} = 5S_3.$$

If $V = \tan^{-1}\left(\frac{xy}{x^2 + y^2}\right)$, prove that

$$x \frac{\partial V}{\partial x} + y \frac{\partial V}{\partial y} = 0.$$

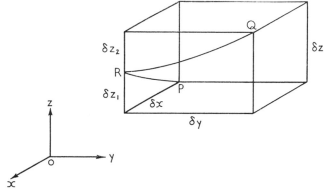

Fig. 15.5

Refer to Fig. 15.5. Let P, Q, R be neighbouring points on the surface $z = f(x, y)$, PR being in a plane parallel to xOz (say $y = a$) and QR in a plane parallel to yOz (say $x = b$), and let the coordinates of P, Q be (x, y, z) and $(x + \delta x, y + \delta y, z + \delta z)$ respectively.

δz is the sum of the increase in z from P to R, say δz_1, and that from R to Q, say δz_2. Now PR is in the plane $y = a$, so that

$$\frac{\delta z_1}{\delta x} \approx \frac{\partial z}{\partial x},$$

i.e. the rate of increase of z with respect to x, keeping y constant. Thus

$$\delta z_1 \approx \frac{\partial z}{\partial x} \delta x. \tag{1}$$

We next proceed from R to Q keeping x constant. $\dfrac{\delta z_2}{\delta y}$ is approximately equal to the gradient, *at the point R*, of z with respect to y (x kept constant).

Assuming that this, in turn, is approximately equal to the corresponding gradient at the neighbouring point P, we have

$$\frac{\delta z_2}{\delta y} \approx \frac{\partial z}{\partial y}$$

or

$$\delta z_2 \approx \frac{\partial z}{\partial y} \delta y. \qquad (2)$$

From (1) and (2) we have

$$\delta z = \delta z_1 + \delta z_2 \approx \frac{\partial z}{\partial x} \delta x + \frac{\partial z}{\partial y} \delta y.$$

More generally, we assert that if z is a function of several variables $x_1, x_2, \ldots, x_n$, then

$$\delta z \approx \frac{\partial z}{\partial x_1} \delta x_1 + \frac{\partial z}{\partial x_2} \delta x_2 + \ldots + \frac{\partial z}{\partial x_n} \delta x_n.$$

EXAMPLE 15.3 *Two sides and the included angle of a triangle are measured to be 6 m, 10 m, and 60°. Show that, if the measurements are 0·01 m, 0·01 m and 10′ too great respectively, the error in calculating the area is approximately*

$$\left(\frac{4\sqrt{3}}{100} + \frac{\pi}{72}\right) m^2.$$

$\Delta = \frac{1}{2}bc \sin A$, where $b = 6$, $c = 10$, $A = \dfrac{\pi}{3}$ (if we are to differentiate

we, of course, put the angle into radian measure). Now if b is measured 0·01 m too long, the error $\delta b = 0·01$. Similarly $\delta c = 0·01$, $\delta A = \dfrac{10}{60} \cdot \dfrac{\pi}{180}$, (converting 10′ to radians).

Now

$$\delta \Delta \approx \frac{\partial \Delta}{\partial b} \delta b + \frac{\partial \Delta}{\partial c} \delta c + \frac{\partial \Delta}{\partial A} \delta A$$

$$= \tfrac{1}{2}c \sin A \, \delta b + \tfrac{1}{2}b \sin A \, \delta c + \tfrac{1}{2}bc \cos A \, \delta A$$

$$= \left(\frac{1}{2} . 10 . \frac{\sqrt{3}}{2} . 0·01\right) + \left(\frac{1}{2} . 6 . \frac{\sqrt{3}}{2} . 0·01\right) + \left(\frac{1}{2} . 6 . 10 . \frac{1}{2} . \frac{10\pi}{60 \times 180}\right)$$

$$= (\sqrt{3} . 0·04) + \frac{\pi}{72} .$$

Thus the error is approximately $\left(\dfrac{4\sqrt{3}}{100} + \dfrac{\pi}{72}\right) m^2.$

It is sometimes necessary to be careful over the meanings of partial derivatives. For example, $\dfrac{\partial V}{\partial x}$ would mean that V was a function of x and y (and possibly z) and that y (and z) were to be treated as constants, but $\dfrac{\partial V}{\partial r}$ would imply that polar coordinates were being used, so that $V = f(r, \theta)$, this time θ being treated as constant.

EXAMPLE 15.4   *If* $x = r \cos \theta$, $y = r \sin \theta$, *find* (i) $\dfrac{\partial x}{\partial r}$, (ii) $\dfrac{\partial x}{\partial \theta}$, (iii) $\dfrac{\partial r}{\partial x}$, (iv) $\dfrac{\partial \theta}{\partial y}$.

(i) $x = r \cos \theta$; $\dfrac{\partial x}{\partial r} = \cos \theta$.

(ii) $x = r \cos \theta$; $\dfrac{\partial x}{\partial \theta} = -r \sin \theta$.

(iii) $\dfrac{\partial r}{\partial x}$ implies that r is a function of x, y, and y *is to be kept constant.*

We have
$$r^2 = x^2 + y^2,$$
so
$$2r \frac{\partial r}{\partial x} = 2x, \quad \text{or} \quad \frac{\partial r}{\partial x} = \frac{x}{r}.$$

(iv) $\tan \theta = \dfrac{y}{x}$, so $\sec^2 \theta \dfrac{\partial \theta}{\partial y} = \dfrac{1}{x}$.

Thus
$$\frac{\partial \theta}{\partial y} = \frac{1}{x \sec^2 \theta} = \frac{1}{x\left(1 + \dfrac{y^2}{x^2}\right)} = \frac{x}{x^2 + y^2}.$$

From (i), (iii) we have
$$\frac{\partial x}{\partial r} = \cos \theta; \quad \frac{\partial r}{\partial x} = \frac{x}{r}, = \cos \theta. \tag{1}$$

There is no question of a rule such as $\dfrac{\partial x}{\partial r} = 1 \bigg/ \dfrac{\partial r}{\partial x}$.

Equations (1) may be made plausible geometrically (Fig. 15.6):

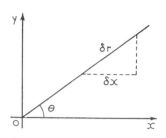
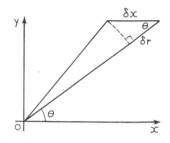

Keeping θ constant,   Keeping y constant,

$$\frac{\partial x}{\partial r} \approx \frac{\delta x}{\delta r} = \cos \theta. \qquad\qquad \frac{\partial r}{\partial x} \approx \frac{\delta r}{\delta x} \approx \cos \theta.$$

Fig. 15.6

If $V = f(x, y)$, $\dfrac{\partial V}{\partial x}$ is another function of x, y and $\dfrac{\partial}{\partial x}\left(\dfrac{\partial V}{\partial x}\right)$ is written as

$\dfrac{\partial^2 V}{\partial x^2}$. Similarly $\dfrac{\partial^2 V}{\partial y^2} = \dfrac{\partial}{\partial y}\left(\dfrac{\partial V}{\partial y}\right)$. We write $\dfrac{\partial}{\partial y}\left(\dfrac{\partial V}{\partial x}\right)$ as $\dfrac{\partial^2 V}{\partial y \partial x}$ and

$\dfrac{\partial}{\partial x}\left(\dfrac{\partial V}{\partial y}\right)$ as $\dfrac{\partial^2 V}{\partial x \partial y}$. Many applications involve these 'second order' partial

derivatives, e.g. (i) the wave equation $\dfrac{\partial^2 y}{\partial x^2} = \dfrac{1}{c^2}\dfrac{\partial^2 y}{\partial t^2}$, where $y = f(x, t)$, and

(ii) Laplace's equation $\dfrac{\partial^2 V}{\partial x^2} + \dfrac{\partial^2 V}{\partial y^2} + \dfrac{\partial^2 V}{\partial z^2} = 0$, where $V = f(x, y, z)$. These

equations are of great importance in physics, but their interpretation is beyond our present scope.

EXERCISE LXXIII

1. Explain the meaning of the symbol $\partial z / \partial x$, when z is a function of two independent variables x and y.

If $z = x \tan^{-1}\left(\dfrac{y}{x}\right)$, show that

$$x\frac{\partial z}{\partial x} + y\frac{\partial z}{\partial y} = z,$$

and find the value of $\dfrac{\partial}{\partial y}\left(\dfrac{\partial z}{\partial y}\right)$.

Show that when x is increased and y decreased by 1 per cent from $x = y = 1$, the corresponding decrease in z is approximately 0·27 per cent.

2. (i) If $x = r \cos \theta$ and $y = r \sin \theta$, prove that

$$\left(\frac{\partial x}{\partial \theta}\right)^2 + \left(\frac{\partial y}{\partial \theta}\right)^2 = r^2,$$

and
$$\left(\frac{\partial x}{\partial r}\right)^2 + \left(\frac{\partial y}{\partial r}\right)^2 = \left(\frac{\partial r}{\partial x}\right)^2 + \left(\frac{\partial r}{\partial y}\right)^2 = 1.$$

(ii) Sketch sections of the surface
$$z = (x^2 + y^2 - 1)^2$$
cut by the planes $x = 0$ and $z = a$ when a takes the values $0, \frac{1}{4}, 1$, and 4. Find the volume enclosed by the surface and the plane $z = 1$.

3. Given that $(x - y) \tan u = x^3 + y^3$,

   (i) prove that $x \dfrac{\partial u}{\partial x} + y \dfrac{\partial u}{\partial y} = \sin 2u$,

   and (ii) find the increase in u, to the nearest minute of arc, corresponding to increases of $0{\cdot}002$ in both x and y from $x = 1$, $y = 0$.

4. V is defined as a function of two independent variables x, y by means of the relations
$$V = \frac{1}{r}, \qquad r^2 = (x - a)^2 + (y - b)^2,$$
   where a and b are constants. Prove that

   (i) $(x - a) \dfrac{\partial V}{\partial x} + (y - b) \dfrac{\partial V}{\partial y} + V = 0$,

   (ii) $\dfrac{\partial}{\partial x} \left(\dfrac{\partial V}{\partial x}\right) + \dfrac{\partial}{\partial y} \left(\dfrac{\partial V}{\partial y}\right) = V^3$,

   (iii) $\dfrac{\partial}{\partial x} \left(\dfrac{\partial V}{\partial y}\right) = \dfrac{\partial}{\partial y} \left(\dfrac{\partial V}{\partial x}\right)$.

5. Find the value of $\dfrac{\partial^2 f}{\partial x^2} + \dfrac{\partial^2 f}{\partial y^2}$ when $f(x, y) = \tan^{-1}(x/y)$.

6. The base radius r and the semi-vertical angle α of a right circular cone are measured and the volume calculated from the formula $V = \frac{1}{3}\pi r^3 \cot \alpha$. Measured values are $r = 8$ cm, $\alpha = 45°$, and these measurements are liable to errors of $\pm 0{\cdot}04$ cm and $\pm 0{\cdot}25°$ respectively. Show that the greatest error that can occur in the calculated volume is a little less than $2\frac{1}{2}$ per cent.

7. Show that $y = \sin(x + ct) + 5(x - ct)^{\frac{3}{2}}$ is a solution of the 'wave equation'
$$\frac{\partial^2 y}{\partial x^2} = \frac{1}{c^2} \frac{\partial^2 y}{\partial t^2},$$
   and find the most general solution you can.

REVISION EXERCISE III

1. Prove *from first principles* that $\dfrac{d}{dx}(\tan x) = \sec^2 x$.

$$\left[\text{You may assume that } \frac{\sin \theta}{\theta} \to 1 \text{ as } \theta \to 0.\right]$$

Differentiate the following functions of x with respect to x:

   (i) $\sqrt{\{x(1 + x^2)\}}$,   (ii) $\dfrac{\sin x}{(2 + \cos x)^2}$.

Prove that, if $y^2 + ay + b = x$,

then
$$\frac{d^2y}{dx^2} + 2\left(\frac{dy}{dx}\right)^3 = 0.$$

2. Find from first principles the derivative of $\sin x$ with respect to x.
Prove that
$$\frac{d}{dx}(\tan x) = \sec^2 x, \qquad \frac{d}{dx}(\sec x) = \sec x \tan x.$$

Prove that, if
$$y = \sec^n x \quad \text{and} \quad z = \tan^n x,$$

then
$$\frac{dz}{dy} = \sin^{n-2} x.$$

3. Find from first principles, and without assuming the binomial theorem for a negative index, the derivative with respect to x of $1/x^4$.
Differentiate
$$\left(x + \frac{1}{x}\right)^n, \qquad (\cos x + \sec x)^n$$

with respect to x, and find the n-th derivative of
$$\left(x + \frac{1}{x}\right)^2$$

for all positive integral values of n, distinguishing the cases $n \leqslant 2$ and $n > 2$.

4. (i) Differentiate with respect to x the functions
$$x^3(3x + 1)^2, \qquad \frac{\ln x}{x}, \qquad \cos^2(3x - 4).$$

(ii) Find the maximum and minimum values of the function
$$\frac{x}{1 + x + x^2},$$

distinguishing between them.

5. Prove that the function $f(x)$ has a maximum value at $x = a$ when $f'(a) = 0$ and $f''(a)$ is negative.

A straight tree trunk of circular cross-section is 3·6 m long and tapers uniformly in radius from 15 cm at one end to 7·5 cm at the other. Show that the pole of greatest volume with *uniform* circular section which can be cut from the tree trunk is 2·4 m long.

6. Suppose that a sheet of paper is folded along a straight line so as to bring the top left-hand corner on to the right-hand edge of the sheet and so that the crease intersects the top edge. If the length of the crease is x times the width of the paper and the acute angle between the crease and the left-hand edge is θ, show that
$$x = \frac{1}{2\cos^2 \theta \sin \theta}.$$

Determine the minimum value of x as θ varies.

7. A cone of semivertical angle θ is circumscribed about a sphere of given radius a. Prove that the volume of the cone is

$$\tfrac{1}{3}\pi a^3 (1 + \operatorname{cosec} \theta)^3 \tan^2 \theta,$$

and find the value of $\sin \theta$ for which this volume is a *minimum*.
[You may assume that the volume of a cone is $\tfrac{1}{3}\pi r^2 h$.]

8. Prove that, if $f'(a) = 0$ and $f''(a)$ is negative, then the graph of the function $f(x)$ has a maximum at the point whose abscissa is a. Find a condition for a maximum of $f(x)$ when $x = a$ if $F(x)$ is a function such that

$$f'(x) = (x - a)F(x) \quad \text{and} \quad F(a) \neq 0.$$

A variable isosceles triangle has a constant perimeter $2c$. If x denotes one of the equal sides, express the area of the triangle in terms of x and c. Prove that the area is a maximum (and not a minimum) when the triangle is equilateral.

9. P is a variable point on the circumference of a circle of unit radius of which AB is a fixed diameter and O the centre. Prove that, when OP is perpendicular to AB, then the sum of the lengths AP and BP is a maximum and the sum of the cubes $(AP)^3$ and $(BP)^3$ a minimum.

10. A conical tent of semivertical angle θ completely encloses a given rectangular box. The base of the box is a square of side $a\sqrt{2}$ and rests on the circular base of the tent. The height of the box is equal to a and the four top corners of the box just touch the sides of the tent. Prove that the area of the curved surface of the tent is equal to

$$\pi a^2 (1 + \tan \theta)^2 \operatorname{cosec} \theta.$$

Prove also that if θ is varied this surface area is a minimum when $t = \tan \theta$ satisfies

$$2t^3 + t - 1 = 0.$$

By drawing a graph of $y = 2t^3 + t - 1$ for values of t between 0 and 2, or otherwise, obtain the value of θ which makes the surface area of the tent least.

11. Find the values of x between 0 and 2π which make the function

$$2 \sin x - \cos 2x$$

a maximum or minimum, and distinguish between them.

Draw a rough graph of the function in this interval.

12. Find the turning points of the function $x^3 - 2x^2$, distinguishing between maxima and minima, and *sketch* the graph of

$$y = x^3 - 2x^2.$$

The straight line $5x - 4y = 0$ through the origin O cuts this curve at the points A, O and B. Prove that the coordinates of A are $(-\tfrac{1}{2}, -\tfrac{5}{8})$ and find the coordinates of B.

Calculate the area contained between the curve and the line OB.

13. Find the coordinates of the turning-points of the function

$$\frac{3 \cos x}{2 - \sin x}$$

between the values 0 and 2π of x, distinguishing between maximum and minimum values.

Sketch the graph of the function between $x = 0$ and $x = 2\pi$.

Prove that the tangents at the points where the function is zero intersect at the point $(3\pi/4, -3\pi/4)$.

14. A straight line, whose slope can be varied, passes through the fixed point (a, b) in the positive quadrant. It meets the coordinate axes at the points $(p, 0)$ and $(0, q)$, where p and q are positive. Prove that the minimum value of $p + q$ is $(\sqrt{a} + \sqrt{b})^2$.

Find also the minimum value of pq.

15. A particle moves in a plane so that its polar coordinates at time t are

$$r = te^{-\frac{1}{2}t^2}, \qquad \theta = 2t.$$

Prove that the speed of the particle is

$$v = (1 + t^2)e^{-\frac{1}{2}t^2}.$$

Find the maximum and minimum values of v, and sketch the graph of v against t, for both positive and negative values of t.

16. Show that the function

$$px + \cos^{-1}(p \sin x), \text{ where } 0 < p < 1, 0 \leqslant x \leqslant \tfrac{1}{2}\pi,$$

increases as x increases through its range, p being constant.

Find the greatest value of the function if x and p can each assume any value in their respective ranges.

17. (a) Show that the function $y = (1 - x)^4(2x^2 - 2x - 3)$ has stationary values at the points where $x = -\tfrac{1}{2}$, 1, and $\tfrac{5}{3}$. Find the sign of dy/dx (i) when $-\tfrac{1}{2} < x < 1$, (ii) when $1 < x < \tfrac{5}{3}$, and determine which of the stationary values of y is a maximum and which a minimum.

(b) When $y = \cos 5\theta$ and $x = \cos \theta$, express dy/dx in terms of θ and show that

$$(1 - x^2)\frac{d^2y}{dx^2} - x\frac{dy}{dx} + 25y = 0.$$

18. Sketch the curve $x^2y = a^2(a - y)$, where a is a positive constant. On the same diagram sketch the curve $(a - y)x^2 = a^2y$.

The finite area between the curves is rotated (i) about the x-axis, (ii) about the y-axis. Find the volumes generated.

19. Find the length of the curve $6xy = x^4 + 3$ from the point at which y is a minimum to the point in the first quadrant at which the tangent passes through the origin.

20. A segment of the parabola $y^2 = 4ax$ is cut off by the line joining the origin to a variable point on the parabola. Prove that the locus of the centre of mass of the segment is the parabola $y^2 = \tfrac{5}{2}ax$.

21. Show that the curve $y = e^{-x} \sin x$ is contained between the two curves $y = e^{-x}$ and $y = -e^{-x}$. Determine the maxima and minima of $y = e^{-x} \sin x$ and draw a rough graph of the curve for $x \geqslant 0$ marking in, for the interval 0 to 2π, the maximum and minimum and the points where the curve touches the two bounding curves.

Prove that the radii of curvature at successive maximum points form a geometric progression of common ratio $e^{2\pi}$.

22. The equation of a curve is given parametrically by

$$x = f(t), \qquad y = g(t).$$

Prove that the radius of curvature ρ, at the point whose parameter is t, is given by
$$\rho = (\dot{x}^2 + \dot{y}^2)^{\frac{3}{2}}/(\dot{x}\ddot{y} - \ddot{x}\dot{y}),$$
where dots denote differentiation with respect to t.

Obtain the radius of curvature of the cycloid
$$x = a(t - \sin t), \qquad y = a(1 - \cos t),$$
at the point P where $t = \theta$, taking $0 < \theta < \pi$.

Find the equation of the normal to the cycloid at P and prove that it passes through the point Q whose coordinates are
$$a(\theta + \sin \theta), \qquad -a(1 - \cos \theta).$$

Verify that the distance PQ is numerically equal to the radius of curvature at P.

23. Prove that $e^x > 1 + x$ when $x > 0$.

Find the constants a, b, c such that the curve
$$y = ax^2 + bx + c$$
touches the curve $y = e^x$ at $x = 0$ and intersects it at $x = 1$.

Assuming that these curves do not intersect when x lies between 0 and 1, prove that
$$e^x < 1 + x + (e - 2)x^2$$
when $0 < x < 1$.

24. (i) Use Maclaurin's theorem to obtain the expansion of $\sin ax$, where a is a constant, in ascending powers of x.

(ii) Prove that, if
$$y = \sin^{-1}(mx) = a_0 + a_1 x + \frac{a_2}{2!}x^2 + \dots,$$
then
$$(1 - m^2 x^2)\frac{d^2 y}{dx^2} = m^2 x \frac{dy}{dx},$$
and
$$a_{n+2} = m^2 n^2 a_n.$$

Prove that a_{2n} is zero, and find the value of
$$a_{2n+1}/(2n + 1)!$$

25. State Taylor's theorem for the expansion of the function $f(x + h)$ in a series of ascending powers of h.

If a is an approximation to a root of the equation $f(x) = 0$, show that
$$a - \frac{f(a)}{f'(a)}$$
is likely to be a closer approximation to the root.

A root of the equation
$$2x^3 + 5\tan^2 x = 6$$
is known to lie close to $\frac{\pi}{4}$. Obtain the value of this root correct to 3 significant figures. [Take $\pi = 3 \cdot 142$.]

26. Find the limit, as x tends to 0, of
$$\frac{\sin kx}{\tan x}; \qquad \frac{\sin(\tan^{-1} px)}{\tan(\sin^{-1} qx)}; \qquad \frac{(e^{nx} - 1)^2}{\sec x - 1}.$$

27. Prove that, if
$$y = \sin(m \sin^{-1} x),$$

then
$$(1 - x^2)\left(\frac{dy}{dx}\right)^2 + m^2y^2 = m^2,$$

and
$$(1 - x^2)\frac{d^2y}{dx^2} - x\frac{dy}{dx} + m^2y = 0.$$

State Maclaurin's theorem for the expansion of a function $f(x)$ in a series of ascending powers of x, and show that the first two terms in the expansion of the principal value of y in ascending powers of x are
$$mx + \tfrac{1}{6}m(1 - m^2)x^3.$$

28. (i) State Maclaurin's theorem and use it to expand
$$y = \sin(e^x - 1)$$
in ascending powers of x, up to and including the term in x^4.

(ii) Differentiate $x \sin x - \tfrac{1}{2}\sin^2 x$ with respect to x. From your result prove that, for $0 < x < \tfrac{1}{2}\pi$,
$$0 < x \sin x - \tfrac{1}{2}\sin^2 x < \tfrac{1}{2}(\pi - 1).$$

29. Given that $y = \tfrac{1}{2}(\sin^{-1} x)^2$, where $-\tfrac{1}{2}\pi \leqslant \sin^{-1} x \leqslant \tfrac{1}{2}\pi$, prove that $(1 - x^2)y_2 - xy_1 - 1 = 0$, where $y_n = \dfrac{d^n y}{dx^n}$.

Differentiate this equation n times* and deduce that
$$(1 - x^2)y_{n+2} - (2n + 1)xy_{n+1} - n^2y_n = 0,$$
and that, when $x = 0$,
$$y_{2n-1} = 0, \qquad y_{2n} = \{2^{n-1}(n - 1)!\}^2.$$

30. State Leibniz's theorem on the n-th derivative with respect to x of a product of two functions of x.

State Maclaurin's theorem on the expansion of a function of x as a power series in x.

If $y = \sin(\sinh^{-1} x)$, prove that
$$(1 + x^2)\frac{d^2y}{dx^2} + x\frac{dy}{dx} + y = 0.$$

Hence, or otherwise, prove that
$$y = \sum_{n=0}^{\infty} \frac{(-1)^n(1^2 + 1)(3^2 + 1)\dots[(2n - 1)^2 + 1]x^{2n+1}}{(2n + 1)!}.$$

31. Integrate the following functions with respect to x:

(i) $\dfrac{x + 6}{x^2 + 6x + 8}$;   (ii) $x \ln x$;   (iii) $\sin^{-1} x$.

Evaluate
$$\int_{\frac{1}{4}}^{1} \frac{(1 + x^{\frac{3}{2}})\, dx}{x\sqrt{\{x(1 - x)\}}}$$
by means of the substitution $x = \cos^2 \phi$.

32. Give a geometrical interpretation of the integral
$$I = \int_a^b f(x)\, dx \qquad (b > a).$$

---

* This is examination jargon.

Without attempting to evaluate them, determine whether the following integrals are positive, negative, or zero:

$$\int_0^1 x^3(1 - x^2)^2 \, dx; \qquad \int_0^\pi \sin^3 x \cos^3 x \, dx; \qquad \int_{\frac{1}{2}}^1 e^{-x} \ln x \, dx.$$

33. (i) Evaluate the integrals

$$\int_1^e x \ln x \, dx, \qquad \int_{\frac{1}{2}}^2 (x^3 + x)^{-1} \, dx.$$

(ii) Evaluate the integral

$$\int_0^{\frac{\pi}{2}} (1 + \sin x)^{-1} \, dx,$$

and hence or otherwise evaluate the integral

$$\int_0^1 (1 + x)^{-\frac{3}{2}}(1 - x)^{-\frac{1}{2}} \, dx.$$

34. (i) Prove that

$$2 \sinh^2 x = \cosh 2x - 1$$

and use this relation to integrate $\sinh^4 x$ with respect to x.

(ii) Evaluate $\int_0^{\frac{a}{2}} x^2 \sqrt{a^2 - x^2} \, dx$ and $\int_0^2 xe^x \, dx$.

35. (i) By means of the substitution $y = a - x$, or otherwise, prove that

$$\int_0^a f(x) \, dx = \int_0^a f(a - x) \, dx.$$

Hence prove that

$$\int_0^\pi \frac{x \sin x}{1 + \cos^2 x} \, dx = \int_0^\pi \frac{(\pi - x) \sin x}{1 + \cos^2 x} \, dx = \frac{\pi^2}{4}.$$

(ii) ABC is a triangular lamina in which $AB = AC$ and the distance of A from BC is h; the density of a thin strip parallel to BC and at a distance x from A is px. Prove that the centre of gravity of the lamina is at a distance $\frac{3}{4}h$ from A.

36. (i) Find the following integrals:

$$\int \frac{x}{x^2 + 2x - 3} \, dx, \qquad \int \tan^{-1} x \, dx.$$

(ii) Use the substitution $x = \tan \theta$ to show that

$$\int_0^{1\cdot} \frac{1}{(x^2 + 1)^2} \, dx = \frac{1}{8}\pi + \frac{1}{4},$$

and that $\qquad \displaystyle\int_0^{\frac{\pi}{4}} \frac{1}{\tan^2 \theta + 3} \, d\theta = \frac{\pi}{4}\left(\frac{1}{2} - \frac{1}{3\sqrt{3}}\right).$

37. Prove that

$$\int_0^{\frac{\pi}{4}} \frac{d\theta}{\sin^4 \theta + \cos^4 \theta} = \int_{\frac{\pi}{4}}^{\frac{\pi}{2}} \frac{d\theta}{\sin^4 \theta + \cos^4 \theta}.$$

By means of the substitution $\tan 2\theta = t$, or otherwise, evaluate

$$\int_0^{\frac{\pi}{2}} \frac{d\theta}{\sin^4 \theta + \cos^4 \theta}.$$

38. (i) Prove that, if I_n denotes

$$\int_0^{\frac{\pi}{2}} \cos^n x \sin^3 x \, dx,$$

then $$I_n = \frac{2}{(n + 1)(n + 3)}.$$

For what values of n is this result true? Give a reason for your answer.
Deduce that, when p is a positive integer,

$$I_n + I_{n+2} + \ldots + I_{n+2p} = \frac{1}{n + 1} - \frac{1}{n + 2p + 3}.$$

(ii) Prove by the methods of the integral calculus that the centre of gravity
of a thin uniform hemispherical shell of radius a is at a distance $\frac{1}{2}a$ from the plane
of the rim.

39. Prove that

$$\int_{-a}^a f(x) \, dx = \int_0^a \{f(x) + f(-x)\} \, dx.$$

Hence show that

$$\int_{-\frac{\pi}{4}}^{\pi} \frac{1}{1 + \sin x} \, dx = 2 \int_0^{\frac{\pi}{4}} \sec^2 x \, dx,$$

and evaluate this integral.

Use the first result to evaluate

$$\int_{-1}^1 \frac{1}{1 + e^{-x}} \, dx.$$

40. (i) Evaluate

$$\int_1^3 \frac{x^2 + x + 1}{x(x + 1)} \, dx \quad \text{and} \quad \int_0^{\frac{\pi}{2}} \frac{d\theta}{1 + 2 \cos \theta}.$$

(ii) By means of the substitution $x = \sin^2 \theta$, or otherwise, determine

$$\int \sqrt{\left(\frac{x}{1 - x}\right)} \, dx,$$

expressing your answer as a function of x.

41. The coordinates of three points on the curve

$$y = a + bx + cx^2$$

are (x_1, y_1), (x_2, y_2), and (x_3, y_3), where $x_3 - x_2 = x_2 - x_1 = h$. Prove that the
area under the curve between the lines $x = x_1$ and $x = x_3$ is equal to

$$\tfrac{1}{3}h(y_1 + 4y_2 + y_3).$$

Deduce Simpson's rule for five ordinates.

The arc cut off from the curve $y = x(2 - x)$ by the x-axis is revolved through four right angles about the x-axis. Find by exact integration the volume enclosed. What is the numerical value for the volume given by Simpson's rule for *five* ordinates?

42. (i) If $I_m = \displaystyle\int_0^{\frac{\pi}{2}} \sin^m \theta \, d\theta$, $(m > 1)$, and n is a positive integer, prove that

$$I_{2n} = \frac{(2n - 1)(2n - 3)\ldots 1}{2n(2n - 2)\ldots 2} \cdot \frac{\pi}{2},$$

$$I_{2n+1} = \frac{2n(2n - 2)\ldots 2}{(2n + 1)(2n - 1)\ldots 3}.$$

(ii) Prove that

$$\int_0^{\frac{\pi}{2}} (2 \sin^5 \theta - 3 \sin^7 \theta) \cos^2 \theta \, d\theta = 0.$$

(iii) By means of the substitution $x = \sqrt{(1 - \sin \theta)}$, or otherwise, prove that

$$\int_0^1 x^2(1 - x^2)^5 \sqrt{(2 - x^2)} \, dx = \tfrac{4}{105}.$$

43. By means of the substitution

$$x = \alpha \cos^2 \theta + \beta \sin^2 \theta,$$

or otherwise, prove that

$$\int_\alpha^\beta \frac{dx}{\sqrt{\{(x - \alpha)(\beta - x)\}}} = \pi.$$

Prove also that

$$\int_\alpha^\beta \{(x - \alpha)(\beta - x)\}^{n-\frac{1}{2}} \, dx = \left(\frac{\beta - \alpha}{2}\right)^{2n} \int_0^\pi \sin^{2n} \phi \, d\phi.$$

44. (i) $F(p) = \displaystyle\int_0^\infty e^{-pt} f(t) \, dt$ is known as the *Laplace transform* of $f(t)$. Find the Laplace transform of (a) 1, (b) t, (c) e^{-at}, (d) $\cos at$.

(ii) (P) Find out what you can about the use of Laplace transforms in solving differential equations.*

45. (a) If $I_n = \displaystyle\int_0^{\frac{\pi}{4}} \tan^n\theta \, d\theta$, find the relation between I_n and I_{n-2}, where $n \geqslant 2$.

(b) Show that

$$\int_0^{\frac{\pi}{2}} f(\sin \theta) \, d\theta = \int_0^{\frac{\pi}{2}} f(\cos \theta) \, d\theta,$$

and evaluate

$$\int_0^{\frac{\pi}{2}} \frac{\sin \theta}{\sin \theta + \cos \theta} \, d\theta.$$

---

* See, for example, M. R. Spiegel, *Laplace Transforms* (Schaum, 1965).

46. Evaluate the integral

$$f(k) = \int_{-\frac{\pi}{2}}^{\frac{\pi}{2}} \frac{1 - k \cos \theta}{1 - 2k \cos \theta + k^2} \, d\theta.$$

Draw a rough sketch to show how $f(k)$ varies with k. Evaluate also

$$\int_{-\pi}^{\pi} \frac{1 - k \cos \theta}{1 - 2k \cos \theta + k^2} \, d\theta$$

in the cases (i) $|k| < 1$, (ii) $|k| > 1$.

47. It is known that the mutual attraction of two given particles is k/x^2, where k is a constant and x is the distance between the particles. Two particles A, B are connected to a fixed point O by rods of lengths a, b $(a > b)$. If A is fixed and B describes a complete revolution about O, show that the average value of the mutual attraction is

(i) $\dfrac{k}{a^2 - b^2}$ if the angle of rotation is regarded as the variable,

(ii) $\dfrac{k}{2ab} \ln \dfrac{a + b}{a - b}$ if the projection of OB on OA is regarded as the variable,

(iii) $\dfrac{k(a^2 + b^2)}{ab(a^2 - b^2)} \tan^{-1} \dfrac{b}{a}$ if the projection of OB on a line perpendicular to OA is regarded as the variable.

48. Prove that $\displaystyle\int_0^{\pi} \cos mx \cos nx \, dx = 0$ if m and n are integers and $m \neq n$.

Sum the series

$$\cos x + \cos 3x + \ldots + \cos (2n - 1)x,$$

$$\cos 2x + \cos 4x + \ldots + \cos 2nx,$$

and hence evaluate

$$\int_0^{\pi} \left(\frac{\sin nx}{\sin x}\right)^2 dx$$

for all positive and negative integral values of n.

49. Evaluate $\displaystyle\int \frac{(1 - x^2) \, dx}{(1 + x^2)\sqrt{(1 + x^2 + x^4)}}.$

50. (a) Find the general solution of the differential equation

$$y(x - 1)\frac{dy}{dx} = y^2 + 1.$$

(b) Given that y satisfies the differential equation

$$(t + 1)\frac{dy}{dt} + y = t \sin t,$$

and that $y = 1$ when $t = 0$, find y in terms of t.

51. Solve the differential equations:

(ii) $x\dfrac{dy}{dx} = y^2 - 3y + 2$;

(ii) $\dfrac{d^2y}{dx^2} + 9y = 3$.

52. Solve the differential equations:

(i) $\sec x\dfrac{dy}{dx} = \sqrt{(1 - y^2)}$,

(ii) $\dfrac{d^2y}{dx^2} + 4y = 1$, given that $y = 0$, $\dfrac{dy}{dx} = \dfrac{1}{2}$ at $x = 0$.

53. Solve the differential equations:

(i) $(x + y)\dfrac{dy}{dx} = x - y$, using the substitution $y = vx$;

(ii) $(x^2 + x)\dfrac{dy}{dx} + y = x + 1$.

54. Given that $y = \dfrac{ax + b}{cx + d}$, where a, b, c, d are constants and $bc \neq ad$, prove that

$$\frac{d^2y}{dx^2} \cdot \frac{d^2x}{dy^2} = \frac{4c^2}{bc - ad}.$$

Prove conversely that, if $\dfrac{d^2y}{dx^2} \cdot \dfrac{d^2x}{dy^2}$ is constant, then $y = \dfrac{\alpha x + \beta}{\gamma x + \delta}$, where α, β, γ, δ are constants.

55. (a) If

$$e^{-x}\frac{dy}{dx} = (1 - y)^2$$

and $y = 0$ when $x = 0$, express y in terms of x.

(b) When a mixture of two liquids is being reduced by boiling it is found that the ratio of the rates at which the liquids are separately decreasing at any instant is proportional to the ratio of the amounts x and y remaining at that instant. Express this fact in the form of a differential equation and solve the equation to obtain the relation between x and y.

56. If the normal at any point P of a curve cuts the x-axis in G and if the length of the radius of curvature at P is twice PG, prove that

$$\frac{2p}{1 + p^2}\frac{dp}{dy} = \frac{1}{y},$$

where $p = dy/dx$. Hence show that $p = (cy - 1)^{\frac{1}{2}}$, where c is a constant. Find the equation of the curve and show that it is a parabola whose axis is parallel to the y-axis.

57. (a) The curve $y = f(x)$ passes through the point $(3, 1)$, and its gradient at the point (x, y) is given by the differential equation

$$\frac{dy}{dx} = \frac{1}{2}\left(1 - \frac{x}{y}\right).$$

Find $f(x)$.

(b) By substituting $y = z^{-\frac{1}{2}}$, or otherwise, solve the differential equation

$$\frac{dy}{dx} = y - xe^{-2x}y^3.$$

58. Explain the meaning of the symbols $\partial z/\partial x$ and $\partial z/\partial y$, where z is a function of the two independent variables x and y.

If $z = (x + y) \ln (x/y)$, prove that

$$x\frac{\partial z}{\partial x} + y\frac{\partial z}{\partial y} = z.$$

If x and y are each increased by 1 per cent from the values $x = 2$, $y = 1$, prove that z is also increased by 1 per cent.

59. (i) If $u = e^{x+y} \cos (x - y)$, prove that

$$\frac{\partial u}{\partial x} + \frac{\partial u}{\partial y} = 2u,$$

$$\frac{\partial}{\partial x}\left(\frac{\partial u}{\partial x}\right) + \frac{\partial}{\partial y}\left(\frac{\partial u}{\partial y}\right) = 0.$$

(ii) A rectangular tank, open on top, is to have a capacity of 32 m³. Prove that, if x m is its length, y m its width, and A m² the area of its outer surface, then

$$A = xy + 64\left(\frac{1}{x} + \frac{1}{y}\right).$$

Prove that, if y remains fixed while x varies, the minimum value of A is $16\sqrt{y} + 64/y$.

If now y is allowed to vary find the minimum value of A.

60. Prove that

(i) $\dfrac{d}{dx}(e^x \sin x) = 2^{\frac{1}{2}}e^x \sin (x + \frac{1}{4}\pi)$,

(ii) $\displaystyle\int e^x \sin x \, dx = 2^{-\frac{1}{2}}e^x \sin (x - \frac{1}{4}\pi)$.

Prove carefully by induction that

$$\frac{d^n}{dx^n}(e^x \sin x) = 2^{\frac{n}{2}} e^x \sin (x + \frac{1}{4}n\pi).$$

61. The functions $f(x)$ and $g(x)$ are real for real values of x and satisfy the equations

$$f(x + y) = f(x)g(y) + g(x)f(y),$$
$$g(x + y) = g(x)g(y) - f(x)f(y),$$

for all values of x and y. Given that $f(x)$ does not vanish for all values of x, prove that

(i) $f(0) = 0$,   (ii) $g(0) = 1$.

If, further, $f'(0) = 1$ and $g'(0) = 0$, prove that

$$f'(x) = g(x), \qquad g'(x) = -f(x).$$

Determine the functions $f(x)$ and $g(x)$.

62. By means of a graph, or otherwise, prove that, if $f'(x)$ is defined and is positive at all points of the interval $a \leqslant x \leqslant b$, then $f(x)$ is an increasing function of x in the interval.

Prove also that, if in addition $f(a) = 0$, then $f(x)$ is positive when $a < x \leqslant b$.

By repeated applications of these results, prove that, when $x > 0$,

$$x - \frac{x^3}{3!} < \sin x < x - \frac{x^3}{3!} + \frac{x^5}{5!}.$$

Criticize the following argument:

'$\sec^2 x - 1 \geqslant 0$ always; hence $\tan x - x$, which vanishes when $x = 0$, is positive for all positive values of x; putting $x = \frac{3}{4}\pi$, we have $-1 - \frac{3}{4}\pi > 0$.'

63. A tank holds 40 litres of brine initially. While water is added at the rate of one litre a minute, the resulting mixture (supposed uniform) is drawn off at the rate of two litres a minute. Find how long it would take for the salt content of the tank to be reduced to one quarter of its initial value.

Papers Q to X

Q

Q1. Find the derivative of $\tan^{-1} x$ with respect to x.

Find an approximate value for $\int_0^1 \frac{dx}{1 + x^2}$ by using Simpson's rule with five ordinates and hence compute the value of π correct to three decimal places.

Q2. A particle moves in a plane so that its position at time t, referred to fixed rectangular axes, is given by the formulae

$$x = te^t \cos t, \qquad y = te^t \sin t.$$

Prove that its speed at time t is

$$e^t \sqrt{(1 + 2t + 2t^2)}$$

in a direction making an angle $\psi + t$ with the x-axis, where

$$\tan \psi = t/(1 + t).$$

Prove also that the acceleration at time t is

$$2e^t \sqrt{(2 + 2t + t^2)}.$$

Q3. Solve the differential equations:

(i) $\dfrac{dy}{dx} = \sin x \tan y$;

(ii) $xy \dfrac{dy}{dx} = y^2 + x^2 e^{y/x}$, by means of the substitution $y = vx$.

Q4. Integrate with respect to x:

(i) $\cos x \operatorname{cosec}^3 x$;   (ii) $\dfrac{x + 3}{\sqrt{(7 - 6x - x^2)}}$.

By making the substitution $x = a \cos^2 \theta + b \sin^2 \theta$, prove that

$$\int_a^b \frac{x\,dx}{\sqrt{\{(x-a)(b-x)\}}} = \tfrac{1}{2}\pi(a+b).$$

Q5. (i) Find the equation connecting s and ψ, where s is the length of arc OP and $\tan \psi$ is the slope of the tangent at P, if P is any point on the cycloid

$$x = a(\theta + \sin \theta), \qquad y = a(1 - \cos \theta).$$

Hence or otherwise show that the radius of curvature at P is twice the distance PN where N is the point of intersection of the line $y = 2a$ with the normal at P.

(ii) A particle slides on a smooth cycloid placed in a vertical plane with Oy vertically upwards. Show that the motion of the particle is given by an equation of the form

$$\ddot{s} = -\kappa^2 s,$$

and find the period of a small oscillation.

R

R1. Prove that the function $f(x)$ has a maximum value at $x = a$ when $f'(a) = 0$ and $f''(a)$ is negative.

P, Q are two points on a circle of centre C and radius a, the angle PCQ being 2θ. Prove that the radius of the circle inscribed in the triangle CPQ is $a \tan \theta(1 - \sin \theta)$, and that this has a maximum value when

$$\sin \theta = \frac{\sqrt{5} - 1}{2}.$$

R2. Find the area enclosed by the parabolas

$$y^2 = 4a(x + a) \quad \text{and} \quad y^2 = 4b(b - x).$$

If this area is rotated through two right angles about the x-axis, find the volume of the solid figure formed.

R3. If $y = \ln(1 + \sin x)$, prove that $y_1 y_2 + y_3 = 0$, where $y_r = d^r y/dx^r$.

Prove, by using Maclaurin's theorem or otherwise, that the expansion of y starts with the terms

$$x - \tfrac{1}{2}x^2 + \tfrac{1}{6}x^3 - \tfrac{1}{12}x^4 + \tfrac{1}{24}x^5.$$

R4. (i) Defining I_n by $I_n = \displaystyle\int_0^\pi \frac{\sin nx}{\sin x}\,dx \ (n \geqslant 2)$, prove, by considering $I_n - I_{n-2}$ or otherwise, that $I_n = 0$ if n is an even integer and that $I_n = \pi$ if n is odd.

By considering $J_n - J_{n-1}$, or otherwise, prove that

$$J_n = \int_0^\pi \frac{\sin^2 nx}{\sin^2 x}\,dx$$

is equal to $n\pi$ for all positive integral values of n.

(ii) If $u_n = \displaystyle\int_0^\infty e^{-px} \sin^n x\,dx \ (p > 0, n \geqslant 2)$, prove that

$$u_n = \frac{n(n-1)}{n^2 + p^2}\,u_{n-2}.$$

R5. A particle moves in a straight line on a smooth horizontal plane in a medium whose resistance is kv^2 per unit mass, where v is the velocity and k is a constant. If the initial velocity of the particle is V, prove that its displacement at time t is

$$\frac{1}{k}\ln(1 + kVt).$$

If the particle is projected vertically upwards and v is its velocity at height x after time t, prove that

$$v\frac{dv}{dx} = \frac{dv}{dt} = -kv^2 - g.$$

If the velocity of projection is $\sqrt{(g/k)}$ and the particle comes to rest at height H after time T, prove that

$$2kH = \ln 2, \qquad 16gkT^2 = \pi^2.$$

S

S1. (i) Integrate with respect to x

$$\sqrt{(a^2 - x^2)} \quad \text{and} \quad x^3 e^{-2x}.$$

(ii) If $f(2a - x) = f(x)$, prove that

$$\int_0^{2a} f(x)\,dx = 2\int_0^a f(x)\,dx$$

and that, if $f(2a - x) = -f(x)$,

$$\int_0^{2a} f(x)\,dx = 0.$$

Evaluate

$$\int_0^\pi \sin^6 x \cos^3 x\,dx \quad \text{and} \quad \int_0^\pi \sin^3 x \cos^2 x\,dx.$$

S2. Prove that, in general, if h is an approximation to a root of the equation $f(x) = 0$, then a better approximation is

$$h - \frac{f(h)}{f'(h)}.$$

Illustrate your result by drawing a diagram, showing a curve $y = f(x)$ and the length which is represented by $h - f(h)/f'(h)$.
What is the exceptional case?
Verify that $x = \frac{1}{3}\pi$ is an approximate solution of the equation

$$\cos x = \frac{1}{2}x,$$

and show that a better approximation is 1·03.

S3. State Maclaurin's theorem for the expansion of a function $f(x)$ as a series of ascending powers of x.
Find the n-th derivatives of the functions $1/(1 + x)$ and $\sin x$. Use Maclaurin's theorem to obtain the expansions of these two functions, giving the general term in each expansion.

S4. (i) The displacement y of a particle at time t satisfies the differential equation

$$\frac{d^2y}{dt^2} + 4\frac{dy}{dt} + 13y = 10 \cos 2t.$$

Solve this differential equation and prove that when t is large the motion of the particle is approximately simple harmonic with amplitude $2\sqrt{(145)}/29$ and period π.

(ii) Solve the differential equation

$$\frac{dy}{dx} + \frac{2xy}{1 + x^2} = \cos x,$$

given that $y = 2$ when $x = 0$.

S5. (i) Find the centre of gravity of a uniform lamina bounded by the curve $r = a(1 + \cos \theta)$.

(ii) By the methods of the integral calculus, find the centre of gravity of a uniform solid hemisphere, and deduce that the centre of gravity of an octant of a sphere, whose radius is a, is distant $\frac{3}{8}\sqrt{3}a$ from the centre of the sphere.

T

T1. (a) Evaluate the integral

$$\int_0^1 \tan^{-1} x \, dx.$$

(b) If

$$I_n = \int_0^1 \frac{dx}{(1 + x^2)^n},$$

show that

$$2nI_{n+1} = 2^{-n} + (2n - 1)I_n.$$

Deduce the value of

$$\int_0^1 \frac{dx}{(1 + x^2)^3}.$$

T2. A rod AB of length a m is hinged to a horizontal table at A. The rod is inclined to the vertical at an angle θ and there is a luminous point at a height h m vertically above A ($h > a$). Prove that the length, in metres, of the shadow of the rod on the table is $ah \sin \theta/(h - a \cos \theta)$.

If the rod rotates in a vertical plane with constant angular velocity ω, find in terms of θ the rate at which the length of the shadow alters.

Prove that the maximum length of the shadow is $ah/(h^2 - a^2)^{\frac{1}{2}}$ m.

T3. Prove that, if

$$y = \sin (\ln x) \qquad (x > 0),$$

then

$$x^2\frac{d^2y}{dx^2} + x\frac{dy}{dx} + y = 0.$$

By induction, or otherwise, prove that

$$x^2\frac{d^{n+2}y}{dx^{n+2}} + (2n + 1)x\frac{d^{n+1}y}{dx^{n+1}} + (n^2 + 1)\frac{d^ny}{dx^n} = 0,$$

where n is a positive integer.

T4. Prove that the radius of gyration of a uniform circular lamina of radius a about a perpendicular axis through its centre is $a/\sqrt{2}$.

Find the volume of the solid of uniform density formed by rotating the portion of the curve $y = a \sin x$ between $x = 0$ and $x = \pi$ through four right angles about the axis of x. Find also the radius of gyration of this solid about the axis of x.

T5. A particle is projected in a medium which offers a resistance to motion equal to kv per unit mass, k being a constant. Prove that in the subsequent motion

$$\frac{dv}{d\psi} = v \tan \psi + \frac{k}{g} v^2 \sec \psi.$$

By means of the integrating factor $\sec \psi/v^2$, solve this differential equation to obtain the relation between v and ψ.

If the particle is projected with velocity $10\sqrt{3g/k}$ at an angle of elevation $60°$, prove that the velocity at the highest point of its path is $5\sqrt{3g/16k}$.

U

U1. (i) Use the method of integration by parts to evaluate

$$\int_1^e \ln x \, dx \quad \text{and} \quad \int_0^{\frac{\pi}{2}} x^2 \cos x \, dx.$$

(ii) Find that solution of the differential equation

$$\frac{dy}{dx} + 2\frac{y}{x} = x$$

which assumes the value zero when $x = 1$.

U2. (i) By considering the sign of the derivative near $x = 0$, or otherwise, prove that the function $x \sin 2x - 2 \sin^2 x$ has a maximum at $x = 0$. [It may be assumed that $\tan x > x$ when $0 < x < \frac{1}{2}\pi$.]

(ii) Prove that the numerically least value of the radius of curvature at points on the curve $y = \ln x$ is $\frac{3}{2}\sqrt{3}$.

U3. Given that V is a function of two independent variables x and y, discuss the meaning of the statement

$$\delta V \approx \frac{\partial V}{\partial x} \delta x + \frac{\partial V}{\partial y} \delta y,$$

where δx, δy and δV denote small increments in x, y and V.

The volume V of a right circular cone is given in terms of its semi-vertical angle α and the radius r of its base by the formula

$$V = \frac{1}{3}\pi r^3 \cot \alpha.$$

The radius r and the angle α are found by measurement to be 6 cm and $45°$, but these measurements are liable to errors of ± 0.05 cm and $\pm\frac{1}{2}°$ respectively; find (to one place of decimals) the greatest percentage error which can occur in the calculated value of the volume. [Take π as approx. $3\frac{1}{7}$.]

U4. State Taylor's theorem for the expansion of a function $f(x)$ near the point $x = a$.

Expand tan $(\frac{1}{4}\pi + z)$ as a power series in z, up to and including the term involving z^4.

Hence prove that if z^3 and higher powers of z are neglected

$$\cos \{\tfrac{1}{2}\pi \tan (\tfrac{1}{4}\pi + z)\} = -\pi(z + z^2).$$

U5. Prove that

$$\int \frac{dx}{\sqrt{(x^2 + a^2)}} = \ln \{x + \sqrt{(x^2 + a^2)}\} + \text{const.},$$

and find the indefinite integral of $\sqrt{(x^2 + a^2)}$.

[*No credit will be given for merely quoting the answer.*]

A lamina is bounded by the parabola $y^2 = 4ax$ and the straight line $x = a$. Prove that the total length of the boundary is $2a \{\ln (1 + \sqrt{2}) + 2 + \sqrt{2}\}$.

V

V1. Differentiate the following functions with respect to x, giving the results in the simplest possible form:

(i) $\tan^{-1} \dfrac{2x}{1 - x^2}$;

(ii) $\ln [x + \sqrt{(x^2 - 1)}]$;

(iii) $\sin (\frac{1}{2} \cos^{-1} x)$, where $0 \leqslant \cos^{-1} x \leqslant \frac{1}{2}\pi$.

V2. Evaluate the integrals

$$\int_0^{\frac{\pi}{2}} \frac{dx}{2 - \cos x}, \qquad \int_2^4 \frac{dx}{x\sqrt{(x^2 - 4)}}, \qquad \int_1^2 \frac{2x - 3}{x(3 - x)}\, dx,$$

$$\int x \sin^2 x \, dx.$$

V3. (i) Find the maximum and minimum values of the function $y = \dfrac{m^2}{x + 1} - \dfrac{n^2}{x - 1}$ and also the values of x at which they occur.

(ii) Prove that the radius of curvature at the point (x, y) on the curve $y = \ln \sec x$ is $\sec x$. By interchanging x and y, or otherwise, prove that the radius of curvature at the point (x, y) on the curve $y = \cos^{-1} (e^x)$ is e^x.

V4. If $y = \sin \ln (1 + x)$, prove that

$$(1 + x^2)\frac{d^2y}{dx^2} + (1 + x)\frac{dy}{dx} + y = 0.$$

Hence prove that if terms of degree 5 or more are ignored,

$$y = x - \tfrac{1}{2}x^2 + \tfrac{1}{6}x^3.$$

Verify that this agrees with the result obtained by using the expansions of $\ln (1 + x)$ and $\sin x$.

V5. A particle of unit mass moves on the axis of x under a force $\lambda^2 x$ directed towards the origin and a resisting force μ multiplied by the speed; initially the particle is at O moving with speed V. Write down the equation of motion of the particle.

Prove that, if $4\lambda^2 > \mu^2$, the position of the particle at time t is given by a formula of the type

$$x = Ae^{kt} \sin \omega t$$

and determine the constants A, k and ω.

Find the corresponding formula for x when $\mu = 2\lambda$.

W

W1. Find whether the following functions of x have a maximum, a minimum, or an inflexion when $x = 0$.

(i) $\tan x - x$,
(ii) $\cos x - \sqrt{(1 - x^2)}$,
(iii) $x^{\frac{4}{3}}$.

W2. Evaluate

$$\int_0^{\frac{1}{2}} \frac{dx}{(1 - x^2)\sqrt{(1 + x^2)}}.$$

W3. (i) V is a function of two independent variables x and y. Interpret V, $\dfrac{\partial V}{\partial x}$ and $\dfrac{\partial V}{\partial y}$ geometrically.

Explain briefly the method of representing a surface on a plane by means of a contour map.

(ii) If $f(x, y) = \dfrac{\cos y}{\sin x}$ and $g(x, y) = \dfrac{\sin y}{\cos x}$, prove that $\dfrac{\partial f}{\partial x} \bigg/ \dfrac{\partial f}{\partial y} = f/g$ and that $\dfrac{\partial g}{\partial x} \bigg/ \dfrac{\partial g}{\partial y} = g/f$.

Prove also that $\dfrac{\partial}{\partial x}\left(\dfrac{\partial f}{\partial x}\right) - \dfrac{\partial}{\partial y}\left(\dfrac{\partial f}{\partial y}\right) = 2f \operatorname{cosec}^2 x$.

W4. Prove that the moment of inertia of a uniform circular disc about an axis through its centre and perpendicular to its plane is $\pi x^4 \sigma/2$, where x is the radius and σ is the mass per unit area.

Prove that the moment of inertia of a uniform solid sphere about a diameter is $8\pi a^5 \rho/15$, where a is its radius and ρ its density.

W5. (i) Find $y = f(x)$, given that $f(0) = 0$, $f(1) = \sinh(1)$ and $\dfrac{d^2y}{dx^2} = y$.

(ii) Solve the equation $\dfrac{dy}{dx} + y \sec x = \dfrac{\cos x}{1 + \sin x}$,

given that $y = \dfrac{\pi}{2(\sqrt{2} + 1)}$ when $x = \frac{1}{4}\pi$.

X

X1. Evaluate the following limits

(i) $\lim\limits_{x \to \infty} \left\{ \sqrt{(x^2 + x)} - \sqrt{(x^2 - x)} \right\}$,

(ii) $\lim\limits_{x \to 0} \left\{ \dfrac{\cos x - \cos 2x}{\cos 2x - \cos 3x} \right\}$,

(iii) $\lim\limits_{x \to \frac{1}{4}\pi} \dfrac{1 - \sin x}{\cos x}$.

X2. (i) Evaluate

$$\int_0^\pi \frac{dx}{\alpha - \cos x}, \qquad \text{for } \alpha > 1.$$

(ii) Prove that

$$\int_0^\infty x^n e^{-ax} \cos bx \, dx = n! \, (a^2 + b^2)^{-(n+1)/2} \cos \left\{ (n + 1) \tan^{-1} \left(\frac{b}{a} \right) \right\}.$$

X3. A light elastic spring of natural length l and modulus of elasticity λ hangs in a vertical plane with one end fixed. From the other end is suspended a particle P of mass m. The particle is displaced from its position of equilibrium O and then released. There is a frictional resistance to motion proportional to the velocity at any instant, the factor of proportionality being mk. If $OP = x$ at time t, prove that

$$\frac{d^2x}{dt^2} + k\frac{dx}{dt} + n^2x = 0,$$

where $n^2 = \lambda/ml$.

Obtain the general solution of this equation when $k^2 < 4n^2$.

If $n = 8$ and $k = \frac{1}{4}$ and an additional force

$$m \, (\cos 4t + 48 \sin 4t)$$

is applied vertically downwards to the particle, prove that after a sufficient lapse of time the displacement is approximately

$$x = \sin 4t.$$

X4.(i) Find the solution of the equation

$$\left(1 - \frac{dy}{dx} \right) \sin x = y \cos x$$

for which $y = \sqrt{2} - 1$ when $x = \pi/4$.

(ii) Prove that the radius of curvature at any point on the curve $x = x(t)$, $y = y(t)$ is given by

$$\rho = \frac{\left[\left(\dfrac{dx}{dt} \right)^2 + \left(\dfrac{dy}{dt} \right)^2 \right]^{\frac{3}{2}}}{\dfrac{dx}{dt}\dfrac{d^2y}{dt^2} - \dfrac{dy}{dt}\dfrac{d^2x}{dt^2}}$$

Hence, or otherwise, find the curvature of the curve

$$x = 3t^2, \qquad y = 3t - t^3$$

at the origin.

X5. For the curve

$$x^4 - y^4 + 16x^2 - 30y^2 - 225 = 0,$$

(i) find the (real) points at which it meets the axes of coordinates;
(ii) state what symmetry it possesses;
(iii) find the equations of the asymptotes;
(iv) sketch the curve.
Sketch the curve

$$x^2 - y^2 = 9$$

in a separate diagram, showing that the two curves are similar in appearance.

The areas bounded by portions of the two curves respectively and the line $x = 7\frac{1}{2}$ are rotated through four right angles about the x-axis. Prove that the volumes generated are

$$\frac{2293\pi}{24} \quad \text{and} \quad \frac{729\pi}{8}.$$

Appendix 1

THE PRINCIPLE OF MATHEMATICAL INDUCTION

The 'Principle of Mathematical Induction' is a method for proving general propositions. We shall be concerned with results involving a positive integer, say n, as for example 'if n is any positive integer, then $1 + 2 + 3 + \ldots + n = \frac{1}{2}n(n + 1)$'. The proofs consist of two parts:

(a) We show that *if* the result holds for any value of n, say $n = k$, then it follows that it also holds when $n = k + 1$;

(b) we then verify that the result does in fact hold for some particular value of n, say n_0 (which is often 1).

The combination of (a) and (b) yields the proof that the proposition holds for all positive integers $n \geqslant n_0$.

The procedure is best illustrated by a number of examples.

EXAMPLE A1   *To prove that* $1^2 + 2^2 + 3^2 + \ldots + n^2 = \frac{1}{6}n(n + 1)(2n + 1)$.

We assume that the proposition is true for some value of n, say $n = k$, i.e.

$$1^2 + 2^2 + \ldots + k^2 = \tfrac{1}{6}k(k + 1)(2k + 1) \tag{1}$$

and attempt to deduce from this that the result would then necessarily follow if k were replaced by $k + 1$, i.e.

$$1^2 + 2^2 + \ldots + k^2 + (k + 1)^2 = \tfrac{1}{6}(k + 1)(k + 2)(2k + 3). \tag{2}$$

Assuming the truth of (1), adding $(k + 1)^2$ to both sides gives

$$\begin{aligned}
1^2 + 2^2 + \ldots + k^2 + (k + 1)^2 &= \tfrac{1}{6}k(k + 1)(2k + 1) + (k + 1)^2 \\
&= \tfrac{1}{6}(k + 1)\{k(2k + 1) + 6(k + 1)\} \\
&= \tfrac{1}{6}(k + 1)(2k^2 + 7k + 6) \\
&= \tfrac{1}{6}(k + 1)(k + 2)(2k + 3).
\end{aligned}$$

Thus (1) implies (2), i.e. if the proposition is true for $n = k$ it is also true for $n = k + 1$.

But it is true when $n = 1$, since $1^2 = \frac{1}{6} \cdot 1(1 + 1)(2 + 1)$.

Therefore, by the Principle of Mathematical Induction,

$$1^2 + 2^2 + \ldots + n^2 = \tfrac{1}{6}n(n + 1)(2n + 1)$$

for all positive integers n.

We may write $1^2 + 2^2 + \ldots + n^2$ more shortly as $\Sigma_{r=1}^{n} r^2$. 'Σ' means *sum*, and $\Sigma_{r=a}^{b} f(r)$ means the sum when r is successively put equal to $a, a + 1$, $a + 2, \ldots, b$, that is $f(a) + f(a + 1) + \ldots + f(b)$.

EXAMPLE A2   *To show that* $3 \cdot 2^n + 2 \cdot 9^{n+1}$ *is divisible by 7 for all* n.

We assume that $u_n = 3 \cdot 2^n + 2 \cdot 9^{n+1}$ is divisible by 7 for some value $n = k$, i.e.

$$u_k = 3 \cdot 2^k + 2 \cdot 9^{k+1} = 7p, \text{ where } p \text{ is an integer.}$$

Now
$$u_{k+1} = 3 \cdot 2^{k+1} + 2 \cdot 9^{k+2}$$
and
$$u_k = 3 \cdot 2^k + 2 \cdot 9^{k+1}$$
so
$$u_{k+1} - 2u_k = 2 \cdot 9^{k+2} - 4 \cdot 9^{k+1}$$
$$= 2 \cdot 9^{k+1}(9 - 2)$$
$$= 14 \cdot 9^{k+1}.$$
Thus
$$u_{k+1} = 2u_k + 14 \cdot 9^{k+1}$$
$$= 14(p + 9^{k+1})$$
$$= 7q, \text{ where } q \text{ is an integer.}$$

Thus if u_k is divisible by 7, so is u_{k+1}.

But u_1 is divisible by 7 ($u_1 = 3 \cdot 2^1 + 2 \cdot 9^2 = 168 = 7 \cdot 24$), and so by the Principle of Mathematical Induction $u_n = 3 \cdot 2^n + 2 \cdot 9^{n+1}$ is divisible by 7 for all positive integers n.

EXAMPLE A3   *The Binomial Theorem.*

A fuller discussion of this theorem is given in books on algebra. We remind the reader that $n!$ means the product $1 \cdot 2 \cdot 3 \cdot \ldots \cdot n$ (e.g. $5! = 1 \cdot 2 \cdot 3 \cdot 4 \cdot 5 = 120$) and that $_nC_r$ (the number of ways of choosing r things from n) is equal to $\dfrac{n!}{r!(n-r)!}$. We shall prove (1) $_{n+1}C_r = {_nC_r} + {_nC_{r-1}}$, and (2) (The Binomial Theorem)

$$(a + b)^n = a^n + {_nC_1}a^{n-1}b + \ldots + {_nC_r}a^{n-r}b^r + \ldots + b^n.$$

(1)
$$_nC_r + {_nC_{r-1}} = \frac{n!}{r!(n-r)!} + \frac{n!}{(r-1)!(n-r+1)!}$$
$$= \frac{n!\{n-r+1+r\}}{r!(n-r+1)!}$$
$$= \frac{(n+1)!}{r!(n+1-r)!}$$
$$= {_{n+1}C_r}.$$

(2) Assume the result to be true for some particular value of n, say $n = k$, i.e.

$$(a + b)^k = a^k + {}_kC_1a^{k-1}b + \ldots + {}_kC_ra^{k-r}b^r + \ldots + b^k.$$

Multiplying both sides by $(a + b)$ gives

$$(a + b)^{k+1} = (a^k + {}_kC_1a^{k-1}b + \ldots + {}_kC_{r-1}a^{k-r+1}b^{r-1}$$
$$+ {}_kC_ra^{k-r}b^r + \ldots + b^k)(a + b)$$
$$= a^{k+1} + \ldots + ({}_kC_r + {}_kC_{r-1})a^{k+1-r}b^r + \ldots + b^{k+1}.$$

Now, from (1) above, ${}_kC_r + {}_kC_{r-1} = {}_{k+1}C_r$ and so

$$(a + b)^{k+1} = a^{k+1} + \ldots + {}_{k+1}C_ra^{k+1-r}b^r + \ldots + b^{k+1}.$$

Thus if the proposition is true for $n = k$ it is also true for $n = k + 1$. But it is trivially true when $n = 1$, and so by the Principle of Mathematical Induction the Binomial Theorem is established for all positive integers n.

$[{}_nC_r$ is also written as ${}^nC_r$ or $\begin{pmatrix} n \\ r \end{pmatrix}.]$

EXERCISE AI

Use the Principle of Mathematical Induction to prove the following results.

1. $1 + 2 + 3 + \ldots + n = \frac{1}{2}n(n + 1)$.

2. $\Sigma_{r=1}^n r^3 = \frac{1}{4}n^2(n + 1)^2$.

3. $a + ar + ar^2 + \ldots + ar^{n-1} = \dfrac{a(1 - r^n)}{1 - r}$.

4. The sum of the angles of a convex polygon with n sides is $2n - 4$ right angles.

5. $3^{n+1} + 5 \cdot 2^{4n+1}$ is divisible by 13 for all positive integers n.

6. $\dfrac{1}{2} + \dfrac{2}{2^2} + \dfrac{3}{2^3} + \ldots + \dfrac{n}{2^n} = 2 - \dfrac{n + 2}{2^n}$.

7. (after p. 82) If n is a positive integer, $\dfrac{d}{dx}(x^n) = nx^{n-1}$.

8. (see p. 83) The general 'product' rule for differentiation.

9. (cf. p. 152) $\dfrac{d^n}{dx^n}(\sin bx) = b^n \sin \left(bx + \dfrac{n\pi}{2}\right)$.

Appendix 2

SOME LIMITS

This appendix concerns certain results on limits that have been quoted in the text. As usual, the symbol n is used to denote a positive integer and x, y, z to denote real numbers.

(1) $\lim\limits_{n \to \infty} \dfrac{x^n}{n!} = 0$

We first restrict x to be positive. To prove the result, it is necessary to show that, given *any* positive number ε (however small), $\dfrac{x^n}{n!} < \varepsilon$ for *all* n greater than some value n_0. (The reader should refer to pp. 10 and 21.)

Denoting $\dfrac{x^n}{n!}$ by u_n, we have

$$\frac{u_{p+1}}{u_p} = \frac{x^{p+1}}{(p+1)!} \times \frac{p!}{x^p} = \frac{x}{p+1}.$$

Now $\dfrac{x}{n+1} < \frac{1}{2}$ for all $n \geqslant p$, if p is so chosen that it is greater than the larger of $(2x - 1)$ and 0. We then have $u_{p+1} < \frac{1}{2}u_p$, $u_{p+2} < \frac{1}{2}u_{p+1} < (\frac{1}{2})^2 u_p$, $\ldots$, $u_{p+r} < (\frac{1}{2})^r u_p$. Writing $p + r = n$, $u_n < (\frac{1}{2})^r u_p$.

Now, given $\varepsilon > 0$, we can choose r_0 so that $u_p < 2^{r_0}\varepsilon$, i.e. $(\frac{1}{2})^{r_0}u_p < \varepsilon$, so that, for $r > r_0$, $\qquad u_n < (\frac{1}{2})^r u_p < (\frac{1}{2})^{r_0}u_p < \varepsilon$.

We have thus found a number $n_0 (= p + r_0)$ such that $\dfrac{x^n}{n!} < \varepsilon$ for all $n > n_0$, so that the result follows.

Writing $y = -x$, so that $y < 0$, we have $\left|\dfrac{y^n}{n!}\right| = \left|\dfrac{x^n}{n!}\right| < \varepsilon$ for $n > n_0$ and

so $\lim\limits_{n \to \infty} \dfrac{y^n}{n!} = 0$ (cf. p. 10). Thus, for all real x, $\lim\limits_{n \to \infty} \dfrac{x^n}{n!} = 0$.

(2) *If $n \to \infty$, then* $\dfrac{x^n}{n} \to 0$ *if* $0 < x \leqslant 1$, *and* $\dfrac{x^n}{n} \to \infty$ *if* $x > 1$.

(i) If $0 < x \leqslant 1$, $x^n \leqslant 1$, so $\dfrac{x^n}{n} \leqslant \dfrac{1}{n} \to 0$ as $n \to \infty$.

(ii) If $x > 1$, we have $\ln\left(\dfrac{x^n}{n}\right) = n \ln x - \ln n$

$$= n\left(\ln x - \frac{\ln n}{n}\right).$$

Now $\ln x > 0$ and $\dfrac{\ln n}{n} \to 0$ as $n \to \infty$ (p. 158). Thus $\ln \dfrac{x^n}{n} \to \infty$ as $n \to \infty$ and so $\dfrac{x^n}{n} \to \infty$.

Before discussing the remaining limits, it is necessary to consider the meaning of a^λ, where a is a positive number and λ is irrational; we shall proceed intuitively. If x is restricted to being rational, a^x is a continuous function at any (rational) point x_0 in the sense that $a^x \to a^{x_0}$, as $x \to x_0$ *taking up rational values only*. Now it is possible for x to tend to an *irrational* limit while still restricted to rational values itself; if this (irrational) limit is λ, we define a^λ as the limit as $x \to \lambda$ through rational values of a^x. We shall assume that $\dfrac{d}{dx}(x^\alpha) = \alpha x^{\alpha-1}$ if $x > 0$ for *all* α. (For α rational, a proof is given on p. 85 above; for α irrational, the reader is referred, for example, to Courant, *Differential and Integral Calculus*, Blackie, Vol. I, 2nd edition, p. 130.)

(3) $\dfrac{x}{e^x} \to 0$ *as* $x \to \infty$.

$e^x = 1 + x + \dfrac{x^2}{2!} + \ldots > 1 + x + \dfrac{x^2}{2!}$, if $x > 0$. Thus

$$0 < \frac{x}{e^x} < \frac{x}{1 + x + \dfrac{x^2}{2!}} \to 0$$

as $x \to \infty$, and the result follows.

(4) $\dfrac{x^n}{e^x} \to 0$ *as* $x \to \infty$.

The proof of (3) can be adapted.

(5) $\dfrac{\ln y}{y} \to 0$ *as* $y \to \infty$.

Putting $x = \ln y$ in (3) gives the required result.

(6) $z \ln z \to 0$ *as* $z \to 0$.

Putting $y = \dfrac{1}{z}$ in (5) above gives $z \ln \dfrac{1}{z} \to 0$ as $z \to 0$, i.e. $-z \ln z \to 0$, so that $z \ln z \to 0$ as required.

EXERCISE AII

1. If k is any positive number not an integer, prove that $\dfrac{x^k}{e^x} \to 0$ as $x \to \infty$. [Consider

the integer n such that $n < k < n + 1$, so that $\dfrac{x^n}{e^x} < \dfrac{x^k}{e^x} < \dfrac{x^{n+1}}{e^x}$ if $x > 1$.]

2. Prove that $\dfrac{\ln x}{x^k} \to 0$ as $x \to \infty$ for *any* positive number k.

3. Prove that $x^k \ln x \to 0$ as $x \to 0$ for *any* positive number k.

Appendix 3

TAYLOR'S THEOREM

As a first taste of 'analysis', a proof is given below of Taylor's theorem. *If $f^{(n-1)}(x)$ is continuous for $|x - a| \leqslant |h|$ and $f^{(n)}(x)$ exists for $|x - a| < |h|$,*

then $\qquad f(a + h) = f(a) + hf'(a) + \ldots + \dfrac{h^{n-1}}{(n-1)!} f^{(n-1)}(a) + R_n,$

where $R_n = \dfrac{h^n}{n!} f^{(n)}(a + \theta h)$, θ *being some number such that* $0 < \theta < 1$.

If, further, $R_n \to 0$ as $n \to \infty$, then $f(a + h) = f(a) + hf'(a) + \dfrac{h^2}{2!} f''(a) + \ldots$

We first extend the definition of continuity (p. 22). Let $f(x)$ be defined over the closed interval $[a, b]$.* If $f(a + h) \to f(a)$ as $h \searrow 0$, we shall say that $f(x)$ is continuous at a; similarly if $f(b + h) \to f(b)$ as $h \nearrow 0$, then $f(x)$ is continuous at b.

If within a certain interval there is a number M such that no value of $f(x)$ exceeds M but any number less than M is exceeded by at least one value of $f(x)$, then M is the (*least*) *upper bound* of $f(x)$ in that interval. We assume the truth of the following theorem:

A function which is continuous over a closed interval attains its upper bound.

We can now prove

Rolle's Theorem. If $f(x)$ is continuous in $[a, b]$ and differentiable in (a, b), and $f(a) = f(b)$, then there is a number ξ ($a < \xi < b$) such that $f'(\xi) = 0$.

We may suppose that $f(a) = f(b) = 0$ (if $f(a) = f(b) = k$, we could consider instead $\phi(x) = f(x) - k$).

(i) If $f(x) = 0$ throughout, $f'(x) = 0$ throughout (a, b).

(ii) If not, let us suppose that $f'(x)$ is somewhere positive; then $f(x)$ has an upper bound M and $f(x) = M$ for some ξ ($a < \xi < b$). If $f'(\xi)$ were positive or negative, there would be values of x near ξ such that $f(x) > M$, giving a contradiction. Hence $f'(\xi) = 0$, so that the theorem is proved.

In case the reader has been tempted to read the above quickly (and perhaps even drawn a diagram and dismissed Rolle's theorem as 'obvious'), he should now ask himself the following questions.†

---

* I.e., for $a \leqslant x \leqslant b$, cf. p. 91.   † He should refer back to Chapter 2 only if in difficulties.

1. (a) What is a *function*?

(b) Which of the following graphs could represent a function?

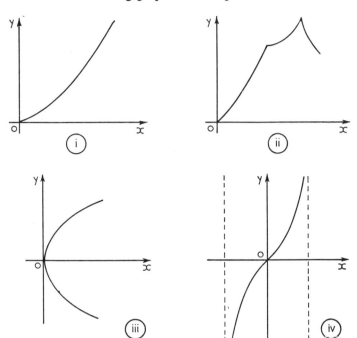

2. (a) What is meant by *continuity*?

(b) Is it possible for a function to be continuous at a point without being differentiable there?

(c) Is it possible for a function to be differentiable at a point without being continuous there?

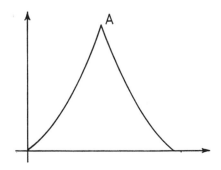

3. Is the function represented by the above graph differentiable at A?

4. If $f(x)$ is continuous in the open interval (a, b), does it necessarily attain its upper bound?

We now have the necessary equipment to prove Taylor's theorem, but to show how the pattern is built up, we first prove

The First Mean-Value Theorem (cf. p. 45). *If $f(x)$ is continuous in $[a, b]$ and differentiable in (a, b), then* $\exists \xi$, $a < \xi < b$, (read 'there exists a number ξ such that $a < \xi < b$') *such that $f(b) - f(a) = (b - a)f'(\xi)$.*

Consider the function*

$$F(x) = f(b) - f(x) - \left(\frac{b - x}{b - a}\right)\{f(b) - f(a)\}.$$

Since $F(a) = F(b) = 0$, by Rolle's theorem we have

$$-f'(\xi) + \frac{1}{b - a}\{f(b) - f(a)\} = 0$$

or $\qquad f(b) - f(a) = (b - a)f'(\xi)$ for some ξ in (a, b).

Writing $b = a + h$, this result can be written

$$f(a + h) = f(a) + hf'(a + \theta h), \text{ where } 0 < \theta < 1.$$

The Second Mean-Value Theorem. *If $f'(x)$ is continuous in $[a, b]$ and $f''(x)$ exists in (a, b), then*

$$f(b) = f(a) + (b - a)f'(a) + \tfrac{1}{2}(b - a)^2 f''(\xi), \text{ where } a < \xi < b.$$

Consider the function

$$G(x) = f(b) - f(x) - (b - x)f'(x) - \left(\frac{b - x}{b - a}\right)^2 \{f(b) - f(a) - (b - a)f'(a)\}.$$

Since $G(a) = G(b) = 0$, by Rolle's theorem we have

$$f(b) = f(a) + (b - a)f'(a) + \tfrac{1}{2}(b - a)^2 f''(\xi) \text{ where } a < \xi < b,$$

or, writing $b = a + h$,

$$f(a + h) = f(a) + hf'(a) + \tfrac{1}{2}h^2 f''(a + \theta h), 0 < \theta < 1.$$

---

* This may seem rather 'out of the blue', but the idea is to build up from Rolle's theorem. The function f doesn't obey the conditions of that theorem, since $f(a)$ is not necessarily equal to $f(b)$; it is therefore necessary to construct a function F, simply related to f, which does obey them.

We now generalize the last two results, obtaining *Taylor's Theorem*. We shall prove:

If $f^{(n-1)}(x)$ is continuous in $[a, b]$ and $f^{(n)}(x)$ exists in (a, b), then

$$f(b) = f(a) + (b - a)f'(a) + \frac{(b - a)^2}{2!}f''(a) + \cdots$$

$$+ \frac{(b - a)^{n-1}}{(n - 1)!}f^{(n-1)}(a) + \frac{(b - a)^n}{n!}f^{(n)}(\xi),$$

where $a < \xi < b$.

Consider the function

$$H(x) = f(b) - f(x) - (b - x)f'(x) - \frac{(b - x)^2}{2!}f''(x) - \cdots$$

$$- \frac{(b - x)^{n-1}}{(n - 1)!}f^{(n-1)}(x) - \left(\frac{b - x}{b - a}\right)^n \left\{f(b) - f(a) - (b - a)f'(a) - \cdots\right.$$

$$\left. - \frac{(b - a)^{n-1}}{(n - 1)!}f^{(n-1)}(a)\right\}.$$

Since $H(a) = H(b) = 0$, by Rolle's theorem we have

$$-f'(\xi) + f'(\xi) - (b - \xi)f''(\xi) + (b - \xi)f''(\xi) - \cdots - \frac{(b - \xi)^{n-1}}{(n - 1)!}f^n(\xi)$$

$$+ \frac{n(b - \xi)^{n-1}}{(b - a)^n} \left\{f(b) - f(a) - \cdots - \frac{(b - a)^{n-1}}{(n - 1)!}f^{(n-1)}(a)\right\} = 0$$

which, on cancelling out the first $(n - 2)$ pairs of terms and then dividing by $\dfrac{(b - \xi)^{n-1}}{n(b - a)^n}$ reduces to

$$f(b) = f(a) + (b - a)f'(a) + \frac{(b - a)^2}{2!}f''(a) + \cdots$$

$$+ \frac{(b - a)^{n-1}}{(n - 1)!}f^{(n-1)}(a) + \frac{(b - a)^n}{n!}f^{(n)}(\xi),$$

where $a < \xi < b$, as required, or, putting $b = a + h$,

$$f(a + h) = f(a) + hf'(a) + \cdots + \frac{h^{n-1}}{(n - 1)!}f^{(n-1)}(a) + \frac{h^n}{n!}f^{(n)}(a + \theta h),$$

where $0 < \theta < 1$.

The argument above holds if $b < a$, i.e. if the interval concerned is $[b, a]$. The reader should verify that the first part of the statement on p. 328 is thus established.

From the above, if $f(x)$ has derivatives of all orders in the interval $(a - \eta, a + \eta)$, then for $|h| < \eta$ we have, for all n,

$$f(a + h) = f(a) + hf'(a) + \ldots + \frac{h^{n-1}}{(n-1)!} f^{(n-1)}(a) + R_n,$$

$$R_n = \frac{h^n}{n!} f^{(n)}(a + \theta h), \qquad 0 < \theta < 1. \tag{1}$$

If $R_n \to 0$ as $n \to \infty$, then

$$f(a + h) = f(a) + hf'(a) + \frac{h^2}{2!} f''(a) + \ldots \tag{2}$$

Thus the proof of Taylor's theorem is completed.

Setting $a = 0$ in (1) and (2) gives Maclaurin's theorem (cf. p. 137), e.g. (2) yields

$$f(h) = f(0) + hf'(0) + \frac{h^2}{2!} f''(0) + \ldots.$$

The two theorems are thus seen to be essentially equivalent.

*   *   *

Recommendations for the School Library

Men of Mathematics by E. T. Bell (S and S Press, U.S.A.).

History of Calculus by C. B. Boyer (Dover).

Origins of the Infinitesimal Calculus by M. E. Baron (Pergamon).

Mathematical Models (2nd edition) by H. M. Cundy and A. P. Rollett (O.U.P.).

Mathematics and the Imagination by E. Kasner and J. Newman (Bell).

Riddles in Mathematics by E. P. Northrop (Penguin).

Fallacies in Mathematics by E. A. Maxwell (C.U.P.).

Mathematical Snapshots by H. Steinhaus (O.U.P.).

A Book of Curves by E. H. Lockwood (C.U.P.).

Mathematical Analysis by K. G. Binmore (C.U.P.).

Calculus (2 vols) by T. M. Apostol (Blaisdell).

Selected Papers on Calculus by T. M. Apostol *et al* (Math. Assn, U.S.A.).

Answers

A few answers have been omitted, which may be found useful for test purposes.

Exercise I. **2.** (i) $4 \cdot 5$ m, (ii) $2 \cdot 5$ m, (iii) $2 \cdot 5$ m s^{-1}, (iv) (a) $2 \cdot 2(5)$, (b) $2 \cdot 0(5)$, (c) 2. **4.** $0 \cdot 2(5)$, $0 \cdot 7(5)$, $1 \cdot 2(5)$, $1 \cdot 7(5)$, $2 \cdot 2(5)$, 5.

Exercise II. **2.** (a) $\frac{5}{8}$, (b) $-\frac{1}{2}$, (c) $\frac{1}{4}$, (d) $\frac{9}{8}$, (e) $-\frac{4}{9}$. **7.** $y - 2 = 4(x - 6), y - 4 = 7(x + 2), y + 3 = -\frac{1}{2}(x - 2)$. **9.** (a) $\frac{12}{5}$, (b) $-\frac{7}{3}$, (c) 1, (d) $\dfrac{s - q}{r - p}$. **11.** $y - 4 = 2(x - 3), y - 7 = \frac{1}{3}(x + 4), y + 1 = -\frac{2}{5}(x - 3)$.

Exercise III. **4.** (iii) 9, (v) $6 \cdot 6$, (vi) $6 \cdot 3$, (vii) $6 \cdot 03$. **5.** (ii) $13 \cdot 61$, (iv) $13 \cdot 006$. **7.** (a) 8, (b) 0. **8.** (i) $(0, 0), (4, 0)$, (ii) 0. **10.** (ii) $27, 27$. **12.** $-\dfrac{1}{x^2}$. **16.** (a) 7, (c) $\frac{4}{3}$, (e) $-\frac{4}{3}$. **17.** Gradient at P, 12. **18.** Gradient at P, 5. **19.** (i) $x^2 + 2x \cdot \delta x + (\delta x)^2$, (iii) $x \delta x + (\delta x)^2$, (v) $16x^3 + 48x^2 \cdot \delta x + 36x(\delta x)^2$, (vii) $x^3 + 3x^2 \cdot \delta x + 3x(\delta x)^2 + (\delta x)^3$. **22.** (ii) $5x^4$. **23.** (b) $10x - 3$. **24.** $-\dfrac{1}{(x + 1)^2}$.

Exercise IV. **1.** (i) -2, (iii) 13, (iv) $a^2 - 4a + 1$, (v) $(a + h)^2 - 4(a + h) + 1$. **2.** (i) -4, (ii) $10\frac{1}{2}$. **3.** (i) 1, (ii) $y - \dfrac{2}{y}$, (iv) $x + h - \dfrac{2}{x + h}$, (vi) $-2\frac{1}{3}$. **4.** E.g. x^2. **5.** $f(2x) = x(2x + 1)$. **6.** (a) D all x, R all values, (b) D all x, R all values $\geqslant 1$, (c) D all x except $x = 2$, R all values except 0, (d) $D x \leqslant 3$, R all values $\geqslant 0$, (e) D $x \leqslant 2$ and $x \geqslant 5$, R all values $\geqslant 0$, (f) D all x except $x = 0$, $x = 3$, R all values.

Exercise V. **1.** $8x - 7$. **2.** (i) $3x^2$, (ii) $6x$, (iii) $-\dfrac{1}{(x + 2)^2}$, (iv) $2 + \dfrac{1}{(x - 4)^2}$.

Exercise VI. **2.** (a) 46, (b) 6, (c) 27. **3.** (a) $3 \cdot 99 < x < 4 \cdot 01 \, (x \neq 4)$. **4.** (b) $x = -1$, (c) $x = -1$, (g) $x = -1$, (h) $x = 2, 4$, (i) $x = 2, 4$, (j) $x = 3$.

Exercise VIII. **1.** $8x^7$. **2.** $56x^6$. **4.** $99x^{98}$. **6.** $24x^2 + 12x - 37$. **7.** $2x + 4$. **8.** $1029x^2 - 882x + 189$. **9.** $\frac{1}{2} - \frac{1}{2}x^{-\frac{1}{2}}$. **10.** $24x^5$. **13.** $4x - 5$. **15.** $-21x^{-\frac{1}{4}} + 49$. **16.** $\frac{1}{2}$. **17.** 4. **18.** 4. **20.** $2 \cdot 30, -1 \cdot 30$. **22.** $3\frac{1}{2}$. **23.** (i) $2, 2 \cdot 48, -7, 3x^2 - 10x + 3, -4$, (ii) $\frac{1}{3}, 3$. Three real roots. **24.** (ii) na^{n-1}. **25.** (i) $y = 6x - 23$. **26.** (iv) $-\frac{1}{4}$. **28.** (b) $-(x - 2)^{-2}$. **31.** (i) -6. **35.** $(0 \cdot 2, 0 \cdot 3)$ min, $(-10 \cdot 2, 1090)$ max.

Exercise IX. **1.** (a) $y = 6x - 9$, $6y + x = 57$, (c) $y - 6x = 1$, $6y + x = 6$, (e) $4x = y + 8$, $4y + x + 15 = 0$. **4.** $y = 24x + 61$.

Exercise X. **1.** (a) $x > -\frac{1}{4}$, (b) $x < -2$ and $x > 1$, (c) $-2 < x < 3$, (d) all x except $x = 1$. **2.** (a) $2\frac{7}{8}$, $x = -\frac{1}{4}$, min, (b) 24, $x = -2$, max, -3, $x = 1$, min, (d) 0, $x = 1$, inflexion. **3.** $x > 4$, $-2 < x < 1$. **5.** (a) Min at $(\frac{5}{2}, -\frac{1}{4})$, (c) Max at $(0\cdot10, 1\cdot05)$, min at $(3\cdot23, -14\cdot23)$, (d) none, (e) Max at $(1, 0)$, min at $(\frac{5}{3}, -\frac{4}{27})$, (f) Inflexion at $(1, 0)$, (h) Max at $(0\cdot77, 0\cdot19)$, inflexion at $(0, 0)$, min at $(-0\cdot77, -0\cdot19)$. **6.** The gradient does not exist at the origin.

Exercise XI. **2.** 32 m. **3.** (ii) 2 cm. **5.** $\frac{1}{3}h$. **7.** $\frac{1}{2}l$. **8.** $12\frac{1}{2}$ cm^2.

Exercise XII. **2.** $t = \frac{3}{2}$, -1, (i) 18, (ii) 0. **3.** (iii) -8, (iv) $1\frac{1}{4}$, $31\frac{1}{4}$. **4.** $t = 1$, $x = 4$; $t = 3$, $x = 0$; -6 m s^{-2}, 6 m s^{-2}; -3 m s^{-1}.

Exercise XIII. **1.** 19·45. **2.** $-1\cdot2$. **3.** $1\cdot2$ m^2. **4.** ± 13 cm^2. **6.** $\frac{1}{2}\sqrt{5}(\approx 1\cdot12)$. **7.** 1·8. **9.** 0·7. **10.** $y = 3x + 2$, $y = 3x + 6$. **11.** 8·56. **12.** (ii) $6\frac{3}{4}$ m. **13.** Each 12. **14.** $y = 13x - 59$, $y = x - 17$. **15.** 0·7 m s^{-1}. **17.** 6 cm. **18.** (ii) $(2, 4)$, $(4, -8)$. **19.** (i) 0, -4 m s^{-1}, (ii) $1\frac{1}{2}$ s, 18 m. **20.** 2·7. **21.** 1·39 cm. **22.** $Q(-\frac{2}{3}, \frac{1}{9})$. **23.** $\dfrac{2x}{3}$. **24.** Inflexion at $(1, 0)$, max at $(4, 27)$. **25.** 576 boxes. **26.** $4h$, $y - k = 4h(x - h)$, $y = \pm 12x$. **27.** -2, $-0\cdot9$. **28.** $180t - 135t^2$, $180 - 270t$, $1\frac{1}{3}$ hr, $53\frac{1}{3}$ miles, 60 m.p.h.

Exercise XVI. **1.** (i) $\frac{1}{5}x^5 + c$, (ii) $\frac{1}{4}x^4 - \frac{3}{2}x^2 + x + c$. **2.** (i) 576·4, (ii) $44\frac{1}{4}$, (iii) $3\frac{5}{6}$, (iv) $13\frac{3}{4}$. **3.** 12. **4.** 116. **5.** $47\frac{1}{3}$.

Exercise XVII. **1.** $7\frac{1}{2}$. **3.** $16\frac{1}{2}$. **4.** $-4\frac{1}{2}$. **6.** $27\frac{11}{12}$. **7.** 0. **8.** $\frac{1}{3}x^3 + 4x - 9\frac{2}{3}$. **9.** $y = \frac{1}{4}x^4 - x^3 + x - 32\frac{1}{4}$. **10.** (a) Even, $\frac{1}{5}, \frac{1}{5}, \frac{2}{5}$, (b) Even, (c) Odd, $\frac{1}{4}, -\frac{1}{4}, 0$, (d) Odd, (e) Neither, $\frac{11}{12}, \frac{5}{12}, \frac{4}{3}$. **11.** (i) 0, (ii) 14. **12.** 3. **13.** $\frac{1}{3}$. **14.** $20\frac{5}{6}$. **15.** (ii) $42x - y = 35$. **17.** 0, $\frac{1}{4}, -\frac{1}{4}$. **18.** 6·4. **19.** $y = (1 - x^2)^2$, ± 1. **20.** (i) 43·2, (ii) 86·4, (iii) 86·4. **21.** (i) $y = 3x$, (ii) $-\frac{8}{3}$. **22.** $\frac{1}{3}$. **24.** Value when $x = 4$ is 95. **25.** Area cut off is $15\frac{3}{4}$.

Exercise XVIII. **1.** 64. **2.** $2\frac{1}{6}$. **3.** $(-) 1\frac{1}{3}$. **4.** 9·6. **5.** $\frac{1}{6}$. **6.** $121\frac{1}{2}$. **7.** $\frac{1}{12}$.

Exercise XIX. **2.** 18π. **3.** $29\frac{1}{5}\pi$. **4.** $67\frac{11}{15}\pi$. **5.** $\frac{3}{5}\pi$. **6.** (i) $6\frac{2}{5}\pi$, (ii) 8π. **7.** $18\frac{2}{3}\pi$. **8.** $\frac{1}{30}\pi$. **9.** $0\cdot08\pi$. **10.** $\dfrac{1792\pi}{9}$.

Exercise XX. **2.** $(\frac{4}{5}, \frac{2}{7})$. **3.** To 2 dec. pl., $(0\cdot69, 2\cdot31)$. **4.** $(2\cdot5, 6\cdot4)$. **5.** $(2\frac{4}{7}, 67\frac{3}{11})$. **6.** $(1\cdot59, 0)$. **7.** $(0, \frac{5}{8})$. **8.** $(\frac{1}{2}, 0)$.

Exercise XXI. **1.** 27 m, $85\frac{1}{3}$ m, 8. **2.** 16 m s^{-2}, $42\frac{2}{3}$ m, $85\frac{1}{3}$ m. **3.** 3 m s^{-2}, -1 m s^{-2}; $11\frac{1}{3}$ m; 1, 6. **4.** 20 m s^{-1}, $466\frac{2}{3}$ m. **5.** (i) $(t^2 - 4t + 3)$ m s^{-1}, (ii) 1, (iii) $\frac{4}{3}$ m.

Exercise XXII. **1.** $1 \cdot 15$. **2.** (i) $68, 0 \cdot 64$, (ii) $64, 0 \cdot 66$. **3.** $7 \cdot (7)$. **4.** $0 \cdot 93$. **5.** $1 \cdot 1$.
7. $4\frac{1}{2}$. **8.** (iii) 825 cm. **9.** $\dfrac{81\pi}{10}$. **10.** $8\frac{1}{6}$, $9\frac{1}{3}$. **11.** Area is $10\frac{2}{3}$. **13.** 35 m s^{-1}, 20 m s^{-2}, 16 m s^{-1}, 16 m. **14.** 134π. **15.** $\frac{8}{3}$, $(\frac{3}{2}, \frac{6}{5})$. **16.** $1 \cdot 2(2)$. **17.** $4\frac{1}{2}$ m, 5 s. **18.** x-coordinate of C.G. is $\frac{1}{5}$.

Exercise XXIII. **2.** $3x^2 + 2x - 3$. **3.** $4x^3 - 24x^2 + 44x - 23$. **4.** $(x + 1)^{-2}$. **5.** $5(x + 2)^{-2}$. **6.** $(4 - 2x)(x^2 - 4x + 1)^{-2}$. **7.** $(-x^4 + 6x^3 - 12x^2 + 4x - 6) \times (x^3 + 2)^{-2}$.

The arbitrary constant of integration is generally suppressed from now on, to save space. The reader should make a habit of inserting it in all indefinite integrations.

Exercise XXIV. **1.** $\frac{1}{2}x^{-\frac{1}{2}}$. **2.** $-\frac{3}{2}x^{-\frac{5}{2}}$. **3.** $1 - \dfrac{1}{x^2}$. **4.** $-7x^{-8}$. **5.** $-\frac{2}{3}x^{-\frac{5}{3}}$.
6. $4(x + 1)^3$. **7.** $14(3x - 2)(3x^2 - 4x + 2)^6$. **8.** $\frac{3}{2}x^2(x^3 + 1)^{-\frac{1}{2}}$. **9.** $-2(x + 3)^{-3}$.
12. $-3(5 - x)^2$. **13.** $(x - 2)^2(x + 1)(5x - 1)$. **14.** $3(3x + 1)(4 - x)^2(7 - 5x)$.
15. $-\dfrac{\sqrt{3}}{6}x^{-\frac{3}{2}}$. **16.** $\frac{15}{2}(1 - 5x)^{-\frac{5}{2}}$. **17.** $(1 + 8x - x^2)(x^2 + 1)^{-2}$. **18.** $(x - 4)^2 \times (3 + 16x - x^2)(x^2 + 1)^{-3}$. **19.** $2(11 + 34x + 5x^2 - 15x^3)(3x - 4)^2(5x^2 + 1)^{-2}$.
20. $(6x^2 + 2x + 3)(x^2 + 1)^{-\frac{1}{2}}$. **21.** $(2x - 3)(3x + 2)^{-2}(x^2 + 1)^{-\frac{1}{2}}$. **22.** $-(x + 1)^{-\frac{1}{2}} \times (x - 1)^{-\frac{3}{2}}$. **23.** $(1 - 2x)(x - 2)^{-2}(x + 1)^{-2}$. **24.** $-\frac{1}{2}(24x^2 + 30x - 25)(2x + 1)^{-2} \times (7 - 3x)^{-\frac{1}{2}} + 2(2x + 1)^{-2}$. **25.** $3\frac{1}{2}$. **26.** $\frac{2}{3}x^{\frac{3}{2}}$. **27.** $-\dfrac{1}{x}$. **28.** $\frac{9}{5}x^{\frac{5}{3}}$. **29.** $\frac{3}{2}(x + 1)^{\frac{4}{3}}$.
30. $\frac{1}{20}(5x - 1)^4$. **31.** $-\frac{1}{20}(1 - 5x)^4$. **32.** $2x^2 - 7x - x^{-1}$. **33.** $1 \cdot 36$.

Exercise XXV. **3.** $3 \cos 3x$. **4.** $2 \cos 2x - 4 \sin 4x$. **5.** $-\cos\left(\dfrac{\pi}{4} - x\right)$.
6. $-\operatorname{cosec} x \cot x$. **7.** $-\operatorname{cosec}^2 x$. **8.** $\sec x \tan x$. **9.** $\sin x(2 \cos^2 x - \sin^2 x)$.
10. $3 \sin^2 x \cos x$. **11.** $\sec x(\tan x + \sec x)$. **12.** $\sec x(1 + 2 \tan^2 x)$.
13. $4 \cos 4x \cos 3x - 3 \sin 4x \sin 3x$. **14.** $-3 \tan x \sec^2 x(1 - 3 \tan^2 x)^{-\frac{1}{2}}$.
15. $\sin x + x \cos x$. **16.** $\dfrac{\pi}{180} \cos \dfrac{\pi x}{180} \left[= \dfrac{\pi}{180} \cos (x^\circ)\right]$. **17.** $\frac{1}{2} \sin 2x$.
18. $-\frac{1}{5} \cos (5x + 1)$. **19.** $\sin x - \cos x$. **20.** $\frac{1}{2}x - \frac{1}{4} \sin 2x$. **21.** $\frac{1}{2}x + \frac{1}{4} \sin 2x$.
22. $-\frac{1}{14} \cos 7x - \frac{1}{2} \cos x$. **23.** $\frac{1}{24} \sin 12x + \frac{1}{4} \sin 2x$. **24.** $\tan x$. **25.** $\tan x - x$.
26. $-\cot x$. **27.** 1. **28.** 1.

Exercise XXVI. **1.** $-\dfrac{1}{1 + x^2}$. **3.** (a) $3(1 - 9x^2)^{-\frac{1}{2}}$, (b) $-4(1 - 16x^2)^{-\frac{1}{2}}$,

$(c) -(1 - x^2)^{-\frac{1}{2}}$. **5.** (i) $\dfrac{\pi}{4}$, (ii) $\dfrac{\pi}{6}$. **6.** (i) $\dfrac{1}{12} \tan^{-1}\left(\dfrac{4x}{3}\right)$. (ii) $\dfrac{\sqrt{3}}{6} \tan^{-1}\left(\dfrac{\sqrt{3}}{2} x\right)$,

(iii) $\frac{1}{3} \sin^{-1} 3x$, (iv) $\dfrac{1}{\sqrt{2}} \sin^{-1}\left(\sqrt{\dfrac{2}{3}}\, x\right)$. **7.** $x^{-1}(x^2 - 1)^{-\frac{1}{2}}$. **8.** $(3 - 8x - 3x^2)(x^4 +$
$11x^2 + 24x + 17)^{-1}$.

Exercise XXVII. **2.** $-\frac{1}{2}$. **3.** (i) $\cos x \operatorname{cosec} y$, (ii) $3(x^2 - y)(3x - 2y)^{-1}$. **4.** $-x^3 \times$
$(1 - x^4)^{-\frac{1}{2}}$. **5.** $-\frac{4}{5}$. **6.** $\frac{4}{3}$.

Exercise XXVIII. **2.** (i) $3x^2 - 8x + 3$, $6x - 8$, 6, (ii) $5x^4 - x^{-2}$, $20x^3 + 2x^{-3}$,
$60x^2 - 6x^{-4}$, (iii) $-(1 + x)^{-2}$, $2(1 + x)^{-3}$, $-6(1 + x)^{-4}$, (iv) $\cos x - 2 \sin 2x$,
$-\sin x - 4 \cos 2x$, $-\cos x + 8 \sin 2x$. **4.** $\frac{3}{2}$. **5.** $(-0.87, 11.8)$ max, $(1.54, 5.41)$
min.

Exercise XXIX. **2.** (ii) Non-existent.

Revision Exercise I. **2.** $\frac{5}{2}x^{\frac{3}{2}} - x^{-\frac{1}{2}}$. **4.** $6(3x - 2)$. **5.** $(3 + 8x - 3x^2)(x^2 + 1)^{-2}$.
6. $3 \cos x - 2 \sin 2x$. **7.** $-2(2x^2 - 6x + 3)(x - 2)^{-2}(x - 3)^{-2}$. **8.** $\cos^2 x \times$
$(4 \cos 4x \cos x - 3 \sin 4x \sin x)$. **9.** $4(1 - 16x^2)^{-\frac{1}{2}} - (2x^2 + 2x + 1)^{-1}$.
10. $-\frac{1}{2}(1 - x)^{-1}(-x)^{-\frac{1}{2}}$ [note $x < 0$]. **11.** $m \cos x(1 - m^2 \sin^2 x)^{-\frac{1}{2}}$. **17.** $2 \cos 2x +$
$6(3x - 1)$, $-\frac{1}{2} \cos 2x + \frac{1}{9}(3x - 1)^3 + c$. **19.** $\frac{3}{4}x^4 + \dfrac{1}{x}$. **20.** $\sin x + \frac{4}{3} \cos 3x$.

21. $-\frac{1}{48}(3 \cos 8x + 4 \cos 6x)$. **22.** $\dfrac{1}{\sqrt{7}} \tan^{-1}\left(\dfrac{x}{\sqrt{7}}\right)$. **23.** $\frac{1}{5} \sin^{-1}\left(\dfrac{5x}{2}\right)$. **24.** (i) 0,
(ii) $\dfrac{\pi}{2}$. **25.** 0 if $|p| \neq |q|$, $\dfrac{\pi}{2}$ if $|p| = |q|$. **26.** (i) $x(1 + x^2)^{-\frac{1}{2}}$, (ii) $(1 + x^2)^{\frac{1}{2}}$.
27. $(\sin x \cos h - \sin x)$ is evidently omitted on the grounds that it tends to 0 as
$h \to 0$, but then so does $\cos x \sin h$. The argument is worthless. **58.** $5x - y = 14$,
$x + 3y = 6$, $1\frac{3}{5}$. **59.** (i) $4x^3 - 12x$, $1 + 2x^{-3}$, (ii) $9x - y = 16$, $(-4, -52)$.
61. $(\frac{9}{16}, -\frac{5}{4})$. **62.** $4y + x = 9$, $y = 4x - 2$. **63.** $x \sin^3 t + y \cos^3 t = \sin^3 t \cos^3 t$,
$x \cos^3 t - y \sin^3 t = \cos 2t(\cos^4 t + \sin^4 t)$. **64.** $xt(t^3 - 2) - y(2t^3 - 1) + t^2 = 0$.
65. $axx_1 + h(xy_1 + yx_1) + byy_1 + g(x + x_1) + f(y + y_1) + c = 0$. **66.** 1. **67.** 3.
68. 0. **70.** m. **71.** (i) $\frac{1}{2}$, (ii) 1, (iii) 1, (iv) $\frac{5}{3}$. **72.** (i) yes, (ii) no. **73.** $g(x)$ is not even
continuous at $\dfrac{\pi}{2}$. **74.** (i) $x = \dfrac{(2n + 1)\pi}{2}$. **75.** (a) 0, (b) non-existent, (ii) this
argument falsely assumes the continuity of f' at $x = 0$. **76.** 32 m s^{-1}, 8 s, -8 m s^{-2},
$170\frac{2}{3}$ m. **77.** When $\upsilon = 2$, acceleration is zero. **78.** (i) 5 m s^{-2}. **79.** (i) -15 m s^{-1},
108 m s^{-2}, (ii) ± 13. **80.** (i) $-5 + 8x - 3x^2$, $6 + 4x^{-2}$, (ii) $-1 < x < 3$. **82.** (i)
$3x^2 + 4x + 3$, $2x - 18x^{-3}$, (ii) $(1, -1)$ min, $(-2, 26)$ max. **83.** (a) (i) $(1 -$
$2x)(x^2 - x + 1)^{-2}$, (ii) $x^2 \cos x$, (b) $(0, 1)$ min, $(-2, -3)$ max. **85.** 48 m^2. **86.** 5 cm,
6 cm. **87.** 2, $\frac{1}{2}$. **88.** (i) $\frac{1}{4}x^4 + 2x^{\frac{3}{2}} + x^{-1} + c$, (ii) 36. **89.** $42\frac{2}{3}$ sq. units, 128π cu.
units. **90.** $\dfrac{2\pi}{3}$. **91.** $y = x(x - 2)^2$, $\frac{4}{3}$, $\frac{128}{105}\pi$. **92.** (i) 8π, (ii) $1\frac{1}{8}$. **93.** 0 max,

2 min; -1. **94.** Area between CP, CQ and curve is 45. **95.** (2, 20) max, (4, 16) min;

36. 96. $4x - 3y = 1$, $-\frac{1}{9}$. **97.** 2, 1; $\frac{1}{3}$. **98.** ± 1, 'infinite'; $(\frac{1}{2}\sqrt{2}, \pm\frac{1}{2})$, $(-\frac{1}{2}\sqrt{2},$

$\pm\frac{1}{2})$, $\dfrac{4\pi}{15}$. **99.** 4·003. **100.** 0·058. **101.** $-0\cdot6$, $-1\cdot5$, $-2\cdot4$. **102.** 1·17. **107.** $\frac{1}{2}$.

Papers A to H. **A1.** (i) $\frac{1}{2}(3x^{-\frac{1}{2}} + 5x^{-\frac{3}{2}})$, $\frac{1}{3}x^3 - 3x + 2x^{-1} + c$, (ii) max 2, min -2.

A2. $6x - y = 20$, $x + 6y = 65$, (3, 2). **A3.** $\frac{2}{3}, \frac{2}{5}, \frac{1}{2}$. **A4.** $y = -4$, $-2, 0$; $\dfrac{dy}{dx} =$

0, -3, 0; $\dfrac{d^2y}{dx^2} = 6, 0, -6$, (i) min, (ii) point of inflexion, (iii) max. **A5.** (i) $y +$

$9x = 27$, $y + 9x + 5 = 0$, (ii) 2:1. **B1.** (i) 9, (ii) $\frac{1}{4}x^4 + x + c$, $1\frac{1}{4}$. **B2.** (i) $\frac{1}{3}x^{-\frac{1}{2}}$,

$\frac{1}{2}x^2 + \frac{4}{3}x^{\frac{3}{2}} + x + c$, (ii) 4 m, 20 m, $1\frac{2}{3}$ m s^{-1}. **B3.** $6\frac{2}{3}$, 10, 4 cm. **B4.** 40. **B5.** $\dfrac{\pi}{20}$,

$(\frac{2}{3}, 0)$. **C1.** (i) $1 + \frac{1}{2}x^{-\frac{1}{2}} - \frac{1}{2}x^{-\frac{3}{2}} - x^{-2}$, (ii) -22, -23. **C2.** 3 m s^{-1}, 1 m s^{-2}. **C3.** Min (3, 2·32), max (1, 2·39). **C4.** $4x + 20y = 145$, 5·5. **C5.** $\frac{1}{12}a^3(2b - a)$, 1:2. **D1.** $-3x^{-4}$, (i) $(1 - x)^{-2}$, (ii) $\frac{1}{2}x^{-\frac{1}{2}}(1 - x^{\frac{1}{2}})^{-2}$, (iii) 3 sin 6x, (iv) $2(1 + x^2)^{-1}$. **D2.** 12 m s^{-1}, 8 m s^{-2}, $3\frac{5}{9}$ m. **D3.** $2\pi h^2$ cm^3, 14 sec. **D4.** $y = 3x$, $y = 3x + 4$;

$2\frac{1}{4}$. **D5.** 100 m by $\dfrac{200}{\pi}$ ($\approx 63\cdot7$) m. **E1.** $(32 - 12x - x^3)x^{-3}(x + 4)^{-2}$,

$2(\tan^3 x - x \sec^2 x)$; 0·186. **E2.** 15π cm^3 s^{-1}, 0·104 cm s^{-1}. **E3.** 1·1. **E4.** $y = x$, $y = -x$; (2, 2); $1\frac{1}{3}$. **E5.** a. **F1.** $x(x - 2)(x - 1)^{-2}$, $4x - 2(2x^2 + 1)(1 + x^2)^{-\frac{1}{2}}$.

F2. 1·2. **F3.** 2:1, yes. **F4.** $\frac{1}{6}\pi(4 + 6\sqrt{2} + 5\sqrt{6})$. **F5.** (i) $t = \dfrac{2\pi}{3p}$, $x = -\dfrac{3a}{4}$,

(ii) $t = \dfrac{\pi}{p}$, $x = -\dfrac{a}{2}$, (iii) $t = \dfrac{4\pi}{3p}$, $x = -\dfrac{3a}{4}$. **G1.** $-(6x^2 - 4x + 3)(2x^2 - 1)^{-2}$,

$\sin^2 x(2 + \sec^2 x)$, $\frac{1}{2}a\{x(a - x)^3\}^{-\frac{1}{2}}$. **G2.** $y = 2 + 3x - x^3$, $\frac{14}{15}$. **G3.** $\frac{4}{3}$; (i) $\frac{1}{2}\pi$,

(ii) $\frac{8}{15}\pi$. **G4.** $-\dfrac{q^2}{p^2}\delta p$; $-1:4$. **G5.** $2a^2$, $\sqrt{3}a^2$. **H1.** (i) 3·5. **H2.** (i) $1 - x^{-2}$,

(ii) $-3x(a^2 - x^2)^{\frac{1}{2}}$, (iii) $(1 + x)^{-1}(2x + 1)^{-\frac{1}{2}}$. **H3.** $2a(1 + 4 \sin \theta - \sin^2 \theta - 4 \sin^3 \theta)$. **H4.** $t^2 \cos t - 4t \sin t - 6 \cos t$; $(2k + \frac{1}{2})\pi$ min, $(2k + \frac{3}{2})\pi$ max.

Exercise XXX. **1.** $\frac{1}{4}x^4 - 2x^3 + 2x^2 + 2x + c$. **2.** $\frac{1}{6}(x + 1)^5$. **3.** $\frac{1}{32}(4x + 3)^8$.
4. $\frac{1}{6}(1 - 2x)^{-3}$. **5.** $\frac{2}{9}(3x + 1)^{\frac{3}{2}}$. **6.** $-\frac{3}{35}(6 - 7x)^{\frac{5}{3}}$. **7.** $-\frac{1}{3} \cos 3x$. **8.** $\frac{1}{4} \sin (4x + 1)$.
9. $-\frac{1}{4} \cot 4x$. **10.** $\frac{1}{4} \tan 4x - x$. **11.** $\frac{1}{3} \sec 3x$. **12.** $\frac{1}{2}x + \frac{1}{4} \sin 2x$. **13.** $\frac{3}{8}x +$

$\frac{1}{4} \sin 2x + \frac{1}{32} \sin 4x$. **14.** $\sin^{-1} x$. **16.** $\frac{1}{2} \sin^{-1}\left(\dfrac{2x}{3}\right)$. **17.** $\dfrac{1}{\sqrt{2}} \sin^{-1} (\sqrt{\frac{2}{3}}x)$.

20. $\frac{1}{6} \tan^{-1} (\frac{2}{3}x)$. **21.** $\dfrac{1}{\sqrt{6}} \tan^{-1} (\sqrt{\frac{2}{3}}x)$. **24.** $\frac{1}{2} \sin x - \frac{1}{22} \sin 11x$. **26.** $\frac{1}{2}x -$

$\frac{1}{28} \sin 14x$. **27.** $-\frac{1}{6}(1 + x^2)^{-3}$. **30.** $\frac{2}{3}(1 + x^3)^{\frac{1}{2}}$. **31.** $-\frac{1}{6}(1 + 4x + x^3)^{-6}$.
33. $-(7 \sin x + 4 \cos x)^{-1}$. **34.** $\frac{1}{2}(\sin^{-1} x)^2$. **35.** $\sin^{-1} x - (1 - x^2)^{\frac{1}{2}}$. **36.** 0.

38. $\frac{1}{6} \sin^6 x$. **39.** $\frac{1}{1024} (24x - 8 \sin 4x + \sin 8x)$. **41.** $\dfrac{4}{3} - \dfrac{\pi}{4}$.

Exercise XXXI.   **3.** $\frac{1}{18}(x+4)^8(2x-1)$.   **4.** $\frac{1}{504}(3x+1)^7(21x-25)$.
5. $\frac{1}{54}(1-9x)(3x-1)^{-3}$.   **6.** $-\frac{1}{6}(3x^2+4)^{-1}$.   **7.** $\frac{2}{105}(x+1)^{\frac{3}{2}}(15x^2-12x+8)$.
8. $\frac{2}{5}(1+x)^{-\frac{1}{2}}(x^3-2x^2+8x+16)$.   **9.** $\frac{1}{10}\tan^{-1}\left(\dfrac{5x}{2}\right)$.   **10.** $\frac{1}{2}\sin^{-1}(2x)$.

12. $\frac{1}{3}(x^2+2)^{\frac{3}{2}}$.   **13.** $\frac{1}{15}(1+x^2)^{\frac{3}{2}}(3x^2-2)$.   **14.** $\frac{1}{8}(1+x^2)^4$.   **15.** $\frac{1}{2}\{\sin^{-1}x +$
$x\sqrt{(1-x^2)}\}$.   **16.** $\frac{1}{4}\sin^4 x$.   **17.** $\frac{1}{5}\sin^5 x - \frac{1}{7}\sin^7 x$.   **20.** $\sec^{-1}x$.   **21.** $\frac{1}{2}\tan^{-1}x +$
$\frac{1}{2}x(1+x^2)^{-1}$.   **22.** $\sqrt{2}\tan^{-1}\sqrt{\left(\dfrac{1+x}{2}\right)}$.   **23.** $\frac{1}{3}\tan^{-1}\left(\dfrac{x-2}{3}\right)$.   **24.** $\sin^{-1}\left(\dfrac{x+1}{2}\right)$.
25. $\frac{1}{2}\tan^{-1}(\frac{1}{2}\tan x)$.   **26.** $\sin^{-1}(x^{\frac{1}{2}})-x^{\frac{1}{2}}(1-x)^{\frac{1}{2}}$.   **27.** $-\cos^{-1}x-(1-x^2)^{\frac{1}{2}}$.
28. $\frac{1}{3}\tan^{-1}(x^3)$.   **29.** $\dfrac{1}{\sqrt{3}}\tan^{-1}\left(\dfrac{\sin x}{\sqrt{3}}\right)$.   **30.** $-\sin x - \operatorname{cosec} x$.   **31.** $\frac{1}{4}\tan^4 x$.
32. $-\cos x + \frac{2}{3}\cos^3 x - \frac{1}{5}\cos^5 x$.

Exercise XXXII.   **1.** $\frac{1}{156}(5\cdot8^{12}-6\cdot7^{12})$.   **2.** $1\frac{1}{3}$.   **3.** $\dfrac{\pi}{4}$.   **4.** $\frac{1}{18}\pi$.   **5.** $\frac{2}{3}$.   **6.** $\frac{2}{15}$.
7. $\dfrac{\pi}{4\sqrt{3}}$ ($\approx 0{\cdot}45$).

Exercise XXXIII.   **1.** $\sin x - x\cos x$.   **2.** $\frac{1}{2}(3x+1)\sin 2x + \frac{3}{4}\cos 2x$.
3. $(x^2-2)\sin x + 2x\cos x$.   **4.** $x\cos x(6-x^2)+3\sin x(x^2-2)$.   **5.** $x\cos^{-1}x -$
$(1-x^2)^{\frac{1}{2}}$.   **6.** $\frac{1}{8}\sin 2x - \frac{1}{4}x\cos 2x$.   **7.** $\frac{1}{2}x\cos x - \frac{1}{2}\sin x - \frac{1}{10}x\cos 5x +$
$\frac{1}{50}\sin 5x$.   **8.** $\frac{1}{8}(2x^2+2x\sin 2x + \cos 2x)$.   **9.** $\frac{1}{6}x^3 - \frac{1}{4}x^2\sin 2x - \frac{1}{4}x\cos 2x +$
$\frac{1}{8}\sin 2x$.   **10.** $\frac{1}{4}\{x\sqrt{(1-x^2)}+\sin^{-1}x(2x^2-1)\}$.   **11.** $\frac{1}{2}\{(x^2+1)\tan^{-1}x - x\}$.
12. $\frac{1}{4}(x^4-1)\tan^{-1}x - \frac{1}{12}x^3 + \frac{1}{4}x$.   **13.** $\dfrac{\pi}{2}-1$.   **14.** $\frac{1}{48}(3\sqrt{3}-\pi)$, $\approx 0{\cdot}043$.
15. $2 + \dfrac{\pi}{8}$.

Exercise XXXIV.   **1, 2, 5.** 0.   **7.** $\dfrac{\pi}{4}$.   **8.** $\dfrac{2}{a}(a>1)$, $2(-1\leqslant a\leqslant 1)$, $-\dfrac{2}{a}(a<-1)$.

Exercise XXXV.   **2.** $0{\cdot}005$.   **3.** (iii) $-\frac{1}{18}(1+9x^2)^{-1}$.   **4.** (iii) $\frac{1}{8}$.   **5.** $2x(x^4+2x^2+$
$3)^{-1}$.   **7.** (i) $\frac{2}{5}(x+1)^{\frac{5}{2}}-\frac{2}{3}(x+1)^{\frac{3}{2}}$,   (ii) $\dfrac{1}{\sqrt{5}}\tan^{-1}\left(\dfrac{x+2}{\sqrt{5}}\right)$,   (iii) $4\sin^{-1}x -$
$(1-x^2)^{\frac{1}{2}}$, (iv) $\frac{1}{32}(12x-8\sin 2x + \sin 4x)$.   **8.** $1, \dfrac{\pi}{4}$; $0{\cdot}8(9)$.   **9.** $\frac{3}{4}\pi^2-6$, $\approx 1{\cdot}40$.
10. $\frac{1}{6}(1-3x)^{-2}$, $\frac{1}{2}(\tan 2x - 2x)$, $(1+x^2)\tan^{-1}x - x$, $\tan^{-1}(\sin x)$.

Exercise XXXVII.   **1.** $1-3x+6x^2-10x^3$, $|x|<1$.   **2.** $1-\frac{3}{2}x+\frac{3}{8}x^2+\frac{1}{16}x^3$,
$|x|<1$.   **3.** $1-\frac{3}{2}x+\frac{27}{8}x^2-\frac{135}{16}x^3$, $|x|<\frac{1}{3}$.   **4.** $\frac{1}{16}-\frac{1}{8}x+\frac{5}{32}x^2-\frac{5}{32}x^3$,
$|x|<2$.   **5.** $1+2x+3x^2+4x^3$, nx^{n-1}, $|x|<1$.   **6.** $6x - \dfrac{(6x)^3}{3!}+\dfrac{(6x)^5}{5!}-\dfrac{(6x)^7}{7!}$,
$\dfrac{(-1)^n(6x)^{2n+1}}{(2n+1)!}$, all x.   **7.** $1-\dfrac{(5x)^2}{2!}+\dfrac{(5x)^4}{4!}-\dfrac{(5x)^6}{6!}$, $\dfrac{(-1)^n(5x)^{2n}}{(2n)!}$, all x.

8. $\frac{1}{2}\left\{\frac{(2x)^2}{2!} - \frac{(2x)^4}{4!} + \frac{(2x)^6}{6!} - \frac{(2x)^8}{8!}\right\}$, $\frac{1}{2}(-1)^{n+1}\frac{(2x)^{2n}}{(2n)!}$, all x. **9.** General term $\frac{1}{2}(-1)^n, \frac{(9^{2n+1} - 1)x^{2n+1}}{(2n + 1)!}$, all x. **10.** $|h| < |x|$. **11.** $1\cdot00332$, error less than 6×10^{-8}. **12.** $1\cdot02470$. **15.** $\frac{1}{x}\{1 - (1 - x)^{\frac{1}{2}}\}$. **16.** $1, \frac{n}{m}a^{n-m}$. **17.** $\frac{1}{2}$.

Exercise XXXIX. **1.** $3e^{3x}$. **2.** $-2e^{4-2x}$. **3.** $(2x + 1)e^{x^2+x}$. **4.** $-\sin xe^{\cos x}$. **5.** $xe^{5x}(2 + 5x)$. **6.** $e^x(\sin x + \cos x)$. **7.** $e^{-x}(1 - 3x)$. **8.** $\{e^x(1 - x) + 1\} \times (e^x + 1)^{-2}$. **9.** $\frac{1}{2}e^{2x}$. **10.** $2e^{x/2}$. **11.** $-\frac{1}{3}e^{4-3x}$. **12.** $e^x(x - 1)$. **13.** $\frac{1}{2}e^{x^2}$. **14.** $\frac{1}{2}e^x(\sin x - \cos x)$. **15.** $81e^{3x}, a^ne^{ax}$. **16.** $Ae^{3x} + Be^{2x}$. **17.** $y = 2e^x$. **18.** $y = Ae^{4x}$. **19.** $1 + kx + \frac{k^2x^2}{2!} + \frac{k^3x^3}{3!}, \frac{k^nx^n}{n!}$. **20.** (i) $1 + x + \frac{1}{2}x^2 - \frac{1}{8}x^4$, (ii) $x + x^2 + \frac{1}{3}x^3 - \frac{1}{30}x^5$. **21.** (i) $a(PN - b)$, (ii) $\frac{1}{2}\pi a(PN^2 - b^2)$.

Exercise XL. **1.** $2(2x + 1)^{-1}$. **2.** $(x - 3)^{-1}$. **3.** $\cot x$. **4.** $3x^2 \ln x + x^2$. **5.** $(1 - \ln x)x^{-2}$. **6.** $(\ln x - 1)(\ln x)^{-2}$. **7.** x^{-1}. **9.** $(-1)^{n-1}2^n(n - 1)!(1 + 2x)^{-n}$. **10.** $\frac{1}{2}x - \frac{1}{8}x^2 + \frac{1}{24}x^3 - \frac{1}{64}x^4$, $(-2 < x \leqslant 2)$. **11.** $\ln 3 - \frac{1}{3}x - \frac{1}{18}x^2 - \frac{1}{81}x^3$, $(-3 < x \leqslant 3)$. **12.** $\ln 2 + \frac{1}{2}x - \frac{1}{8}x^2 + \frac{1}{192}x^4$, $(x \leqslant 0)$.

Exercise XLI. **1.** $0\cdot4055$. **2.** (i) $(1 + 2x - x^2)(1 - x)^{-2}$, (ii) $(1 + \frac{1}{2}\ln x)x^{-\frac{1}{2}}$, (iii) $x \sec xe^{x \tan x}$. **4.** $2, 0; x = ey$. **6.** $p, p^2, p^3 + p, p^4 + 4p^2$. **7.** $x + x^2 + \frac{5}{6}x^3$. **8.** $x = n\pi + \tan^{-1} a$. **9.** $3\cdot02$. **10.** (i) $(x^2 + 2x - 3)(1 + x)^{-2}$, (ii) $2x\{x^2(1 + x^4)^{-1} + \tan^{-1}(x^2)\}$, (iii) $\sec^2 x(1 + \tan x)^{-1}$.

Exercise XLIII. **3.** $\frac{1}{2}\ln |x|$. **4.** $-\frac{1}{3}\ln |1 - 3x|$. **5.** $\ln |x| - \ln |x + 1|$. **6.** $\frac{1}{2}\ln\left|\frac{x - 1}{x + 1}\right|$. **7.** $\frac{1}{2}\ln |x(x - 2)| - \ln |x - 1|$. **8.** $\ln\left|\frac{x - 1}{x}\right| + x^{-1}$. **9.** $\frac{1}{2}\ln |x - 1| - \frac{1}{2}\tan^{-1} x - \frac{1}{4}\ln (x^2 + 1)$. **10.** $-\frac{1}{5}\ln |x| + \frac{3}{17}\ln |3x - 1| + \frac{2}{85}\ln |2x + 5|$. **11.** $\frac{1}{3}\ln |1 + 2x| - \frac{1}{3}\ln |1 - x|$. **12.** $x \ln |x| - x$. **13.** $\frac{1}{4}x^2(2 \ln |x| - 1)$. **14.** $e^x(x - 1)$. **15.** $x\{(\ln |x|)^2 - 2 \ln |x| + 2\}$. **16.** $\ln |x^2 - 1|$. **17.** $\frac{1}{3}\ln |x^3 + 1|$. **18.** $\ln |7x^3 + 5x^2 - 2x + 1|$. **19.** $\ln |\sin x|$. **20.** $\frac{1}{2}\ln |\sec 2x|$. **21.** $\ln (1 + e^x)$. **22.** $\ln (1 + \sin x)$. **23.** (i) $\sec x$, (ii) $\ln |\sec x + \tan x|$. **24.** $0\cdot347$. **26.** $x^4e^x\{(x + 5) \ln |x| + 1\}$.

Exercise XLIV. **1.** $\sinh (A - B) = \sinh A \cosh B - \cosh A \sinh B$. **2.** $\sinh 2\theta = 2 \sinh \theta \cosh \theta$. **3.** $\cosh (A + B) = \cosh A \cosh B + \sinh A \sinh B$. **4.** $\text{sech}^2 \varphi = 1 - \tanh^2 \varphi$. **6.** $\cosh X - \cosh Y = 2 \sinh \frac{X + Y}{2} \sinh \frac{X - Y}{2}$. **7.** $\sinh 2\theta = \frac{2 \tanh \theta}{1 - \tanh^2 \theta}$. **8.** $\cosh A = \dfrac{1 + \tanh^2\frac{A}{2}}{1 - \tanh^2\frac{A}{2}}$.

10. $1 + \frac{x^2}{2!} + \frac{x^4}{4!} + \ldots + \frac{x^{2n}}{(2n)!} + \ldots; 3x + \frac{(3x)^3}{3!} + \ldots + \frac{(3x)^{2n+1}}{(2n + 1)!} + \ldots$

12. $x = \ln 2$ or $\ln 3$.

The modulus sign is often suppressed from here onwards. When expressions such as
$\ln\left(\dfrac{x-1}{x+3}\right)$ *arise from indefinite integrations, it is assumed that x lies within appropriate limits.*

Exercise XLV. 2. (i) $3\cosh 3x$, (iii) $-\tanh x \operatorname{sech} x$, (iv) $2(1+4x^2)^{-\frac{1}{2}}$,

(v) $2^{\frac{1}{2}}(2x^2+3x+1)^{-\frac{1}{2}}$. 4. $\cosh^{-1} x$. 5. $\dfrac{1}{\sqrt{3}}\cosh^{-1}(\sqrt{\frac{3}{2}}x)$. 6. $\dfrac{1}{\sqrt{3}}\sinh^{-1}(\sqrt{3}x)$.

8. $\frac{1}{4}\sinh^{-1} 2x + \frac{1}{2}x(1+4x^2)^{\frac{1}{2}}$. 9. $\sinh^{-1}(x-1)$. 10. $\sinh^{-1}(x+2)$.

11. $\cosh^{-1}\left(\dfrac{x-3}{\sqrt{2}}\right)$. 12. $x\sinh^{-1} x - (1+x^2)^{\frac{1}{2}}$.

Exercise XLVI. 3. $m = m_0 e^{-at}$, $p = p_0\left(\dfrac{m_0}{m}\right)^b$, $p = p_0 e^{abt}$. 4. (i) $2x - \frac{1}{3}x^3 +$

$\frac{2}{5}x^5 - \frac{1}{2}x^6$, $|x| < 1$. 7. 0.712. 8. $y = pex$. 9. $\dfrac{dx}{dt} = (n-x)U$,

$x = n - (n-m)e^{-Ut}$. 10. $\frac{5}{4}, \frac{25}{24}$. 11. $\left(2n\pi + \dfrac{\pi}{4}, 2^{-\frac{1}{2}}e^{-2n\pi-(\pi/4)}\right)$,

$\left(2n\pi + \dfrac{5\pi}{4}, -2^{-\frac{1}{2}}e^{-2n\pi-(5\pi/4)}\right)$, $\left(n\pi + \dfrac{\pi}{2}, (-1)^n e^{-(n+\frac{1}{2})\pi}\right)$. 12. 2010, 737 per day.

Exercise XLVII. 1. $-\frac{1}{18}(3x+4)^{-6}$. 2. $\frac{1}{375}(25x-13)(5x-2)^5$. 3. $x\sin^{-1} x +$

$(1-x^2)^{\frac{1}{2}}$. 4. $\sin x - x\cos x$. 5. $\frac{1}{2}\ln\left(\dfrac{x-1}{x+1}\right)$. 6. $\ln(x^2-1)$. 7. $\frac{1}{2}e^{x^2}$.

8. $2(x+5)^{\frac{1}{2}}$. 9. $\frac{1}{4}\ln(4x+1)$. 10. $14(x+3)^{\frac{1}{2}} + 38(x+3)^{-\frac{1}{2}}$. 11. $e^x(x^2 +$

$2x-1)$. 12. $x - \ln(x+1)$. 13. $\ln\left(\dfrac{2-x}{1-x}\right)$. 14. $\frac{2}{3}(x-2)(x+1)^{\frac{1}{2}}$.

15. $\dfrac{2}{\sqrt{3}}\tan^{-1}\left(\dfrac{2x+1}{\sqrt{3}}\right)$. 16. $\ln\left(\dfrac{x+2}{x+3}\right)$. 17. $\frac{1}{2}\ln(x^2+2x+7)$.

18. $\frac{3}{2}\ln(x^2+3x+4) - \sqrt{7}\tan^{-1}\left(\dfrac{2x+3}{\sqrt{7}}\right)$. 19. $\frac{1}{2}x^2 + \frac{1}{2}\ln(x^2-1)$.

20. $\frac{1}{2}x^2 - \frac{1}{2}\ln(x^2+1)$. 21. $\frac{1}{2}\ln(x-1) - \frac{1}{4}\ln(x^2+1) - \frac{1}{2}\tan^{-1} x$.

22. $\frac{2}{5}\cdot 7^{\frac{5}{2}}x^{\frac{5}{2}}$. 23. $\frac{1}{4}\ln\left(\dfrac{1+x}{1-x}\right) + \frac{1}{2}\tan^{-1} x$. 24. $-\dfrac{1}{3}\ln(1-x) + \dfrac{1}{6}\ln(1+$

$x+x^2) - \dfrac{1}{\sqrt{3}}\tan^{-1}\left(\dfrac{2x+1}{\sqrt{3}}\right)$. 25. $2\sqrt{(x-1)} - 2\tan^{-1}\{\sqrt{(x-1)}\}$.

Exercise XLVIII. 1. $(x^2-1)^{\frac{1}{2}}$. 2. $\dfrac{1}{\sqrt{2}}\sin^{-1}(\sqrt{2}x)$. 3. $\sin^{-1} x - (1-x^2)^{\frac{1}{2}}$.

4. $\cosh^{-1}\left(\dfrac{x+2}{\sqrt{3}}\right)$. 5. $\sin^{-1}\left(\dfrac{x-2}{\sqrt{5}}\right)$. 6. $\dfrac{1}{\sqrt{7}}\sin^{-1}\left(\dfrac{14x+3}{\sqrt{65}}\right)$. 7. $\sec^{-1} x$.

8. $\frac{1}{2}(x^4 + 1)^{\frac{1}{2}}$.　**9.** $x \ln x - x$.　**10.** $\frac{2}{9}(3x + 4)^{\frac{3}{2}}$.　**11.** $\frac{1}{3}(x^2 + 1)^{\frac{3}{2}}$.
12. $\frac{2}{105}(x + 1)^{\frac{3}{2}}(15x^2 - 12x + 8)$.　**13.** $\frac{1}{9}(5 + 4x - x^2)^{\frac{3}{2}}(2 - x)^{-1}$.
14. $-\frac{1}{4} \ln (1 - x^4)$.　**15.** $-\frac{1}{5}(3x^2 + 4x - 1)^{-\frac{1}{2}}$.

Exercise XLIX.　**1.** $\ln \sin x$.　**2.** $\tan x$.　**3.** $-\cot x$.　**4.** $-\cot x - x$.
5. $-\frac{1}{16} \cos 8x + \frac{1}{4} \cos 2x$.　**6.** $\frac{1}{2}x - \frac{1}{28} \sin 14x$.　**7.** $\frac{1}{2} \sin x - \frac{1}{14} \sin 7x$.
8. $-\frac{1}{6} \cos^6 x$.　**9.** $-\frac{1}{9} \cos^9 x + \frac{1}{11} \cos^{11} x$.　**10.** $\frac{1}{8}x - \frac{1}{32} \sin 4x$.　**11.** $-\tan^{-1}(\cos x)$.
12. x.　**13.** $\dfrac{1}{\sqrt{2}} \tan^{-1}\left(\dfrac{\tan x}{\sqrt{2}}\right)$.　**14.** Same as 13.　**15.** $\ln \tan \dfrac{x}{2}$.

16. $\frac{1}{13} \ln \tan \left(\dfrac{x+\alpha}{2}\right)$, where $\alpha = \cos^{-1} \frac{5}{13}$.

17. $\dfrac{1}{\sqrt{2}} \ln \tan \left(\dfrac{x}{2} + \dfrac{\pi}{8}\right)$.　**18.** $\dfrac{1}{\sqrt{7}} \ln \left(\dfrac{3t + 4 - \sqrt{7}}{3t + 4 + \sqrt{7}}\right)$, where $t = \tan \dfrac{x}{2}$.
19. $\frac{2}{13}x - \frac{3}{13} \ln (2 \sin x + 3 \cos x)$.　**20.** $\frac{1}{2} \sin^{-1} x - \frac{1}{2}x(1 - x^2)^{\frac{1}{2}}$.
21. $-\frac{1}{3} \operatorname{cosec}^3 x$.　**22.** $-\cos^{-1} x - (1 - x^2)^{\frac{1}{2}}$.　**23.** $x(1 + x^2)^{-\frac{1}{2}}$.　**24.** $\frac{1}{3} \sec^3 x$.
25. $\frac{1}{24} \sec^4 x(3 \sec^4 x - 8 \sec^2 x + 6)$.　**26.** $\frac{1}{2} \sec x \tan x + \frac{1}{2} \ln (\sec x + \tan x)$.

Exercise L.　**1.** $-\frac{1}{8} \cos 4x - \frac{1}{4} \cos 2x$.　**3.** $\sin x - x \cos x$.
5. $(x + 3) \ln (x + 3) - x$.　**7.** $\frac{2}{15}(x + 1)^{\frac{1}{2}}(3x^2 - 4x + 8)$.
9. $-\cos x + \cos^2 x - \frac{1}{3} \cos^3 x$.
11. $\dfrac{2}{\sqrt{19}} \tan^{-1}\left(\dfrac{2x + 3}{\sqrt{19}}\right)$.　**13.** $\frac{1}{4} \ln (x - 1) + \frac{3}{4} \ln (x + 3)$.

15. $\dfrac{1}{\sqrt{2}} \tan^{-1}\left(\dfrac{3 \tan \dfrac{x}{2} - 1}{2\sqrt{2}}\right)$.　**17.** $x - 2 \tan^{-1} x$.　**19.** $\frac{1}{13}e^{2x}(3 \sin 3x + 2 \cos 3x.)$
21. $x \tan^{-1} 2x - \frac{1}{4} \ln (1 + 4x^2)$.　**23.** $\frac{1}{3}(1 + x^2)^{\frac{3}{2}}$.　**25.** $\frac{1}{4}x^2 + \frac{1}{4}x \sin 2x + \frac{1}{8} \cos 2x$.
27. $\frac{1}{3} \sin^{-1}\left(\dfrac{3x}{2}\right)$.　**29.** $x \cos^{-1} x - (1 - x^2)^{\frac{1}{2}}$.　**31.** $\frac{1}{60}(x - 1)(12x^4 + 27x^3 +$
$47x^2 + 77x + 137) + \ln (x - 1)$.　**33.** $\sin^{-1}\left(\dfrac{2x + 5}{\sqrt{41}}\right)$.　**35.** $\cosh^{-1}\left(\dfrac{2x - 1}{3}\right)$.
37. $\frac{1}{3} \tan x(\sec^2 x + 2)$.　**39.** $\frac{1}{32}(12x - 8 \sin 2x + \sin 4x)$.
41. $\sin^{-1}(x^{\frac{1}{2}}) - x^{\frac{1}{2}}(1 - x)^{\frac{1}{2}}$.　**43.** $\frac{1}{2}x^2 \sin^{-1} x - \frac{1}{4} \sin^{-1} x + \frac{1}{4}x(1 - x^2)^{\frac{1}{2}}$.
45. $\frac{1}{18} \sin^{18} x$.　**47.** $\frac{1}{2} \sec^{-1} x + \frac{1}{2}x^{-2}(x^2 - 1)^{\frac{1}{2}}$.　**49.** $\ln \left(1 + \tan \dfrac{x}{2}\right)$.

50. $\dfrac{\sqrt{2}}{8}\left\{2 \tan^{-1}\left(\dfrac{x^2 - 1}{\sqrt{2}x}\right) - \ln \left(\dfrac{x^2 - \sqrt{2}x + 1}{x^2 + \sqrt{2}x + 1}\right)\right\}$.　**51.** 1.　**52.** $\frac{1}{5}$.　**53.** $\frac{1}{2} \ln \frac{6}{5}$,
≈ 0.091.　**54.** $\ln \frac{6}{5}$.　**55.** $\ln \frac{3}{2}$, ≈ 0.406.　**56.** 1.　**57.** 1.　**58.** $\frac{1}{12}\pi + \frac{1}{2}\sqrt{3} - 1$,
≈ 0.13.　**59.** 1.444.　**60.** 0.423.

Exercise LI.　**1.** $\frac{1}{2}\pi$, $1 - \frac{1}{4}\pi$, $3(e - e^{-1})$.　**2.** $\ln (2x^2 - x - 3)$, $\frac{1}{3}s^3 - \frac{1}{5}s^5$, $\pi - 2$.
3. $\frac{1}{4} \sin^4 x - \frac{1}{6} \sin^6 x$, $\ln \left(\dfrac{2x + 3}{x + 2}\right)$, $1 - \frac{1}{4}\pi$.

4. $\frac{1}{2}(1 - x^2)^{-1}$, $\frac{1}{2}\tan^2 x$, $2\ln(1 + x) - \frac{1}{2}\ln(1 - 2x)$, $0\cdot264$.   **5.** (i) $\frac{1}{4}\ln\left(\dfrac{x-1}{x+3}\right)$, $\ln(x^2 + 2x + 2) - \tan^{-1}(x + 1)$, $\tan^{-1}(\sin x)$, (ii) $\frac{1}{2}\tan^{-1} 2$, $\frac{1}{2}\tan^{-1}\frac{1}{2}$.   **6.** $-x + 2\ln(1 + x)$, $\frac{2}{15}(3x + 2)(x - 1)^{\frac{3}{2}}$.   **7.** $\frac{1}{2}\ln(x - 1) - \frac{1}{4}\ln(x^2 + 1) - \frac{1}{2}\tan^{-1} x$, $\sin x - \frac{1}{3}\sin^3 x$, $(x - 1)e^x$.   **8.** (i) $\frac{1}{3}x^3 + 2x - x^{-1}$, (ii) $s - \frac{2}{3}s^3$, (iii) $3\ln(x + 3) - \ln(x + 1)$, $\frac{1}{4}\pi - \frac{1}{2}\ln 2$.   **9.** $\ln\left(\dfrac{x}{x+1}\right) + (x + 1)^{-1} + \frac{1}{2}(x + 1)^{-2}$, $(2x^2 - 1)\sin^{-1} x + x(1 - x^2)^{\frac{1}{2}}$.   **10.** $\frac{1}{3}, \frac{1}{2}, 6\frac{2}{3}$.   **11.** (i) $-\frac{1}{5}\cos\left(\dfrac{5x}{2}\right) - \frac{1}{3}\cos\left(\dfrac{3x}{2}\right)$,

(ii) (a) $0\cdot347$, (b) $0\cdot215$.   **12.** $x - (x - 1)^{-1} + 2\ln(x - 1)$, $x\ln(2x) - x$, $\tan^{-1}(\sin x)$; $\frac{5}{2}\ln 3 - \ln 2$.   **13.** (i) $\frac{13}{5}\ln(x - 1) - \frac{1}{10}\ln(2x + 3)$, (ii) $\frac{1}{2}x - \frac{1}{4}\sin 2x$, $\frac{1}{2}\sqrt{19} - 2$.   **14.** (i) $\frac{1}{2}\tan^{-1}\left(\dfrac{x + 2}{2}\right)$, (ii) $\frac{1}{2}\tan^2 x$, (iii) $\frac{2}{3}x\,(\ln x - \frac{2}{3})$.

15. $(2x + 1)^{\frac{1}{2}}$, $x\sin x + \cos x$, $\frac{2}{3}\ln\frac{4}{3}$.   **16.** (i) $s - \frac{2}{3}s^3 = \frac{1}{6}\sin 3x + \frac{1}{2}\sin x$, (ii) $\ln x - \tan^{-1}(\frac{1}{2}x)$, 4.   **17.** (i) $\frac{1}{3}\sin^{-1} 3x$, (ii) $\frac{1}{2}e^{x^2}$, $\frac{1}{5}\ln 6$.   **18.** (i) $\frac{1}{3}(x - 1) \times (2x + 1)^{\frac{1}{2}}$, $\frac{1}{4}(3x + \frac{1}{4}\sin 4x)$, (ii) $\frac{1}{4}(e^2 + 1)$, $\approx 2\cdot10$.   **19.** (ii) $-\frac{1}{2}$, $\frac{1}{16}\pi^2 - \frac{1}{4}$,

(iii) $\frac{1}{2}\ln(x^2 + x + 1) - \dfrac{1}{\sqrt{3}}\tan^{-1}\left(\dfrac{2x + 1}{\sqrt{3}}\right)$.   **20.** $x - 6x^{-1} - 3x^{-3}$, $\frac{2}{15}$, $\frac{1}{2}\ln 2$.

21. $\ln(x - 1) - 2(x - 1)^{-1}$, $\frac{1}{2}x^2\ln x - \frac{1}{4}x^2$, $\frac{1}{3}\sec^3 x$.

22. $(x + 1)^{-1} + \ln\left(\dfrac{x}{x + 1}\right)$, $\ln(x^2 + 2x + 2) + \tan^{-1}(x + 1)$.   **23.** (i) $-s - s^{-1}$, (ii) $\frac{1}{2}e^x(\sin x - \cos x)$, (iii) $2\cdot24$.   **24.** $\frac{1}{6}\ln\left(\dfrac{1 + x^3}{1 - x^3}\right)$, $\sin^{-1}(x - 1)$, $x\tan x + \ln\cos x$, $\frac{1}{3}\tan^3 x - \tan x + x$.   **25.** (i) $\sin^{-1}\left(\dfrac{x + 2}{3}\right)$, (ii) $-\frac{1}{2}(x^2 + 1)e^{-x^2}$, $x + A$, $\ln(a\sin x + b\cos x) + B$.

Exercise LII.   **1.** $(0, 5)$, $(2, 9)$; $34\frac{2}{3}$; $\dfrac{3592\pi}{15}$.   **2.** $\frac{8}{5}a^2$, $(\frac{25}{28}a, \frac{25}{14}a)$.   **3.** $y = \pm x$,

$\frac{1}{2}\pi - 1$, $\frac{5}{3}\pi - \frac{1}{3}\pi^2$.   **4.** 2π.   **5.** $3\frac{1}{5}$, (i) $\dfrac{64\pi}{3}$, (ii) $\dfrac{256\pi}{21}$, $(\frac{40}{21}, \frac{10}{3})$.   **6.** $\frac{1}{2}\pi - 1$; $\dfrac{\pi^2 - 8}{2\pi - 4}$, $\dfrac{\pi^3 - 6\pi}{48\pi - 96}$.   **7.** $a^2(1 - \frac{1}{2}\sqrt{3})$.   **9.** Fraction is approx. $0\cdot87$.

13. $\bar{x} = \dfrac{\pi\sqrt{2} - 4}{4(\sqrt{2} - 1)}$, $\bar{y} = \dfrac{1}{4(\sqrt{2} - 1)}$.   **14.** $x - 2y = 0$.

Exercise LIII.   **1.** $2x - 3ty + t^3 = 0$, $2tx + (1 - t^2)y - (1 + t^2) = 0$.   **2.** $bx = \pm ay$.   **4.** $x = 2$, $x + y + 1 = 0$, $y = x + 1$.   **5.** (ii) $3tx + (t^3 - 2)y = t^3$, $3x + 3y = 1$, (iii) $(0, 0)$, $(0, 1)$; $(\frac{2}{9}, \frac{1}{9})$.   **6.** $x - 3t_1^2 y + t_1^2(2t_1 + 3) = 0$.   **8.** $x \pm y = 0$, $x \pm 2y = 0$.   **11.** (i) $x = 0$, $y = 0$, (ii) $x \pm y = 0$, $x \pm 2 = 0$, (iii) $x \pm 2y = 0$.

Exercise LIV.   **3.** $\frac{1}{27}(85^{\frac{3}{2}} - 40^{\frac{3}{2}}) \approx 19\cdot7$.   **4.** (i) $\frac{1}{2}\pi a$, (ii) $6a$.   **6.** $\frac{1}{6} + \ln\frac{5}{3}$.

7. $\dfrac{3\sqrt{2}}{8}(X^{\frac{4}{3}} + X^{\frac{2}{3}})$.   **8.** (i) $\pm\frac{1}{2}$, (ii) $\frac{64}{3}a^2$, (iii) $2\{2\sqrt{5} + \ln(2 + \sqrt{5})\}a = 2\{2\sqrt{5} + \sinh^{-1} 2\}a$.

Exercise LV. **1.** $\dfrac{4a}{3\pi}$ from bounding diameter, on central radius. **2.** $2\pi^2 a^2 b$.

4. $a\{\sqrt{2} + \ln(1 + \sqrt{2})\}$. **6.** $\left(\dfrac{\pi}{2}, \dfrac{3}{8}\right), \dfrac{\pi^3}{2}$. **7.** $\sqrt{2}(e^t - 1)$.

Exercise LVI. **2.** $\dfrac{8\sqrt{10}}{5} \approx 5 \cdot 06$. **3.** $\tfrac{19}{15}M$.

Exercise LVII. **2.** $\dfrac{\sqrt{5}}{2}a$. **3.** $\tfrac{3}{20}M(r^2 + 4h^2)$. **4.** $\dfrac{25\sqrt{7}}{7} \approx 9 \cdot 4(5)$.

7. $\tfrac{1}{6}M\left(\dfrac{a^2 b^2}{a^2 + b^2}\right)$. **9.** (i) $\pi a^2 \rho_0 (\pi - 2)$, (ii) $\pi a^4 \rho_0 \left(\dfrac{\pi^3}{4} - 6\pi + 12\right)$.

Revision Exercise II. **1.** $-4x(1 + x^2)^{-2}$, $2x \ln x + x$, $6 \sin^2 3x + 3 \tan^2 3x$.

3. (i) $(x^2 + 2x - 1)x^{-2}(1 - x)^{-2}$, (ii) $(1 + 2x^2)e^{x^2}$, (iii) $2 \sec^2 x(1 - \tan^2 x)^{-1}$.

5. (i) $(x^2 + 1)^{-\frac{3}{2}}$, $\sec x$.

7. $2(3x - 4)(3x + 1)^{-3}$, $2 \tan x \sec^2 x \exp(\tan^2 x)$, $-(1 + x^2)^{-1}$.

9. (i) $(2x^2 - 3)x^{-4}(1 - x^2)^{-\frac{1}{2}}$, (ii) $(1 + \cos x)^{-1}$, (iii) $e^{-2x^2}(x^{-1} - 4x \ln 3x)$,

$-\tfrac{5}{6}t^{\frac{1}{6}}(1 + t^2)^{-1}$. **10.** $\tfrac{1}{5}\ln\left(\dfrac{x + 3}{x - 2}\right)$, $\tfrac{1}{2}x^2 \ln x - \tfrac{1}{4}x^2$, $\tfrac{2}{3}$. **12.** $\tfrac{1}{4}\ln\left(\dfrac{x^2 - 1}{x^2 + 1}\right)$,

$x - \ln(x^2 + x + 1) + \dfrac{2}{\sqrt{3}}\tan^{-1}\left(\dfrac{2x + 1}{\sqrt{3}}\right)$, $\tfrac{1}{3}\sec^3 x$, $x \sin^{-1} x + \sqrt{(1 - x^2)}$.

13. (*last part*). $\pi - 2$. **15.** $-(1 - x^2)^{\frac{1}{2}} + 2 \sin^{-1} x$, $x \ln x - x$, $\tfrac{1}{2}\pi$, $\tfrac{3}{2}\pi$.

18. (*last part*). $1 \cdot 484$. **20.** $\tfrac{1}{2}x^2 - x + 2 \ln(x + 1)$, $-x^{-1}(1 + \ln x)$, $\tfrac{1}{4}\pi$.

21. (i) $\ln 2 - \tfrac{1}{3}\ln 5 \approx 0 \cdot 157$, $\tfrac{1}{4}\pi$, (ii) $-x(x^2 - 1)^{-\frac{1}{2}}$. **22.** (i) $1 \cdot 59$, (ii) $3 \cdot 21$.

24. $2y = 2x^3 + 3x^2 - 36x + 6$. **26.** (ii) Max $\dfrac{1}{\sqrt{3}}$, min $-\dfrac{1}{\sqrt{3}}$.

27. (b) $2\pi a^2 \sin 2\theta \cos \theta$, $\sin^{-1}\left(\dfrac{1}{\sqrt{3}}\right) \approx 35° 15'$. **30.** $-\tfrac{16}{21}(x + 5)^{-2} +$

$\tfrac{4}{21}(4x - 1)^{-2}$, $\tfrac{32}{21}\{(x + 5)^{-3} - (4x - 1)^{-3}\}$, max $(1, \tfrac{1}{9})$, min $(-\tfrac{1}{3}, \tfrac{9}{49})$, inflexion

$(2, \tfrac{5}{49})$. **32.** $6\sqrt{3}$. **33.** 8. **36.** (i) (a) $(1 - x^2)^{-\frac{3}{2}}$, (b) $4 \operatorname{cosec} 4x$, (c) $5 \ln 3 \cdot 3^{5x}$,

(ii) $0 \cdot 99942$. **38.** $2 \cdot 99$. **40.** $y + 2 \cos^3 \theta \sin \theta x = a \cos^2 \theta (1 + 2 \sin^2 \theta)$. **41.** $\tfrac{2}{3}a^2, \tfrac{1}{4}\pi^2 a^3$.

42. (ii) 1. **45.** (i) $\dfrac{2}{\sqrt{e}}$, 1, (ii) 250π. **46.** At points of inflexion, $x = \pm a\left(\dfrac{4\pi}{3} + \dfrac{\sqrt{3}}{2}\right)$,

$y = -\tfrac{1}{2}a$; area $= a^2\left(4 + \dfrac{\pi}{2}\right)$. **47.** $B\left(\dfrac{2}{3}a, \dfrac{2\sqrt{3}}{9}a\right)$. **50.** $5\tfrac{1}{3}$, $(4\tfrac{1}{5}, -\tfrac{2}{5})$.

53. (i) $\left(\dfrac{4a}{3\pi}, \dfrac{4b}{3\pi}\right)$, (ii) $\tfrac{2}{3}\pi a^2 b$. **54.** $\dfrac{3a}{2}$, 2. **55.** $(1 + x)\{1 + 2 \ln(1 + x)\}$, $3 +$

$2 \ln(1 + x)$. **57.** (i) $(1 + 2x - x^2)(1 - x)^{-2}$, (ii) $(1 + x^2)^{-\frac{1}{2}}$, (iii) $2(1 - x^2)^{\frac{1}{2}}$,

$y = 1 + 4x + \tfrac{7}{2}x^2 - \tfrac{22}{3}x^3$. **59.** $x + x^2 + \tfrac{1}{3}x^3$, $\dfrac{2^{n/2}}{n!}\sin \tfrac{1}{4}n\pi$. **61.** $1 + x^2 + \tfrac{1}{6}x^4$.

63. $0 \cdot 10$. **64.** (ii) (a) $e^{2x} \cdot 2^{n-3}\{8x^3 + 12(n - 2)x^2 + 6n(n - 5)x + n^3 - 9n^2 +$

$8n + 8\}$, (b) $\tfrac{1}{3}(-1)^n n! \{(x - 1)^{-n-1} - (x + 2)^{-n-1}\}$. **66.** (ii) 15. **68.** (i) $\ln \tfrac{4}{3} -$

$\tfrac{14}{65} \approx 0 \cdot 072$. **69.** (i) $2bh$, (ii) $\dfrac{ah^2}{3b}$, (iii) $\dfrac{h}{\sqrt{3}}$. **70.** $\tfrac{3}{4}\sqrt{(ah)}$, $\tfrac{2}{3}\sqrt{(3ah)}$. **71.** $\tfrac{1}{6}Ma^2$.

72. Maximum.

Papers I to P. **I1.** (i) $3(1 + x)^2 \tan 3x + 3(1 + x)^3 \sec^2 3x$, (ii) $-16c(3 + 5s)^{-2}$, $2\frac{1}{4}$. **I2.** (i) $\frac{1}{5}(1 + x^2)^{\frac{5}{2}}$, (ii) $\frac{1}{2}\ln(1 + 2x) - \ln(1 - 2x)$, (iii) $\frac{1}{9}x^3(3\ln x - 1)$, $\dfrac{9\pi^2 - 4}{144}$. **I3.** $\frac{32}{81}\pi R^3$. **I4.** (i) $\frac{1}{3}\ln(x^3 + 4)$, (ii) $9\ln(x - 3) - 3\ln(x - 1)$,

(iii) $\frac{1}{4}(\sin 2x - 2x\cos 2x)$, $\frac{1}{4}(\pi + 2)$. **I5.** (ii) 20, 20, 5 m. **J1.** $\frac{1}{12}(16\sqrt{2} - 17)$. **J2.** (i) $2\ln(x - 2) - \ln(x - 1)$, (ii) $\frac{1}{3}x^3(\ln x - \frac{1}{3})$, (iii) $\frac{1}{2}\ln(x^2 + 4x + 10)$,

$\frac{1}{13}(3 - 2e^\pi)$. **J4.** $\frac{1}{4}(e^{2b} - e^{-2b})$, $\dfrac{\pi}{64}(e^{4b} - 8e^{2b} + 24b + 8e^{-2b} - e^{-4b})$.

K1. (i) $(1 - x)^4(1 - 4x - 7x^2)$, $4x(x^4 - 1)^{-1}$, $(\sin x + \cos x)^{-2}$,

(ii) $125e^{3x}\sin(4x + 3\beta)$, where $\beta = \tan^{-1}\frac{4}{3}$. **K2.** $\dfrac{3\cot B}{5\pi}$, $-\dfrac{3}{5\pi}(\cot B + \cot C)$.

K3. Max 0, min 1, (i) $\frac{1}{2}d^2$, (ii) $\dfrac{\sqrt{3}}{4}d^2$. **K4.** (i) $\frac{1}{2}\ln\left(1 - \dfrac{2}{x}\right)$, (ii) $(x^2 - 2)\sin x +$

$2x\cos x$, (iii) $\frac{1}{2}\tan^2 x - \ln\sec x$, $\dfrac{\pi\sqrt{3}}{9}$. **K5.** $\bar{x} = \dfrac{4a}{3\pi}$, $\bar{y} = 0$.

L1. (ii) $2\sqrt{(1 + \tan x)}$, $\frac{1}{2}(e^{2x} - e^{-2x}) + 2x$. **L3.** $0{\cdot}7854, 3{\cdot}1416$. **L5.** $\dot{x} = -2a \times$ $(\sin t + \sin 2t)$, $\dot{y} = 2a(\cos t + \cos 2t)$, $\ddot{x} = -2a(\cos t + 2\cos 2t)$, $\ddot{y} = -2a \times$ $(\sin t + 2\sin 2t)$, $4a\,|\cos\frac{1}{2}t|$. **M1.** $A = \frac{1}{2}$, $B = \frac{1}{2}$. **M2.** $\frac{1}{4} + \dfrac{\sqrt{3}}{2}x + \frac{1}{2}x^2 - \dfrac{\sqrt{3}}{3}x^3$,

$0{\cdot}2730$. **M3.** 5:1. **M4.** (i) $\frac{1}{5}x^5 - 2x^2 - 4x^{-1}$, (ii) $\frac{1}{4}\cos 2x - \frac{1}{16}\cos 8x$,

(iii) $2\ln x - \ln(1 + x^2)$, $\frac{1}{3}$. **M5.** (i) $\frac{8}{5}$, $\frac{12}{7}$, (ii) $\dfrac{2}{\sqrt{15}}$, $\dfrac{48\pi}{5}$. **N1.** (i) (a) $(1 + x^2)^{-\frac{1}{2}}$,

(b) $-2xe^{-x^2}$, (ii) $80\ \mathrm{cm^2/s^{-1}}$. **N2.** (i) $4{\cdot}33$. **N4.** $\dfrac{3\pi}{8}$. **O1.** $2a\,|\sin\frac{1}{2}t|$, $\sqrt{(2ay)}$.

O2. (i) $x + \ln(x^2 + 4) - \frac{3}{2}\tan^{-1}\dfrac{x}{2}$, $\frac{1}{3}\sec^3\theta$, $\frac{1}{9}x^3(3\ln x - 1)$, (ii) $\frac{2}{3}\ln 2$, $\frac{5}{12}\sqrt{2}$. **O3.** $\dfrac{(15 + \sqrt{3})\pi^2}{324\ln 2}$ **O4.** $4\pi a^2\sin\theta\cos^2\theta$. **O5.** $\frac{3}{4}l$, $\dfrac{2a + bl^2}{3a + bl^2}$, $k^2 = \frac{1}{5}l^2$, $\dfrac{5a + 3bl^2}{3a + bl^2}$.

P1. $-3x^{-4}$, $-\frac{3}{2}x^{-5}$, (i) $\tan x$, (ii) $(1 - 2x^2)(1 - x^2)^{-\frac{1}{2}}$. **P2.** $\dfrac{5a}{8}$. **P3.** 2, 0, $8y - 4 = 11(x - \frac{1}{4}\pi)$. **P4.** 1, $2a$, $3a^2 - 1$, $4a^3 - 4a$, $5a^4 - 10a^2 + 1$.

Exercise LVIII. 1. (i) $\dfrac{(9x^4 + 6x^2 + 2)^{\frac{3}{2}}}{6x}$, (ii) $-(1 + \cos^2 x)^{\frac{3}{2}}\csc x$, (iii) $-\csc x$.

2. (i) 2, (ii) 1. 3. $-3a\,|\sin t\cos t|$. 4. $-2a(\sin t + \sin 2t)$, $2a(\cos t + \cos 2t)$, $4a\,|\cos\frac{1}{2}t|$, $\rho = \frac{8}{3}v$. 5. $\frac{3}{2}$, $\frac{3}{4}$. 7. $\dfrac{t^2 + 2t}{t + 1}$. 8. $x = 1$, $y = \frac{1}{4}$.

Exercise LIX. 1. $x = \dfrac{c(3t^4 + 1)}{2t^3}$, $y = \dfrac{c(t^4 + 3)}{2t}$. 2. (i) $x^2 + y^2 = a^2$, (ii) $\dfrac{x^2}{a^2} + \dfrac{y^2}{b^2} = 1$. 3. $27ay^2 = 4(x - 2a)^3$. 5. $x = a(\theta + \sin\theta)$, $y = a(\cos\theta - 1)$. 7. $\dfrac{(9x + 32a)^{\frac{3}{2}}x^{\frac{1}{2}}}{48a}$, $(-\frac{17}{4}a, \frac{28}{3}a)$.

Exercise LXI. **1.** $\frac{1}{16}\pi a^2$. **2.** $a \sec \alpha (e^{2\pi \cot \alpha} - 1)$. **3.** $\frac{1}{2}$. **4.** $\pi + 3\sqrt{3} \approx 8.34$.

5. $a^2(\pi - \frac{1}{2}\tan^{-1} 2 - \frac{7}{5}) \approx 1.19a^2$. **6.** $\int_0^{2\pi} \sqrt{(41 + 40\cos\theta)}\, d\theta$, no.

7. $2\pi a^2\left(1 - \dfrac{1}{\sqrt{2}}\right)$.

Exercise LXII. **3.** $\frac{2}{3}ma^2$. **4.** $\dfrac{3at}{1+t^3}, \dfrac{3at^2}{1+t^3}; \dfrac{3a}{2}, \dfrac{3a}{2}$. **5.** $\dfrac{\pi\sqrt{5}}{10}$. **7.** $a\sqrt{2} +$

$a \ln (1 + \sqrt{2})$. **8.** (i) $\frac{2}{15}k\pi a^5$, (ii) $\frac{4}{15}k\pi a^5$. **9.** $\dfrac{18a^2}{5}, \dfrac{81a^2}{160}$. **11.** πab.

Exercise LXIV. **1.** Non-existent. **2.** $\frac{1}{12}\pi$. **3.** $\frac{1}{12}\ln 5$. **4, 5.** Non-existent. **6.** 1.

7. $-\frac{1}{4}$. **8.** $\dfrac{\pi}{8}(b-a)^2$.

Exercise LXV. **1.** (i) $\dfrac{11.9.7. \ldots 1}{12.10.8 \ldots 2}, \dfrac{\pi}{2}$, (ii) $\dfrac{12.10.8 \ldots 2}{13.11.9 \ldots 3}$.

2. (i) $(n-1)(I_n - I_{n-2}) = \tan^{n-1} x$, (ii) $J_n + nJ_{n-1} = x(\ln x)^n$. **4.** $\frac{1}{5}\sin^5 x -$

$\frac{1}{7}\sin^7 x, \dfrac{3\pi - 8}{32}$. **5.** 1. **6.** Zero.

Exercise LXVI. **1.** (i) $\frac{1}{3}\ln(x^3 + 4)$, (ii) $9\ln(x-3) - 3\ln(x-1)$,
(iii) $\frac{1}{4}(\sin 2x - 2x\cos 2x)$, $\frac{1}{4}(\pi + 2)$. **2.** (i) $x(L^3 - 3L^2 + 6L - 6)$, $L = \ln x$,
(ii) $\dfrac{2n(2n-2). \ldots 2}{(2n+1). \ldots 3} a^{2n+1}$, (iii) $\frac{1}{6}\tan^{-1}\frac{3}{2}$. **4.** (i) $x + \ln(x-2) - \ln(x+2)$,
(ii) $-(x^2 + 2)e^{-\frac{1}{2}x^2}$, (iii) $\sin^{-1}\left(\dfrac{x-2}{3}\right)$, $\dfrac{\pi\sqrt{6}}{12}$. **5.** $\dfrac{5e^4 - 1}{32}$. **6.** 16·0.

7. $\dfrac{\pi}{2a(a+b)}, \dfrac{\pi}{2b(a+b)}, \dfrac{\pi}{2ab}$. **8.** (ii) $\dfrac{\sqrt{2}}{4}\pi$. **9.** (i) $-(2x-3)^{-\frac{1}{2}}$, (ii) $\sin x -$

$x \cos x$, (iii) $\frac{2}{15}(x-4)^{\frac{3}{2}}(3x+8)$, π. **11.** $\frac{3}{2}$. **12.** $\dfrac{\cosh^{-1}\lambda}{\sqrt{(\lambda^2 - 1)}}$.

Exercise LXVII. **2.** (i) $\dfrac{dy}{dx} = \dfrac{y}{x}$, (ii) $\dfrac{dy}{dx} = \dfrac{y^2 - x^2}{2xy}$.

Exercise LXVIII. **1.** $y = \ln \sec x + A$. **2.** $\ln \sin y = x + A$. **3.** $\ln\left|\dfrac{1+y}{1-y}\right| =$

$2\tan^{-1} x + A$. **4.** $\sin y = A \sin x$. **5.** $\ln y = x - \frac{1}{2}x^2 + A$. **6.** $xy = \sin x -$

$x \cos x + A$. **7.** $\ln\left|\dfrac{y-3}{y-2}\right| = -x^{-1} + A$. **8.** $\tan^{-1}\left(\dfrac{2y+1}{\sqrt{3}}\right) = \dfrac{\sqrt{3}}{4}x^2 + A$.

9. $xy = 1 + (x+2)\{\ln(x+2) + A\}$. **10.** $\sqrt{\left(\dfrac{1+x}{1-x}\right)}\, y = \sin^{-1} x + A$.

11. $\sqrt{2}\tan^{-1}\left(\dfrac{y}{\sqrt{2}x}\right) - \ln(y^2 + 2x^2) = A$.

12. $y = x\{A + \ln|x| - 2\ln|x - y|\}$. **13.** $(y+1)\sqrt{(1-x^2)} = 2$.

14. $y = e^x(1 + x)^{-1}$. **15.** $y = x \ln x + cx$. **16.** $e^{-2y} = 2\sqrt{(1 - x^2)}$.
17. $\sqrt{(x^2 + y^2)} = x(\sqrt{2} + 1) - 1$.

Exercise LXIX. 1. $y = Ae^{3x} + Be^{-2x}$. **2.** $y = Ae^{-\frac{1}{2}(5+\sqrt{33})x} + Be^{-\frac{1}{2}(5-\sqrt{33})x}$.
3. $y = \frac{1}{7}(34e^{2x} + 15e^{-5x})$. **4.** $y = \frac{7}{6}x^3 + Cx + D$. **5.** $y = Ae^{2x} + B - \frac{5}{4}x - \frac{3}{4}x^2$. **6.** $y = Ae^x + Be^{-4x} - \frac{3}{4}x - \frac{17}{16}$. **7.** $y = Ae^x + Be^{-6x} - \frac{1}{27} + \frac{2}{9}x - \frac{1}{6}x^2$.
8. $y = Ae^{3x} + Be^{-x} - \frac{1}{3}e^{2x}$. **9.** $y = Ae^{3x} + Be^{-2x} - \frac{1}{50}(3 \sin x + 21 \cos x)$.
10. $y = \frac{1}{137}(17 \sin x - 56 \cos x) + Ae^{(2+\sqrt{14})x} + Be^{(2-\sqrt{14})x}$.

Exercise LXX. 1. $y = \frac{4}{3} \sin 3x$. **2.** $y = \frac{6}{5} + A \sin (\sqrt{5}x + \varepsilon)$. **3.** $y = Ae^{3x} + Be^{-3x}(= C \cosh 3x + D \sinh 3x)$. **4.** $y = e^{-\frac{1}{2}x}\left\{A \cos \left(\frac{\sqrt{7}}{2} x\right) + B \sin \left(\frac{\sqrt{7}}{2} x\right)\right\}$.
5. $y = e^{-x}(A \cos \sqrt{3}\,x + B \sin \sqrt{3}\,x) + \frac{1}{13}(3 \cos x + 2 \sin x)$. **6.** (i) $y = (A + Bx)e^{2x}$, (ii) $y = (A + Bx)e^{2x} + \frac{1}{4}$. **7.** (i) $x\dfrac{d^2x}{dt^2} = \left(\dfrac{dx}{dt}\right)^2$; (ii) $y = 3 \ln x - x^{-1} + 1$.
8. (i) $Ae^x \sin x$, (ii) $2 + A \sin (3x + B)$. **9.** (i) $y = Ce^{-x} + De^{-3x} + \frac{1}{10}(\sin x - 2 \cos x)$, (ii) $(x + 1)y = e^{2x} + Ae^x$. **10.** (i) $x^2y = e^x(x^2 - 2x + 2) + c$, (ii) $x^2 + y^2 = A \exp \left(2 \tan^{-1}\dfrac{y}{x}\right)$. **11.** (last part). $\frac{1}{3}(\pi - \tan^{-1} \frac{5}{12})$.
13. (ii) $y = c(l - x) - cl \cos nx + \dfrac{c}{n} \sin nx$. **14.** (a) $(x - 1)^2 + (y - 1)^2 = A$,
(b) $\dfrac{dy}{dx} = - \sin x(1 + 4 \cos x)$. **15.** (i) $y = \frac{1}{2}x^2 + \ln (1 + x + y) + c$, (ii) $y = 2x + 4 \cos 2x - \frac{1}{3} \sin 3x - 2 \cos 3x$. **16.** $Ay = 2 + Be^{Ax}, y = C$. **18.** $\frac{1}{10}e^{-3t} \times (3 \cos 3t + 9 \sin 3t) + \frac{1}{10}(6 \sin 2t + 7 \cos 2t)$.

Exercise LXXI. 1. $A\{(p^2 - q^2)^2 + 4k^2q^2\}^{-\frac{1}{2}}, \tan^{-1} \left(\dfrac{2kq}{p^2 - q^2}\right); B \cos (qt - \alpha) + \dfrac{Bq}{k} e^{-kt} \sin \alpha \cosh \{\sqrt{(k^2 - p^2)}t\}$. **5.** $y = C \sin (nt + \varepsilon) + \dfrac{A}{n^2 - p^2} \cos pt$, resonance.
6. $m\ddot{x} = eE - eH\dot{y}, m\ddot{y} = eH\dot{x}, A = -B = \dfrac{mE}{eH^2}$. **7.** $n^2 > k^2$.
8. $\dfrac{1}{2\lambda} \ln \left(1 + \dfrac{\lambda V^2}{g}\right)$.

Exercise LXXII. 4. (i) $e + 1$, (ii) $x, y(x^2 + y^2)^{-\frac{1}{2}}, -y, x(x^2 + y^2)^{-1}$.
5. (ii) $2(r + a \cos \theta), -2ar \sin \theta; -a, 0, a, 0; 0, \frac{1}{2}\pi, \pi, \frac{3}{2}\pi$.

Exercise LXXIII. 1. $-2x^2y(x^2 + y^2)^{-2}$. **2.** (ii) $\frac{4}{3}\pi$. **3.** $10'$. **7.** $y = f(x + ct) + g(x - ct)$.

Revision Exercise III. 1. (i) $\frac{1}{2}(1 + 3x^2)(x + x^3)^{-\frac{1}{2}}$, (ii) $(2 + 2c - c^2)(2 + c)^{-3}$.
4. (i) $3x^2(3x + 1)(5x + 1)$, $(1 - \ln x)x^{-2}$, $-3 \sin (6x - 8)$, (ii) $\frac{1}{3}, -1$. **7.** $\frac{1}{3}$.
8. $F(a) < 0, (c - x)c^{\frac{1}{2}}(2x - c)^{\frac{1}{2}}$. **10.** $30° 30'$. **11.** Max $\dfrac{\pi}{2}, \dfrac{3\pi}{2}$; min $\dfrac{7\pi}{6}, \dfrac{11\pi}{6}$.

13. $(\frac{1}{6}\pi, \sqrt{3})$, max; $(\frac{5}{6}\pi, -\sqrt{3})$, min. **14.** $4ab$. **15.** Max $2e^{-\frac{1}{2}}$, Min 1.

16. $\frac{1}{2}(\pi^2 - 4)^{\frac{1}{2}} + \cos^{-1}\left\{\dfrac{(\pi^2 - 4)^{\frac{1}{2}}}{\pi}\right\}$. **17.** (i) Max at $x = 1$, min at $x = -\frac{1}{2}$, $\frac{5}{3}$.

18. (i) $(\pi - 2)\pi a^3$, (ii) $\pi a^3(2 \ln 2 - 1)$. **19.** $\frac{1}{3}(1 - 2^{-\frac{3}{2}})$. **22.** $-4a \sin\dfrac{\theta}{2}$, $y +$

$(x - a\theta)\tan\dfrac{\theta}{2} = 0$. **23.** $e - 2, 1, 1$. **24.** $\dfrac{1 . 3 . \ldots . (2n - 1)m^{2n+1}}{2 . 4 . \ldots . 2n(2n + 1)}$. **25.** 0.787.

26. $k, \dfrac{p}{q}, 2n^2$. **28.** (i) $x + \frac{1}{2}x^2 - \frac{5}{24}x^4$. **31.** (*last part*). $1 + \dfrac{1}{\sqrt{2}}$. **33.** (i) $\frac{1}{4}(e^2 + 1)$,

$\ln 2$, (ii) $1, 1$. **34.** (i) $\frac{1}{32}\sinh 4x - \frac{1}{4}\sinh 2x + \frac{3}{8}x + c$, (ii) $\dfrac{a^4}{192}(4\pi - 3\sqrt{3})$, $e^2 + 1$.

37. $\dfrac{\pi}{\sqrt{2}}$. **38.** $n > -1$. **40.** (i) $2 + \ln \frac{3}{2}$, $\dfrac{1}{\sqrt{3}}\ln(2 + \sqrt{3})$, (ii) $\sin^{-1}(\sqrt{x}) -$

$\sqrt{(x - x^2)}$. **41.** $\dfrac{16\pi}{15}, \dfrac{13\pi}{12}$. **44.** (a) p^{-1}, (b) p^{-2}, (c) $(p + a)^{-1}$, (d) $p(p^2 + a^2)^{-1}$.

46. $\frac{1}{2}\pi + 2\tan^{-1}\left(\dfrac{1 + k}{1 - k}\right)$; (i) 2π, (ii) 0. **48.** $\dfrac{\sin 2nx}{2 \sin x}$; $\dfrac{\sin(2n + 1)x}{2 \sin x} - \frac{1}{2}$; $n\pi$ for

$n > 0$, $-n\pi$ for $n < 0$. **49.** $\sin^{-1}\left(\dfrac{x}{x^2 + 1}\right)$. **50.** (a) $y^2 + 1 = A(x - 1)^2$,

(b) $(1 + \sin t - t \cos t)(1 + t)^{-1}$. **51.** (i) $x(y - 1) = C(y - 2)$, (ii) $y = \frac{1}{3} +$
$A \sin 3x + B \cos 3x$. **53.** (i)$x^2 - 2xy - y^2 = C$, (ii) $xy = (x + 1)\ln\{C(x + 1)\}$.

55. (a) $1 - e^{-x}$, (b) $\dfrac{dx}{dy} = \dfrac{ky}{x}$, $y = Ax^k$. **57.** (ii) $e^{2x} = y^2(x^2 + A)$. **59.** 48.

61. $\sin x, \cos x$. **62.** No derivative at $\frac{1}{2}\pi$. **63.** 20 minutes.

Papers Q to X. **Q3.** (i) $\ln \sin y = A - \cos x$, (ii) $x(\ln x + c) + e^{-y/x}(x + y) = 0$. **Q4.** (i) $k - \frac{1}{2}\operatorname{cosec}^2 x$, (ii) $k - (7 - 6x - x^2)^{\frac{1}{2}}$. **Q5.** (i) $s = 4a \sin \psi$,

(ii) $4\pi \bigg/ \sqrt{\left(\dfrac{a}{g}\right)}$.

R2. $\frac{8}{3}a^{\frac{1}{2}}b^{\frac{1}{2}}(a + b)$, $2\pi ab(a + b)$. **S1.** (i) $\frac{1}{2}a^2 \sin^{-1}\left(\dfrac{x}{a}\right) + \dfrac{x}{2}(a^2 - x^2)^{\frac{1}{2}}$, $-e^{-2x} \times$

$(\frac{1}{2}x^3 + \frac{3}{4}x^2 + \frac{3}{4}x + \frac{3}{8})$, (ii) $0, \frac{4}{15}$. **S4.** (i) $e^{-2t}(A \cos 3t + B \sin 3t) + \frac{1}{29}(18 \cos 2t +$
$16 \sin 2t)$, (ii) $y(x^2 + 1) = (x^2 - 1)\sin x + 2x \cos x + 2$. **S5.** On Ox, $\frac{5}{6}a$ from O.

T1. (a) $\dfrac{\pi}{4} - \frac{1}{2}\ln 2$, (b) $\frac{1}{4} + \dfrac{3\pi}{32}$. **T2.** $\dfrac{\omega a h(h \cos \theta - a)}{(h - a \cos \theta)^2}$ m s^{-1}. **T4.** $\frac{1}{2}\pi^2 a^2$, $\frac{1}{2}\sqrt{\frac{3}{2}}a$.

U1. (i) $1, \dfrac{\pi^2}{4} - 2$, (ii) $y = \frac{1}{4}(x^2 - x^{-2})$. **U3.** 4.2. **U5.** $\frac{1}{2}x\sqrt{(x^2 + a^2)} + \frac{1}{2}a^2 \times$

$\ln\left\{\dfrac{x + \sqrt{(x^2 + a^2)}}{a}\right\}$. **V1.** (i) $2(1 + x^2)^{-1}$, (ii)$(x^2 - 1)^{-\frac{1}{2}}$, (iii) $-\dfrac{1}{2\sqrt{2}}(1 - x)^{-\frac{1}{2}}$.

V2. $\dfrac{2\pi}{3\sqrt{3}}$, $\frac{1}{6}\pi$, 0; $\frac{1}{4}x^2 - \frac{1}{4}x \sin 2x - \frac{1}{8}\cos 2x + c$. **V3.** Min $\frac{1}{2}(m + n)^2$, where

$x = \dfrac{m - n}{m + n}$; max $\frac{1}{2}(m - n)^2$, where $x = \dfrac{m + n}{m - n}$. **V5.** $\ddot{x} + \mu\dot{x} + \lambda^2 x = 0$;
$2V(4\lambda^2 - \mu^2)^{-\frac{1}{2}}$, $-\frac{1}{2}\mu$, $\frac{1}{2}(4\lambda^2 - \mu^2)^{\frac{1}{2}}$; $x = Vte^{-\frac{1}{2}\mu t}$. **W1.** (i), Inflexion,

(ii), (iii) minimum.   **W2.** $\dfrac{1}{2\sqrt{2}} \ln \left(\dfrac{\sqrt{5} + \sqrt{2}}{\sqrt{5} - \sqrt{2}} \right) \approx 0 \cdot 53$.   **W5.** (i) $\sinh x$,

(ii) $y(\sec x + \tan x) = x + \dfrac{\pi}{4}$.   **X1.** (i) 1, (ii) $\frac{3}{5}$, (iii) 0.   **X2.** $\pi(\alpha^2 - 1)^{-\frac{1}{2}}$.

X3. $x = Ae^{-\frac{1}{2}kt} \cos \{\sqrt{(n^2 - \frac{1}{4}k^2)}t + \alpha\}$.   **X4.** $y = \operatorname{cosec} x - \cot x$; $-\frac{2}{3}$.
X5. (i) $(\pm 3, 0)$, (ii) both axes, (iii) $x \pm y = 0$.

Index